TROPICAL GEOMORPHOLOGY

Tropical Geomorphology

A Morphogenetic Study of Rajasthan

H. S. SHARMA

CONCEPT PUBLISHING COMPANY, NEW DELHI-110015.

This study was conducted by the author in 1980-1983 under the UGC Career Award Scheme.

ISBN 81-7022-041-6

First Published 1987

Hari Shanker **Sharma** (B. 1945-

Published by
Naurang Rai
Concept Publishing Company
H-13, Bali Nagar
NEW DELHI-110015 (India)

Printed at
New Gian Offset Printers
Shazada Bagh
New Delhi-110035

DEDICATED TO

Honourable Guru dev
Baba Swami Ram Pal Das ji
Dayal Ashram, Newai

And

Dr. G.C. Sharma, M.S., F.A.C.S.
Former Principal, S.M.S. Medical College
Jaipur

FOR THEIR BLESSINGS

Foreword

It gives me great pleasure to introduce the book entitled 'Tropical Geomorphology - A Morphogenetic Study of Rajasthan' by Dr. H.S. Sharma. Dr. Sharma deserves our congratulations for this innovative research work in the field of physical Geography which is not been given attention that it deserves in our country. We are aware of the fact that most of our environmental problems and inadequacies of our planning strategies are rooted in our partial understanding of physical processes which articulate the interplay between exogenetic and endogenetic forces.

This work was made possible by a Career Award being made available to Dr. Sharma in 1979. It may be mentioned that the University Grants Commission had initiated a scheme of Career Awards to young scientists in various disciplines. The purpose of this scheme was to identify young talented scientists who have shown a high degree of competence as well as of potential in their areas of specialization and to promote their professional development by enabling them to devote their energies to research and study with minimum encumbrances and teaching responsibilities.

The present study was carried out by Dr. Sharma under this scheme. I am particularly happy that the results of this study are now being made available to geographers as well as decision makers.

The study has been divided into three parts dealing with morphogenetic environment, morphogenetic forms as well as processes and morphogenetic regionalization of the state of Rajasthan. By laying stress on the significance of climate and structure in shaping landforms, Dr. Sharma has been able to throw sufficient light on climatogenetic evolution of this part of a tropical region. He has focussed attention on climatic changes and process of desertification during the quaternary period. Dr. Sharma has discussed at length the origin of interdunal flats, sanddunes, Kankar pan, palaeochennels of the Saraswati as well as Ghaggar, and boulders as well as pebbles in the desertic parts of Rajasthan. He has thus ruled out the possibility of Pleistocene glaciation. Besides, he has also advocated the thesis that Rajasthan has witnessed various phases of desertification and that the former limit of Rajasthan Desert was beyond the Aravallis in the north east. He has also cautioned against the possibility of intensification of desertification processes, if the natural resources are used or exploited beyond the optimum limit.

Other aspects dealt within the book relate to the origin of inselbergs, planer surfaces and analysis of morphometric properties in relation to mean annual rainfall. By stressing the significance of ravine erosion in some of the most important river systems from different parts of Central India, Dr. Sharma has been able to throw light on his major conclusion to the effect that ravine formation is mainly due to neotectonism in the quaternary period and has been accelerated by the unwise use of land as well as climatic changes. The Rejuvenation Dynamic Disequilibrium model of ravine genesis has been a distinct contribution of Dr. Sharma in the field of fluvial geomorphology. He has also been able to develop a number of models of ravine origin.

In the last part of this study, Dr. Sharma has presented a scheme of morphogenetic regions and landsystems which may form the basis for planning strategies in the State of Rajasthan in the years to come. The geomorphological map may also be of great use for planners.

I have no doubt that this study will stimulate interest in physical geography and various findings of Dr. Sharma will lead to detailed studies by other scholars. I am particularly glad that Dr. Sharma has not only stated the various problems of morphogenesis as well as their solution in clear terms but has also provided an adequate discussion of the same, which would be of great interest to scholars. Dr. Sharma has justified his Career Award by virtue of his consistent efforts for bringing out this publication. I have great pleasure in commending the book to all those who are interested in this aspect of geoscience.

Vice-Chancellor
University of Delhi
Delhi-110007
3 September 1987

Moonis Raza
Professor Emeritus
Jawaharlal Nehru University
New Delhi-110067

Acknowledgements

The present work is an outcome of the guidance, encouragement and co-operation of senior scientists and friends in India and abroad. First of all I am indebted to Professor Moonis Raza, former Director National Institute of Educational Planning and Administration, New Delhi, and Vice Chancellor, University of Delhi, Delhi, for his keen interest, guidance and constructive suggestions throughout the preparation of this work.

I am equally grateful to Professor N.B.K. Reddy of Shri Venkateswara University, Tirupati and Professor R. Vaidyanadhan of Andhra University, Waltair, for their suggestions and advice. Professor G.S. Gosal and Professor A.B. Mukerjee of Punjab University, Chandigarh, also deserve my sincere thanks and gratitude for their suggestions in the initial stage of the work.

I would like to thank some of my overseas friends who have sent reprints of their research work on request. I am also indebted to: Professor C.A.M. King of Nottingham University (Retired), Nottingham, U.K. for her comments in finalizing the chapter scheme of this work; to Professor Tricart, Director du C.G.A. 43 rue Goethe, 67083 Strasbory Cedex, France, for his comments and for providing guidelines for the work in climatic geomorphology; to Professor Gardiner of the University of Leicester, U.K., Professor Gellert of the German Democratic Republic and Prof. A.D. Abrahams of the state university of New York at Bufallo, U.S.A. for their valuable suggestions.

I also express my sincere thanks to Professor Daya Krishana, then Pro-Vice-Chancellor of the University of Rajasthan for his keen interest in the project, and to Dr. R.B. Singh, the then Head of the Department of Geography, for allowing me the use of all the facilities in the department for my research work.

My sincere thanks are owing to Professor Indra Pal, Department of Geography of the University of Rajasthan for his encouragement and also to Miss A.E. Dass and Mrs Laxmi Shukla, my teachers and Associate Professors for their blessings and good wishes for the completion of this work.

I am very grateful to Dr. Pratap Singh Dhinwa and Mr Ram Awatar Sharma, Research Fellows in this Project for their fullest co-operation and sincerity and devotion in bringing this work to fruition without their co-operation this work would not have been completed. I would also like to thank Mr. Ram Deo Gurjar, Miss Komal Srivastava and Mr. R.D. Doi for their assistance at the writing stage of this work.

I am also thankful to Mr. P.C. Bakliwal of the Photo-geology Laboratory, Geological Survey of India, Jaipur; Mr. Satya Narain Sharma, Deputy Director, Hydrology Department, Jaipur; Mr. R.K. Sood of the Space Application Centre, Ahmedabad; Dr. D.P. Agrawal, Physical Research Laboratory, Ahmedabad; Dr. Surender Singh of the Central Arid Zone Research Institute; Dr. R.C. Agrawal, Ex-Director, Archaeology Department, Jaipur; Dr. R.C. Banerjee, ex-Director, Indian Meteorology Department, Jaipur, for their help in bringing out this work in its present form.

I also wish to express my sincere thanks to Prof. Pramila Kumar for her suggestions from time to time and Dr. M.L. Sharma of Government College, Sambher for his help during field work.

I am thankful to a number of organizations and institutions that have helped in one way or another by extending to me various facilities in the course of this work: Geological Survey of India for providing me Satellite Imageries and Aerial Photographs for the study in the Photo-geology Laboratory; the Space Application Centre, Ahmedabad for providing facilities for interpretation of Landsat Imageries; the Survey of India and Ministry of Defence for supplying aerial Photographs and defence clearance, respectively; to Indian Meteorology Department, Jaipur, the Hydrology Department, Jaipur, and the State Soil Survey Department for supplying necessary data.

My sincere thanks are owing to Mrs M.K. Singh for shouldering the responsibility of typing the manuscript well in time which was an important factor in this project.

Finally, I am indebted to the University Grants Commission for offering me this prestigious career award and financial assistance for this project and to the University of Rajasthan for granting me special Academic Leave for three years to take up this assignment.

5 June, 1987.

Contents

List of Figures

List of Tables

Introduction

LIMITS OF TROPICAL REGIONS AND GENERAL CHARACTERISTICS

The word 'Tropics' is evidently derived from the 'Tropics' (with a capital T) of Cancer and Capricorn, the parallels of a latitude of 23 1/2 degrees, which indicate the outer limit of the areas where the sun can ever be in zenith. It is generally understood that tropical regions are mainly found between these two lines. They are, therefore the regions of 'Low' latitudes, but the outer limits of these 'Low' latitudes are not easily determined because the Tropics of Cancer and of Capricorn themselves are unsuitable as boundaries. They are too regid; some areas with clear tropical characteristics are found at latitudes of more than 23 1/2 degrees while on the other hand some clearly non-tropical regions are found much closer to the Equator.

The best way to determine the outer limits of the tropical areas is therefore to use climatic characteristics which distinguish these regions from the rest of the world. All geomorphic processes are directly and indirectly related to climatic features. Delimited in this way the Tropics constitute a belt around the Equator varying in width from about 40°S to 40°N latitudes.

Within the tropics lies approximately 40 per cent of the earth's surface, three quarters of which is covered with water. The tropics receive over half of the world's total rainfall. As much as two third of the total rainfall has been estimated to fall between 30°N and 30°S. Annual rainfall varies from 0 to 10,000mm in the tropics. Tropical regions are frost free, have only small variations in the length of day, have seasons defined by water availability rather than temperature, possess some distinctive soil characteristics, more especially high levels of mineral accumulation in soil, combined with rapid leaching of nutrients.

As the temperature is relatively uniform, rainfall, particularly its monthly distribution, determines the vegetation and geomorphic processes of each tropical region. The main characteristic of rainfall is the intensity of the storms, much of the rain falling in a few heavy downpours usually accompanied by strong winds. As the intensity of these storms is often in excess of the infiltration capacity of the soil, much of the rain is lost by run-off, rather than by building up of the soil reserves. The run-off takes with it fertile top soil and causes flash flooding of the streams and rivers. These effects are most marked where the ground water is scanty rather than dense as with forest cover.

The investigations in the last two decades in the tropical regions have indicated that these regions have been subjected to climatic changes especially during quaternary period, giving rise to various morphogenetic features. So, these regions have now received scientific attention in order to formulate new hypotheses of landform development which were otherwise initially based on field investigations pertaining to temperate regions.

The present book aims to study the morphogenetic forms and processes of Rajasthan which lies in the north western outer limits of the Indian tropical region. The present processes are controlled by the monsoonal character of rainfall. (Fig. 0.1)

CLIMATE AND GEOMORPHOLOGY

The influence of climate on morphogenetic processes and the associated landform assemblages has long been recognized and emphasized by various studies in the United States, Asia and Africa. Generally, the early German workers overstressed the erosive action of wind and under estimated the action of water. However, the dominant trend in desert research in the 20th century has been towards studying the action of running water. Throughout this the arid landscape has continued to attract much attention as may be seen from the work of Gautur (1923) on the Sahara desert, Mortensen's (1927) study of the north Chilean desert and the recent valuable research available from the Institute de Recherches Sahariennas (Algiers) and Scientifique cherifien (Rabat). However, since the end of the 1920 interest has been turning more and more to other climatic landform assemblages.

FIG. NO. 0·1 KEY MAP

At the beginning of this century, Koppen (1923) and Thornthwaits (1938) suggested their important classification of the world into climatic zones and regions. Thereafter, studies relating to climate landform assemblages were extended, global morpho-climatic schemes were proposed, and the effects of palaeo-climatic changes estimated. Some of the outstanding works on the subject are of Budal (1948), Peltier (1950), Tricart and Cailleux (1955, 1972), and Birot (1968). Their original contention is that the "normal cycle" is so over-simplified a concept and for existing conditions, either unreal or so really abnormal that "present land forms must be interpreted as a product of complex evolution, as a mixture of newly developing agents and of survivals in course of removal at varving rates" (Tricart and Cailleux, 1955).

The research in climatic geomorphology has been dominated by two great methodological traditions. First, the application of the concept of zonality based essentially on the climatological and ecological principles and second, the inductive definition of climate-process provinces based on assumed general relationship between the efficacy of selected geomorphic processes and standard climatic means. Much recent work in this field has been directed towards resource appraisal so that regions have been defined according to phytological, pedological and geomorphological criteria. Nevertheless, there are signs of a movement towards areal mapping based on long term field measurement and using a framework of system theory as a means of establishing geo-systems as a basis for the synthetic establishment of areal climatogenetic units.

CONCEPTUAL FRAMEWORK

Morphogenetic Classification

The concept of a morphogenetic region is that under a certain set of climatic conditions,a particular geomorphic process will predominate and hence will give to a landscape of a region characteristics that will set it off from those of other areas, developed under different climatic conditions (Thornbury, 1954). Morphogenetic regions are therefore conceptual devices by which a geomorphologist may relate climate, process, landforms and regions. The concept of morphogenetic region dominating processes sub-ordinates the factors of structure and time in geomorphology. As a matter of fact there is abundant room for debate about the relative importance of climatically controlled landforms and structurally controlled landforms. Tricart (1981) has pointed out that climatic geomorphology and structural geomorphology do not exist per se. They are not independent geomorphology. They just focus upon a certain aspect of landform shaping, a certain aspect which is in fact very interdependent. it is an abstraction. Thus climatic geomorphology has no intrinsic reality, it is just a point of view on an aspect of geomorphology as a whole-the study of landforms.

History of the Concept

The first geomorphologist to specifically define climate landform assemblages was Budel (1948). He recognized the existance of five morphogenetic regions in Europe. While peltier (1950) recognized nine morphogenetic regions. However, he emphasized the ideal relationship between climate and process and thus was concerned with climate process systems. Other morphogenetic classification have been proposed by Tanner (1961) and Leopold, Wolman and Miller (1964). Cotton (1958) was the first to state that the actual morphogenetic regions of today are the result of the action of alternating morphogenetic systems during the Pleistocene era. Specific studies on morphogenetic regions have been frequently taken up in recent years, for example Common (1966) applied the concept of morphogenetic regions to the specific problem of slope failure. Of special interest and importance is the ever increasing number of quantitative field studies in the world in which actual correlation is made between climatic factor, magnitude of specific geomorphic process and the resultant nature of the landscape.

While morphogenetic analysis is still rather speculative, it represents a major aspect of morphogenesis. One goal of geomorphology should be the quantification of climate and process. The relationship between morphogenetic systems, climate and region which is one of the basic methodological problems of morphogenetic regionalization, has still not been properly investigated by geomorphologists. Thus morphogenetic classification is subjective, qualitative and artificial. Such classification constitutes a geomorphic model which must be used with care and understanding.

India embraces a wide range of rainfall regions and vegetation communities and there can be little doubt that the balance of geomorphic processes varies greatly from humid rain forest through deciduous woodland and savannas into the margin of deserts. If we add to this the possible effect of plate tectonics, part of the country should have experienced a climate different from the one it is exposed to now. At least some of the landforms and deposits should have recorded evidences of these changes.

GEOGRAPHICAL LOCATION OF THE AREA

Rajasthan has been chosen for the present morphogenetic study. The state can conveniently be considered as an environmental laboratory which encompasses a wide range of environmental conditions. The state of Rajasthan covers an area of about 3,26,127 square kilometres and lies between latitudes 23°03' and 30°10' (north) and Longitudes 69°30' and 78°27' (east). It is bounded on the west and north-west by Pakistan, on the north and north east by Punjab, Haryana and Uttar Pradesh and on the south-south east and south-west by Madhya Pradesh and Gujarat respectively.

Geologically, it is the north western part of the Indian shield which has not been affected by tectonic movement of the orogenic type since the Mesozoic age. However, western Rajasthan has witnessed various phases of transgression and regression of the sea up to the late Tertiary period. the evidences of neotectonism are exhibited in the form of incised meanders and ravines and gullies along the Chambal river and its tributaries (Sharma, 1968, 1976).

The study area is exposed to sub-tropical monsoon climate which may be described as having arid, semi-arid, transitional between arid and semi arid and sub humid zones. The normal annual rainfall in the state decreases from south east to north west. It receives rainfall ranging from between 100 mm and 1650 mm. The rainfall variability is high and evaporation losses are also very high. Almost 90 per cent of rainfall occurs between the months of June and September. During remaining months of the year rainfall is light and sporadic. Annual and daily ranges of temperature are high there are marked seasonal contrasts in humidity, wind speed and radiation.

REVIEW OF GEOMORPHOLOGICAL STUDIES IN RAJASTHAN

A fairly large amount of work has been done at the Central Arid Zone Research Institute at Jodhpur and by a Cambridge-Baroda-Poona team of archaeologists and geomorphologists on the changes in the climate of Rajasthan in the Quaternary and Recent times. While all agree that there has been a change from arid to semi-arid to subhumid climate and a repetition of this cycle, contradictions are to be found in the duration assigned to these climatic changes. Broadly speaking, dry river beds are supposed to be indicative of aridity. There have also been attempts to correlate geomorphic features with the glacial and inter-glacial periods in the Himalaya. It is surmized that sand shields formed during the glacial period were washed out during the much longer fluvial (raining) inter-glacial periods. During these periods in Rajasthan the climate is supposed to have been arid and humid respectively fossil parabola seem to have shaped into longitudinal dunes and later berchans predominated. Major dunes are supposed to have been formed during the arid phase and stabilized during the humid phase. One interesting aspect of these studies is that a chronology could be established because of the occurrence of human implements with these landform. Great emphasis has been laid on the degeneration of drainage in these parts though different authors attribute different reasons for this (tectonic, climate, river piracy, human interference, etc.).

Besides the work done at the Central Arid Zone Research Institute some geomorphic studies of river valleys and basin morphometry have been made by Varma (1960); H.S. Sharma (1968, 1972, 1976 and 1979); Padmaja (1976); M.L. Sharma (1982); and Dhinwa (1982). The work of Verma (1980) regarding the droughts of Rajasthan is also worth mentioning. He has studied rainfall trends and delineated drought regions on the basis of climatic parameters and revenue records.

In spite of considerable work on palaeoclimatic changes and geomorphology in Rajasthan, studies on the morphogenetic-climate-process system are still lacking. A proper understanding of the paleoclimate of different regions of Rajasthan requires extensive field work and a proper appreciation of the role of temperature, rainfall and intensities of processes at work and resultant landforms in different climates. Morphogenetic study as a foundation for regional planning is not only of scientific interest but also has great value for the practical economic activity of the state. In recent years the studies concerning regionalization have been moved to the forefront of Indian Geography and various related problems are receiving the attention of individual geographers and a number of organizations. The Planning Commission and the Indian Statistical Institute are developing physico-economic and resource regionalization. The office of the Registrar General of India is investigating the region-wise distinction in development level and is also concerned with questions of general economic regionalization. The Central Water and Power Commission is surveying the energy resource regions of the country. The National Council of Applied Economic Research conducts studies and offers recommendations on the perspective developments state-wise of the main branches of the economy. Recently, Despande (1982) emphasized the need to re-think the demarcation of the geographical regions of the India. In his scheme of regionalization he has emphasized the role of tradition (historic-cultural) in identifying the limits of geographical regions. The present study of morphogenetic regionalization of Rajasthan is another such attempt in the mainstream of the regionalization process with particular emphasis on morpho-climatic parameters.

METHODOLOGICAL FRAMEWORK

The present study is based on the visual interpretation of landsat imageries for the year 1972-1975 in a variety of forms and formats, topographical sheets on the scale 1:50,000 and one inch to four miles, aerial photographs on the scale 1:25,000 and subsequent field work. The landsat imageries especially have proved to be extremely useful in the study of geomorphology and have helped to identify lineaments, paleo-drainage channels, erosional surfaces and dunal patterns and orientation. The method of survey is based on the assumption that each type of country is represented on landsat imageries by a distinctive pattern. However, field work is very essential in geomorphic studies in order to confirm the features identified on landsat imageries. With this in a view about 20,000 kilometers were traversed in different parts of Rajasthan and more detailed observations made on various morphological features identified on landsat imageries and topographical sheets.

In order to show the relationship between climate and processes, 62 random order basins and 54 third order basins were selected from various climatic regions (Fig. 0.2). Various morphometric parameters were computed and with each other. A simple linear correlation and coefficient have been made to show the significance of the relationship. Morphometric parameters such as drainage density, stream frequency, constant of channel maintenance, elongation ratio, relief ratio, texture ratio, circularity ratio, basin slope and tortuosity index were correlated with mean rainfall. some data of discharge sites (8) and runoff (59) have also been collected and correlated with mean rainfall. The morphometric values were later used as a basis of demarcating boundaries of some of the morphogenetic regions and landsystems.

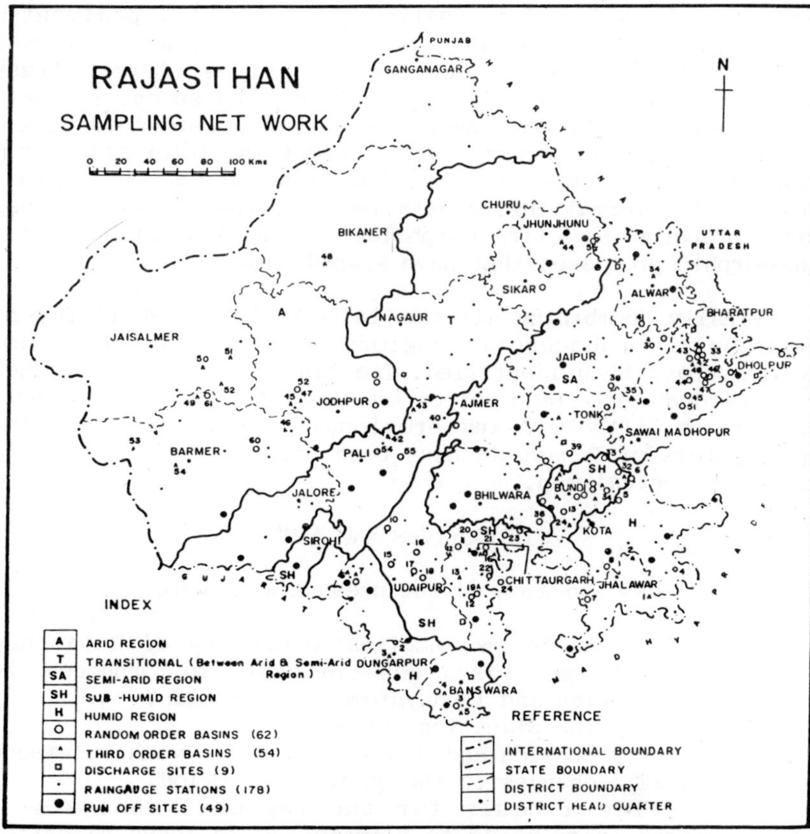

FIGURE. 0.2

The areal pattern of drainage density, stream frequency, dissection index, slope, relative relief and runoff have also been analysed. The data computed from quarter inch toposheets were later reduced to a million inch map and cartographed.

Geological, soil and climatic data were collected from the records of the Geological Survey of India, the State Agricultural Department, the Indian Meteorology Department and the Department of Hydrology at Jaipur.

Aerial photographs on the scale 1:25,000 have also been studied for Alwar, Mount Abu and Jhalawar areas. Soil samples especially from interdunal flats, paleosoils and kankar-pan have been collected and analysed in a soil laboratory. The field investigation was undertaken to identify fossil dunes, playa, saline flats, buried pediments, rocky/gravelly pediments, obstacle dunes, kankar-pans, planer surfaces, unconformities, laterites, ferricretes, river terraces, waterfalls, inselbergs, and so on, the data necessary to trace out the climatogenetic history of Rajasthan since the Quaternary period. Combining information from topographical maps, landsat imageries, aerial photographs and field informations, a detailed geomorphological map has been prepared depicting morphological units and the various geomorphic processes that have shaped them.

Finally, combining all the information from different sources 5 morphogenetic regions of Rajasthan and 45 land systems have been identified. The land systems were further subdivided into land units on the basis of morphometric properties, lithology and erosional characteristics. The landsystems of Rajasthan have been mapped in the form of a geomorphological map.

OBJECTIVES AND AIMS

The objectives and aims of the present work were:

1. To reconstruct the morphoclimatic history of Rajasthan in order to provide a genetic basis to morphogenetic regionalization and landsystems of Rajasthan.
2. To identify the planation surfaces.
3. To identify erosion and deposition phases with reference to climatic changes in the Quaternary period.
4. To provide a basis for the regional analysis of landforms in relation to climate and to focus attention on the relationship between climate and processes.
5. To delineate morphogenetic regions and landsystems on the basis of climatic, pedogenic, biotic, morphometric, erosional peculiarities and stage of evolution.
6. To prepare and compile a geomorphological map of Rajasthan on a 1:1,000,000 scale depicting composite units of landforms. This morphogenetic approach to resource evaluation is of value in the context of the regional planning of the state.

PLAN OF WORK

The study has been divided into three parts, part one deals with the morphogenetic environment including litho-stratigraphic characteristics, lineaments, quaternary deposites, climate and palaeoclimatic changes. Part two deals with morphogenetic forms and processes including aeolian forms and processes, fluvial morphology, basin morphometry, ravine morphology, planation surfaces and inselbergs. In the last part morphogenetic regions and landsystems and morphological units have been delineated and the association between climate, processes, landform and region have been discussed.

1. The humps in eastern Rajasthan is a group of granite domes surrounded by a plain.

2. Flared slopes in Erinpura granite near sendra.

3. A typical hogback ridge near khandar

4. A view of Vindhyan country around kota

5. A view of Berach near Chittorgarh

6. A view of pans near Dhirwas Churt district

7. Views of soil profiles

8. A quartizitic dome exposed in scarpment near Danta Ramgarh

9. Inter-dundal flats and typical dester vegetation

10. A view of arid landscape with small hamlets

11. Boulders and gravels near Jayal

12. A view of hill slopes of Aravalli on the way to Sikar

13. A typical hornhardt inselberg near Ajitgarh

14. Block strewn granitic gneiss bornhardt near Diggi, Tonk district

15. Jointing in Erinpura granite, granite and thor vegetation along joints, near Sendra

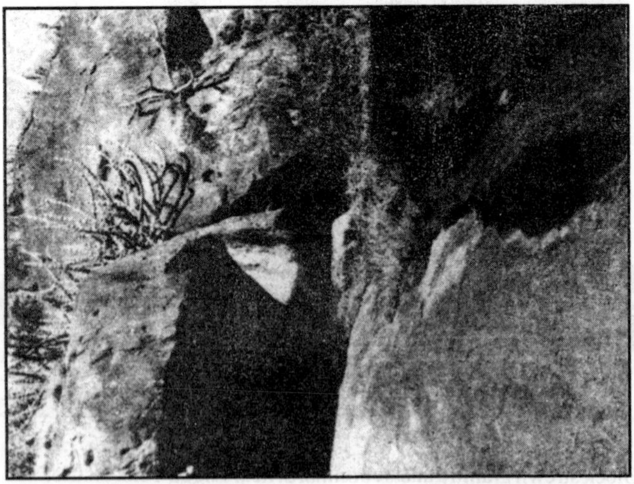

16. Angular block of quartizite resting on quartizite, near Jhunjhunu

PART ONE
MORPHOGENETIC ENVIRONMENT

ONE

Geology

Geological structure forms the foundation of the landforms of any area. Geologically, Rajasthan is made up of diverse rock types ranging from the oldest Archaean metamorphites through sub-recent to recent alluvium and aeolian sand. Three-fourths of the state is covered by aeolian sand and alluvium, the remaining area exposes a wide variety of hard rocks comprising crystalline and sedimentary rocks. The crystalline rocks include various types of metamorphic schists, quartzites, marbles and gneisses of the pre-Cambrian Vindhyan super group and of Permo-Carboniferous to Jurassic, Cretaceous and Tertiary ages. The south-eastern extremity of the state is occupied by a pile of basaltic flows of Deccan traps of the Eocene age (Fig.1.1).

The present analysis of geological lithostratigraphy and structural history of Rajasthan is based on the compilation and synthesis of the data from the records and memoirs of The Geological Survey of India during the last three decades. The revision mapping carried out by the Geological Survey of India since Independence resulted in the acquisition of considerable basic data.

A brief description of the various stratigraphic units are given in Table 1.1.

ARCHAEAN

Bhilwara Super Group

The high-grade metamorphites and gneisses (banded gneissic complex) together with associated low-grade metamorphites (earlier assigned to Aravalli system and Raialo series by Heron) which represent the oldest metasedimentary sequences in Rajasthan were included in the Bhilwara group.

It has been redesignated as Bhilwara super group. This group comprises a complex assemblage of meta sediments like, quartzite, conglomerate, metagraywacke, metaprotoquartzite, marble, dolomite, calc-silicate rocks, calcschist, calcgneiss, paragneisses, paraamphibolite, basic metavolcanic, migmatite, composite gneiss, granite, granite gneiss, grenulitic rocks, granodiorite, charnokite, norite and ultramafics. The lithological assemblage suggest a geosynclinal milieu, though the original fabric has largely

FIGURE 1.1

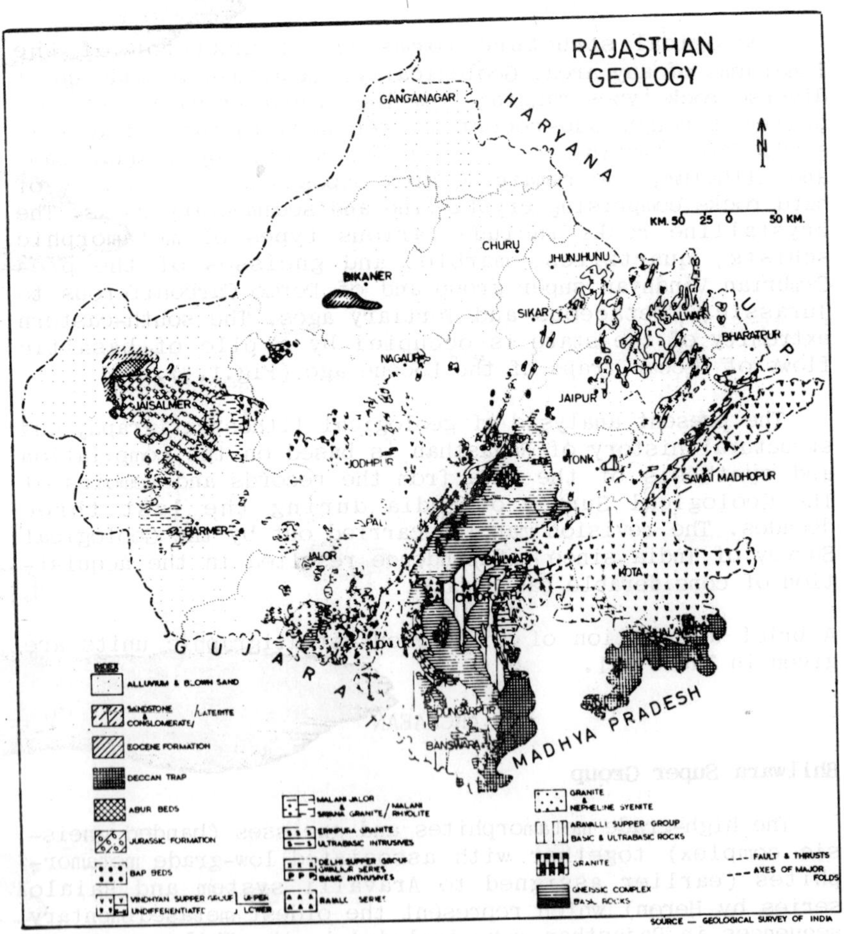

been obliterated by blanket (?) metamorphism and polyphase deformation. The boundaries are defined by erosional unconformities. The conformity is overlain by the Aravalli super group in the west, Delhi super group in the north-west, Vindhyan sedimentaries in the east and Deccan traps in the south-east. The base of the Bhilwara super group is not exposed and is concealed under the Vindhyans.

The Bhilwara super group has been assigned to Archaean II (>2500 m.y.) on the basis of radiometric dating of the late tectonic Berach granite (c.2885 Crawford, 1980; C. 2600 PRL, 1980) and Untala Gingla Granites (c.2950 $\pm$ 15 PRL, 1980).

The Bhilwara super group has been divided into the following : Hindoli group, Mangalwar complex, Sandmata complex, Rajpura-Dariba group, Pur-Banera group, Jahazpur group, Sewar group and Ranthambhor group (Fig.1.2).

Outcrops of the gneissic complex occur in two separate geographical areas. One of these being that of the northern outcrops, covering a stretch of the country eastward between Kisangarh and Antolia, and the other being the area of the southern exposures, also occurring eastward between Nathadwara and Pipal Khunt an oval patch of gneissic rocks between Nathadwara and Kankroli forming the link between these two areas. In the southern areas the rocks are well-exposed and there is undoubted evidence of unconformable relations with the overlying Aravalli sediments. In the northern area the gneissic rocks are overlain by rocks of the Delhi super group. In the intermediate area between Nathawara and Kankroli, a metamorphic discordance between the migrantites and meta-sediments, suggesting a break between the Banded Gneissic Complex and Aravallis has been observed. The correlation of the gneissic rocks in other areas with the Banded Gneissic Complex of the type area is still an open question and remains to be worked out.

The Berach granite and gneiss occur over a crescent-shaped area of about 100 km in length and 30 km in width in central Rajasthan in the vicinity of Chittorgarh (Fig. 1.2). Over a large part of the area the rock is essentially a non-porphyritic granite with no trace of banding, medium and uniform in grain, and varying in colour from reddish-pink to greenish yellow. The felspars-orthoclase and acid plagioclase are much decomposed to kaolin, sericite, quartz, calcite, zoisite and epidote. Microcline is sometimes recognizable westward. There is a gradual transition from a

TABLE 1.1. Lithostratigraphic Units of Rajasthan

Quaternary	Recent		Alluvium and blown sand	Extrusive and intrusive rocks Deccan volcanics
Tertiary	Eocene System	Mandai series	Sandstones	
		Akli, Kapurdhi/ Jogira, Mar/ Bandah/Khuila/ series	Sandstones, bentonitic clays and Fuller's earth	
Mesozoic	Cretaceous system	Abur/Fatehgarh/ Palana series	Sandstone, phosphatic sandstones, limestones, clays and lignite	
	Jurassic system	Parihar series	Felspathic sandstones and quartzitic sandstones	
		Bedesir series	Ferruginous sandstones	
		Baisakhi series	Calcareous sandstones	
		Jaisalmer series	Fossiliferous limestones, sandstones and conglomerates	
		Lathi-Mayakor series	Sandstones, siltstones, conglomerate and schist	
Palaeozoic	Permo-Carboniferous System	Badhaura series	Sandstone and Bap boulder bed	
	Cambrian (?) System	Birmania formation	Yellow cherty limestone, calcareous sandstones, delomitic limestones and cherty limestones phosphorite and phosphatic bands, quartzitic sandstones, variegated shales, calcareous and ferruginous sandstones, grey and yellow cherty limestone	
		Panda formation	Thin bedded sandstones with gritty layers, quartzites and quartzitic sandstones with ferruginous sandstones, silty shales	

TABLE 1.1 Contd.

			Erinpura Granite and Gneiss (c.740 m.y)			Godhra Granite and Gneiss (c. 955 m.y.)		
Lower Palaeozoic to Upper Proterozoic	Vindhyan Super Group	Bhander Group	Upper Bhander shales, limestones sandstones Sirbu shales and limestones	Marwar Super Group	Nagaur Group	Sandstones gypsum and silstones	Malani Jalore Sewana	rhyolites and granites
			Lower Bhander sand-stones, shales					
			Sameria shales and limestones		Bilara group	Limestones and dolomites		
			Lower Bhander lime-stones					
			Ganurgarh shales and sandstones					
		Rewa Group	Upper Rewa sandstones Jhiri shales		Jodhpur group	Sandstones, cherts, shales and boulder beds		
			Lower Rewa sandstones Panna shales					
		Kaimur Group	Sandstones, shales and conglomerates					
		Semri Grou	Tirohan limestones breccia and haematite shales					
			Suket shales and limestones					
			Jhalrapatan sandstones					
			Nimbahera shales					
			Nimbahera limestone					
			Khori-malan conglomerates					
			Jiran sandstones					
			Binota shales and sandstones					
			Sawa sandstones and shales					
			Bhagwanpura limestones, quartzites and shales					
			Khardeola sandstones, shales and grits Kannouj grits and shales Khairmalia flows					

TABLE 1.1 Contd.

	Vindhyan	Super group (platformal)		Godhra Granite and Gneiss · (c.955 m.y)			
Proterozoic II 2000-740 m.y. Delhi Super Group	Punagarh Group	Sowania formation Khambal formation Bombolai formation Sojat formation	Sindreth Group		Goyali formation Angor formation		
	Sirohi Group	Ambeshwar formation Reodar formation Jiyapura formation	Khiwandi formation	Sendra Ambaji Granite and Gneiss (c.800 m.y.)			
			Kishangarh Syenite (c.1490 - 1910 m.y.) Phulad Ophiolite Suite				
	Kumbhalgarh group	Beawar formation Todgarh formation	Kotra formation	Barr formation Rass formation Sendra formation	Kalakot formation Basantgarh formation	Ajabgarh Group	Ajmer sector Ajmer formation
	Gogunda Group	Kelwara formation Antalia formation Richer formation				Alwar Group	Naulakha formation Srinagar formation
Proterzoic I 2500-2000 m.y.	Champaner group	Aravalli supergroup Rajgarh formation Shivrajpur formation Jaban formation Narukot formation Khandia formation Lambia formation					
	Lunavada group	Kadana formation Bhukia formation Chandanwara formation Bhawanpura formation Wagidora formation					

TABLE 3.1 Contd.

Synorogenic Granite and Gneiss (c.2275 m.y.)
Rakhabdev Ultramafic Suite

	Jharol group	Samlaji formation Goran formation	Dovda group	Devthari formation Depti formation	Nathdwara group Haldighati formation	Rama formation Kadmal formation
	Barilake group	Khamor formation Varla formation Sajangarh formation				Kankroli group
	Udaipur group	Udaipur sector	Sarda sector			Sangat formation puthol formation
		Banswara formation Nimachmata formation	Zawar formation			Rajnagar formation
		Balicha formation Eklinggarh formation	Barolmagra formation			Morchana formation
		Sabina formation	Mandli formation			Madra formation
	Debari group	Debari sector	Jaisamand sector	Ghatol sector		Sarda sector
		Jhamarkotra formation Berwas formation	Babarmal formation Dakankotra formation	Jagpura formation Mukandpura formation		Kathalia formation
	Maton Bus group	Jaisamond formation Delwara formation Gurali formation	Jaismond formation Delwara formation	Jaismond formation Delwara formation Gurali formation		Sismagra formation Natharia ki Pal formation Basal formation
Archaren II > 2500 m.y. Bhilwara group		Unclassified granites and basic rocks				
	Barisadri formation	Ranthambhor group Hora formation		Mandalgarh formation		
		Berach granite and gneiss (2585 m.y.)				
	Jahazpur group	Rajpur-dariba group	Pur-banera group	Sawar group		
	Chuleshwarji formation	Satdudhia formation	Samodi formation	Morhi formation		
	Jhirkri formation	Sindesar Formation	Tiranga formation	Ghatiali formation		
	Umar formation	Dariba formation Malikhera formation	Rewara formation Pur formation			Giyangarh-Asind acidic rock

TABLE 1.1 Contd.

	Intrusives	Untala, Gingla Granites (2950 $\pm$ 150 m.y.)			Raipur Jalayan mafic rock	ultra-mafic rock
Hindoli group Nangauli formation Sujanpura formation Bhadesar formation		Potla formation	Mangalwar Complex	Rajmahal formation	Sandmata Complex Baranch formation Badnor formation Sambhugarh formation	
		Lasaria formation	Suwana formation Kekri formation	Sarada formation		

FIGURE 1.2

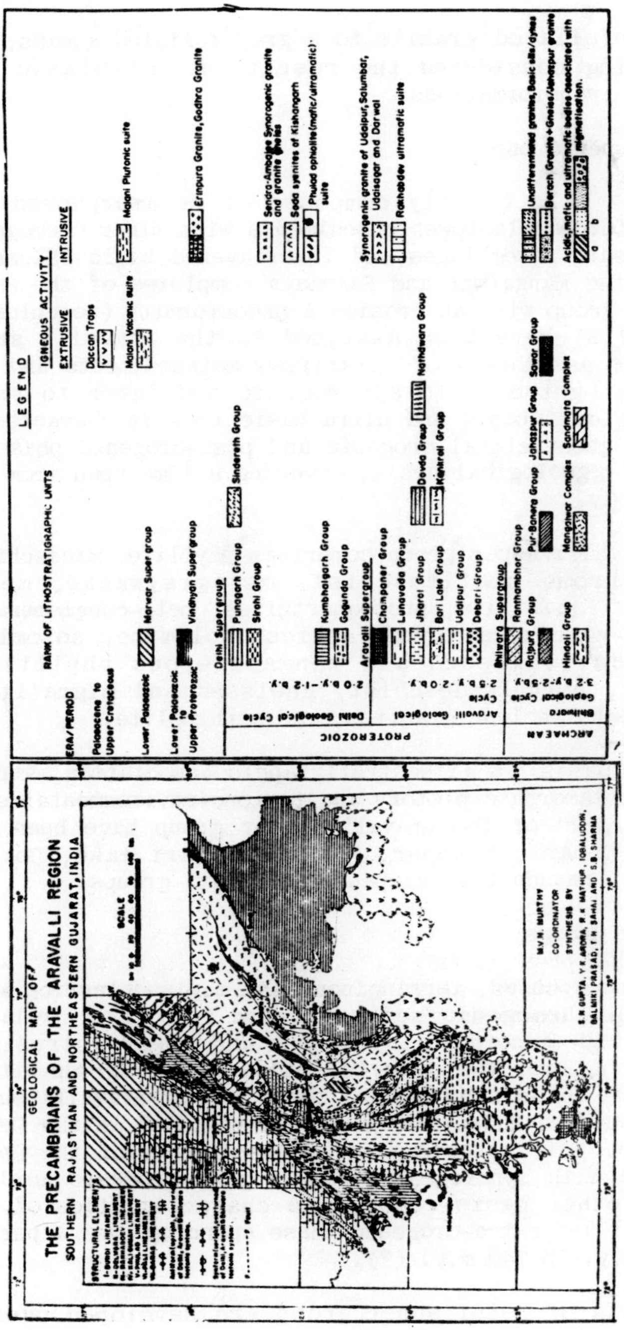
GEOLOGICAL MAP OF
THE PRECAMBRIANS OF THE ARAVALLI REGION
SOUTHERN RAJASTHAN AND NORTHEASTERN GUJARAT, INDIA
SCALE
STRUCTURAL ELEMENTS
M.V.N. MURTHY
CO-ORDINATOR
SYNTHESIS BY
S.N. GUPTA, Y.K. ARORA, R.K. MATHUR, IQBALUDDIN, BALMIKI PRASAD, T.N. SAHAI AND S.B. SHARMA
LEGEND
ERA/PERIOD
RANK OF LITHOSTRATIGRAPHIC UNITS
IGNEOUS ACTIVITY
EXTRUSIVE
INTRUSIVE
Palaeocene to Upper Cretaceous
Deccan Traps
Malani Volcanic suite
Malani Plutonic suite
Lower Palaeozoic
Marwar Super group
Lower Palaeozoic to Upper Proterozoic
Vindhyan Supergroup
Erinpura Granite, Godhra Granite
PROTEROZOIC
Delhi Geological Cycle
2·0 b.y.–1·2 b.y.
Delhi Supergroup
Punagarh Group
Sindreth Group
Sirohi Group
Kumbhalgarh Group
Gogunda Group
Sendra-Ambaji Synorogenic granite and granite gneiss
Soda syenites of Kishangarh
Phulad ophiolite (mafic/ultramafic) suite
Aravalli Geological Cycle
2·5 b.y.–2·0 b.y.
Aravalli Supergroup
Champaner Group
Lunavada Group
Jharol Group
Bari Lake Group
Udaipur Group
Debari Group
Dovda Group
Kankroli Group
Nathdwara Group
Synorogenic granite of Udaipur, Salumbar, Udaisagar and Darwal
Rakhabdev ultramafic suite
ARCHAEAN
Bhilwara Geological Cycle
3·2 b.y.–2·5 b.y.
Bhilwara Supergroup
Ranthambhor Group
Rajpura Group
Hindoli Group
Pur-Banera Group
Mangalwar Complex
Jahazpur Group
Sandmata Complex
Sawar Group
Undifferentiated granites, Basic sills/dykes
Berach Granite + Gneiss/Jahazpur granite
Acidic, mafic and ultramafic bodies associated with migmatisation.

reddish unfoliated granite to a grey foliated gneiss, the latter being considered the result of cataclasic and retrograde transformations.

Aravalli Super Group

A thick pile chiefly composed of metamorphosed and complexly-folded clastogenic sediments with minor chemogenic and oranogenic assemblages and interlayered basic volcanics overlying the Mangalwar and Sandmata complexes of the Bhilwara super group with an erosional unconformity (peneplained pre-Aravallis) have been assigned to the Aravalli super group. This assemblage of stratified metasediments and interlayered together with synorogenic and later to post-orogenic acidic, basic and ultra basic rock is characteristic of the geosynclinal orogenic and post-orogenic phase of the Aravalli geological cycle, covering a time span from 2.5 b.y. to 2 b.y.

The Aravalli super group comprises phyllite, micaschist, garnatiferrous biotiteschist, metagraywacke, metasiltstones, quartzite, protoquartzites, meta-conglomerate, basic meta-volcanics, pyroclastics, dolomite, dolomitic marble, carbonaceous and manganiferous phyllites, calcschist, hornblendeschist, gneisses and migmatites. Plutonic rocks include granites, and amphibolite.

On the basis of lithostratigraphic association, structural and metamorphic history and tectono-environmental setting, the rocks of the Aravalli super group have been assigned to Debari, Udaipur, Kankroli, Bari lake, Dovda, Nathdwara, Lunavada and Champaner (Gujarat) groups.

Delhi Super Groups

The metamorphosed, ferringinous and marine clastics with subordinate chemogenic sediments and temporarily related concordant and discordant intrusives and their intrusives phases resting over the rocks of Bhilwara super group with a first order unconformity, and the rocks of Aravalli super group in southern Rajasthan have been included in the Delhi super group. This assemblage of metasediments of the Delhi super group with synorogenic to late post-orogenic acidic, basic and other basic rocks, are characteristics of the geosynclinal and extra-orogenic phase spanning over a period from 2000 m.y. to 740 m.y. (?).

The rocks of Delhi super group are developed over a

large part of Rajasthan. These rocks extend from Delhi in the north-east to Gujarat in the south-west with their best development in the main Aravalli hill ranges near Ajmer and in western Mewar. North-east of Ajmer, they are best developed around Jaipur and Alwar.

Delhi super group consists dominantly of quartzite, biotite schist, calc schist, calc gneiss, phyllite and marble with subordinate conglomerate and chert. Granites and gneisses, basic meta-volcanics, ultrabasics, hornblende schist and gaboro occur as associated intrusive and extrusives.

The base of the Delhi super group is represented by the Raialo group, which is best developed in north-eastern Rajasthan. The Delhi super group has been divided into Raialo, Alwar, Ajabgarh, Gogunda, Kumbhalgarh, Sirohi, Punagarh and Sindereth groups. Among these the Alwar group is most prominent and is well developed in the hill ranges of the Alwar belt in north-east and in the southern parts of the main Aravalli hill range in the south, but is almost not represented in the middle.

Vindhyan Super Group

The Delhi's succeeded by the Vindhyans after a time interval marked by distinct erosional unconformity. The Vindhyans in Rajasthan were deposited in two separate basins on either side of the Aravalli hill range, constituting the main Vindhyan basin in eastern Rajasthan and Trans-Aravalli Vindhyans, correlated with the former, in western Rajasthan.

Eastern Vindhyans:

The main Vindhyan basin of eastern Rajasthan especially in south-eastern Mewar, the entire Vindhyan succession is well developed. The strata were deposited in a large basin extending from Karauli and Dholpur in the north-east to Nimbhahera and Suket in the south-west. The north-eastern and south-western outcrops are connected by narrow, disconnected patches extending from Sawai Madhopur to Bundi.

In the south-western part of the Vindhyan basin in eastern Rajasthan the formations have reached a maximum thickness of 2400 m, the lower and upper divisions being 1000 m and 1400 m respectively. Arid conditions probably existed during the deposition of the Upper Vindhyans in eastern as well as in western Rajasthan. There are signs that life existed during the Vindhyan period as evidenced by the impressions, considered to be of organic remains, noticed in Suket shales in eastern Rajasthan. The Vindhyans have been affected by overthrusts from the west resulting in a Great Boundary fault of about 800 km strike length, with a very great throw of about 700 m. This has brought the undisturbed upper Vindhyans against the highly folded pre-Aravallis.

Western Vindhyans - Marwar Super Group.

The outcrops of western Rajasthan form two, nearly semi-circular areas, which join near Jodhpur city. The outcrops are much interrupted by aeolian sand and alluvium and are of about 100 m thickness resting on the Malani rhyolites or the Delhi slates. These rocks are grouped into a new super group called the Marwar Super Group. The rock units of this super group are divided into three groups namely, the Jodhpur, Bilara and Nagaur groups, consisting mainly of sandstones, limestones and sandstones with interbedded shales, gypsum and anahydrite. This super group has been provisionally correlated as equivalent to the upper Vindhyans. The lower Vindhyans deposited under marine conditions in eastern Rajasthan are absent in western Rajasthan where volcanic conditions seemed to have existed during this period.

IGNEOUS INTRUSIVES

Ultrabasic Rocks

Ultrabasic intrusives such as periodotite and pyroxite, metamorphosed to talc-serpertine-chlorite rocks, occur in Banswara, Dungarpur, Udaipur, Pali and Ajmer districts. They have been assigned post-Aravalli and post-Delhi ages.

Nepheline-Syenite

Nepheline-syenite is exposed in the vicinity of Kishangarh town and is in the form of sheet-like intrusions into the pre-Aravallis and Banded Gneissic Complex along the margin of the Delhi super group of rocks giving rise to elongated ridges. The pegmatites associated with syenite show large crystals of eleolite and interstitial sodalite (both blue and colourless) and cancrimite. On the basis of radiometric dating (1475 m.y.) this intrusive is considered to be of post-Delhi age.

Post-Delhi Igneous Rocks

In Erinpura granite is the principal intrusive into the

Delhi super group. It occurs as boses and sills of all sizes, although varying in texture and grain size, and having a constant chemical composition, with biotite as the principal ferromagnesium mineral. The outcrops of Erinpura granite, at places isolated, are found extending from Ajmer and Beawar in the north-east to the main belt of the type area of Erinpura and Mount Abu in the south-west. In eastern Mewar, it shows well-developed gneissic structure especially at contact with mica schists.

Malani Suites of Igneous Rocks

The Malani suites of igneous rocks are represented by the Siwana and Jalor granites with their volcanic representatives known as the Malani rhyolites. The Malani rocks are intrusive into Delhi and are later than the Erinpura granite. They show a marked unconformity with the overlying Marwar super group and the Aravallis and hence their age may be post-Delhi, and probably lower Vindhyan. Apart from the Siwana and Jalor granite, a greater part of the Malani suite is composed of rhyolites, felsites, etc. with varying textural gradation, colour and petrological composition.

PALAEOZOIC

There was no sedimentary deposition of significance in Rajasthan between the end of the Vindhyan (uppermost pre-Cambrian or Cambrian) and the beginning of the Jurassic. However, in western Rajasthan in the Jaisalmer district, a sequence of sediments consisting of two formations namely, the Randa and Birmania formations occur over a small area but overlying the Malani suite of igneous rocks. This sequence has been considered to be homotexial with the jodhpur group and Bilara group and this is considered as representing a record of Cambrian sedimentation in Rajasthan.

MESOZOIC AND TERTIARY

From the late Palaeozoic to the early Jurassic period parts of Rajasthan remained as land and this length of large gap permitted these rock formations to undergo extensive erosion leaving only isolated remnants. During the Jurassic age a large part of Rajasthan was covered by a life-supporting sea. Shelf was delineatable into two basins, Jaisalmer and Barmer, which had different litho facies. The Mesozoic period thus witnessed marine transgressions during the lower Jurassic and the upper Cretaceous, the sedimentation having been checked in the Narbada valley, Kutch and Saurashtra by volcanic activity.

With the beginning of the Palaeocene period, tectonic differentiation got intensified and the Palaeocene sea penetrated oblique by, to the Mesozoic strike of beds. The Rajasthan shelf witnessed the Palaeocene sea, the sediments being quite thin and lignite-bearing. The Eocene sediments form a shallow marine sequence, as Marh and Jogira formations in the Palana-Kolayat tectonic embayments, Khuiala and Banda formations in the Jaisalmer basin, and Fatehgarh, Akli, Mandai and Kapurdih formations in the Barmer basin. The Rajasthan shelf area finally got uplifted and remained as land from the middle Eocene onwards. The sequence of rocks of the Mesozoic-Tertiary eras in western Rajasthan is given in Table 1.2.

Deccan Traps

Volcanic activity in the form of fissure eruptions in central and western India marked the close of the Mesozoic era. In Rajasthan the Deccan traps occupy a part of south-eastern Mewar and cover the older formations such as the pre-Aravallis, Aravallis and Vindhyans. The traps are basaltic to doleritic in composition and texture and are uniform over a large area. They are greyish dark to dark green in colour and are commonly composed of plagioclase feldspar and pyroxenes. The rocks generally show spheroidal weathering and exfoliation giving rise to rounded boulders.

RECENT TO SUBRECENT

Alluvium, aeolian sand, kankar, evaporite deposits and soils belong to this group. Alluvium is fairly common and widespread in Bharatpur, Sawai Madhopur and parts of Tonk and Jaipur districts and also along the valleys of the Banas and Luni, Berach and Chambal rivers. A large part comprising most of western Rajasthan is occupied by thick accumulations of aeolian sand constituting the marusthali. These sands cover an irregular rock floor below but local prominences and ridges of Aravalli, Delhi, Trans-Aravalli Vindhyans, Malani, rhyolite and granite are wide spread. Mesozoic and Tertiary rocks rise at places above the aeolian sand. The low-lying ground on the leeward side of the dunes is often marked by calcareous kankar and tufa resulting from the deposition of lime by capillary action.

TABLE 1.2 **The Stratigraphic Sequence in Western Rajasthan**

Age	Jaisalmer Basin	Palana Basin	Barmer Basin
Lower Middle Eocene: Kirthar	Banda formation (75 m)	Jogira formation (121 m)	Kapudi formation (30 m)
Palaeocene-Lower Eocene: Laki	Khuila formation (100 m)	Marh formation (200 m)	Mandai formation (100 m)
Palaeocene Ranikot	Sarur formation (?)	Palana formation (?)	Akil formation (265 m)
Upper Cretaceous	-	-	Fatahgarh formation (50 m)
Lower Middle Cretaceous	Abur formation (60 m)	-	-
Lower Cretaceous	Parihar formation (305 m)	-	-
Upper Jurassic	Bedesir formation (61 m)	-	-
	Baisakhi formation (150 m)	-	-
Middle Jurassic	Jaisalmer formation (130 m)	-	Jaisalmer formation
Lower Jurassic	Lathi formation (330 m)	-	Lathi formation
Permian	Bhadura formation (30 m)	-	-
Upper Carboniferous	Bap boulders spread (G.)	-	-

STRUCTURE

Rajasthan provides a classic example of the pre-Cambrian supracrustal evolution. The sedimentation, magmatism and deformation have followed a characteristic rhythm. The pattern has persisted through space and time. The pre-Cambrian exhibit widely varying structural patterns resulting from repeated periods of deformation and accompanying emplacement of igneous rocks. The regional strike of foliation of the rocks of the pre-Aravalli and Delhi super groups is dominantly to NNE-SSW to NE–SW, varying to N–S and NW–SE in the Aravallis and partly in the pre-Aravallis of southern Rajasthan. The regional structure is defined by the presence of a series of isoclinal, overturned and at places as in the Delhi of the Alwar region, by normal upright folds. The fold axis in the pre-Aravallis of the Pur-Banera Jahajpur-Zawar belt of eastern Rajasthan trends dominantly NE–SW swerving to NS in the Rajpur Dariba and Bhinder belts of southern Rajasthan. On the other hand, the fold axis in the Aravallis of Udaipur-Banswara region in southern Rajasthan lies NW–SE swerving gradually to the NS and NE–SW near Udaipur city and Nathdwara. There are a series of culminations and depressions as well as other structural complications along the Aravalli belt, as for example, the anticlinal culminations in the Sarara dome and the Hammer-head syncline of Kankroli. In comparison with these divergent structural trends in the pre-Aravallis and Aravallis, the fold exis in the Delhi super group along the main Aravalli mountain range and the Alwar-Jaipur-Sikar region trends consistently in the NNE–SSW and NE–SW directions.

Detailed structural studies of the three super groups of pre-Cambrians in Rajasthan suggest that they have been subjected to at least three periods of superposed deformation resulting in varying structural trends. The earliest folds are tight isoclinal folds plunging at about 20°–40° towards WNW. The second deformation which is more intense and prominent has given rise to open to tight upright folds with their plunge varying from 10°–50° towards SW or SSW in the southern parts and NE or NNE in the northern parts. The third deformation is developed over the second generation and exhibits a NNE-ESE trend. A study of these structural elements in different parts of Rajasthan clearly indicates interrelationship in the structural history of different rock groups. The absence of first generation folds between the Delhi's in the Delhi super group masks a major structural break and the Bhilwara/Aravalli super groups of Rajasthan. The other tectonic element in these rocks is the presence of a series of sub-parallel NE-NSW trending faults. The Erinpura granite of Mount Abu region is presumably connected with some of these syn and a post-tectonic lineaments along which they have been emplaced.

In contrast to the complex deformation history of the above super groups, the rocks of the lower Vindhyans extending from Chittaurgarh to Kota and Dholpur on the south-east and eastern Rajasthan are hardly disturbed and are composed of low dipping strata being dominantly towards the south and SSE. Locally, however, they are marked by north plunging open upright folds and flexures as in Rana Pratap Sagar and other areas. The northern boundary of the Vindhyans is also marked by a major fault of importance, namely, the Great Boundary Fault, which has brought the less disturbed Vindhyans against the highly folded pre-Aravalli rocks in Bhilwara, Tonk, Kota, Sawai Madhopur and Bharatpur districts. The Vindhyan rocks along this fault zone are highly folded and locally metamorphosed.

The rocks of the Marwar super group in Jodhpur, Pali, Jaisalmer and Bikaner districts of western Rajasthan are practically undisturbed except for the presence of a north plunging main, basinal, synclinal axis trending NS to NNW-SSE from Sojat city to Bilara and Nagaur.

The rocks of the Mesozoic and Tertiary groups of western Rajasthan are completely underformed and constitute very low dipping strata marked at places by gravity faults.

Lineaments

Rajasthan exhibits a spectacular array of lineaments of varied dimensions and directions which control the morphogeny of Rajasthan to a great extent.

The term lineament is used to indicate linear features of any length on a landsat imagery or aerial photograph. Landsat imageries have proved to be very effective tools for deciphering lineaments. However, a few studies have been made on the relationship of lineaments and drainage. The studies of Vaidyanadhan (1971) and Arunachalam (1981) on the relation between geomorphic lineaments and drainage are worth mentioning.

A regional analysis of the spectral signatures obtained by landsat imageries has revealed numerous lineaments of varying length, intensity and azimuth (Fig.1.3). About 1500 lineaments have been deciphered in various parts of Rajasthan. These have been divided into major, intermediate and minor lineaments based on their spatial extent of more than 300 km, 100 km and less than 100 km respectively.

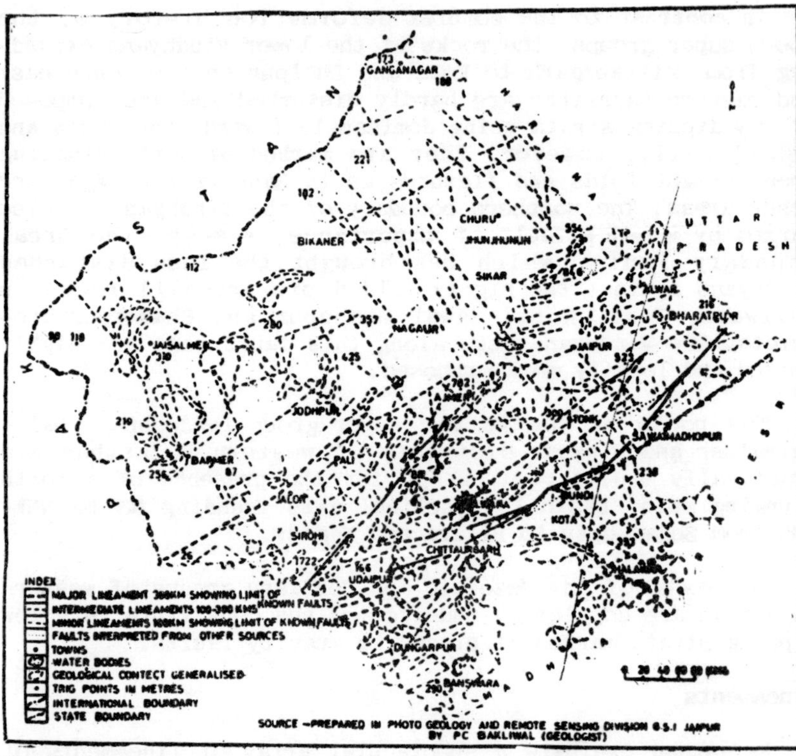

FIGURE 1.3 *Lineament map of Rajasthan—prepared from landsat-IMSS Imagery.*

Four major lineaments of more than 300 km length have been identified in Rajasthan as:

1. **Trans Aravalli lineament** extends over a distance of about 1000 kilometers from the Rann of Kutch beyond Dholpur in a ENE direction, passing along the eastern portion of the river Chambal.
2. **West Aravalli lineament** which extends from Palanpur in the south to Ajmer in the north over a distance of about 500 km in the NNE-SSW direction.
3. **Todabhim-Kakor lineament** which passes through the rocks of the Bhilwara super group in a NE-SW direction over a

length of about 350 km. It has been picked up in a flat terrain which otherwise could not be noticed in the field.

4. **Luni-Sukri lineament** which has been named after the Luni and Sukri rivers. It extends from the east of Jodhpur in the north to beyond Kutch into Pakistan in the south, over a distance of 540 km and lies in the EEN-^{N}WWS to NE-^{N}SW direction.

<u>Intermediate lineaments</u> : There are numerous intermediate lineaments which are mostly seen controlling the drainage pattern in Rajasthan. A few intermediate lineaments between 100 km and 300 km in length are the Chambal, Rikhabdev, Zawar, Agacha, Khetri, Sirohi and Jaipur (Fig.1.3) (Katli, Sota, Mahi, Morel, Berach, Kothari, Banganga and Sabi rivers).

<u>Minor Lineaments</u> : Numerous minor lineaments have been plotted. Most of them show evidence of faulting, the major fracture systems. Quite a few are aligned along the straight courses of the major rivers. Their frequency and intensity are high in southern Rajasthan and very low in western and northern Rajasthan, their intensity has a direct bearing on the structural disturbances in the area.

The existence of lineaments in a south-central Rajasthan in the Udaipur-Banswara area signifies the presence of neotectonics as indicated by the tilting of quaternary river terraces, the deep seated nature of the few linears associated with the emplacement of ultramafics and carbonatite intrusive bodies which have also been confirmed by the geophysical profiles run across the Trans-Aravalli lineament near Palanpur.

TWO

The Nature of Quaternary Deposits

Soils are the product of bio-chemical and physical processes operating upon earth materials under various topographic and climatic conditions. They reflect as much as do landforms, the climatic and geomorphic history of the region in which they involved. Palaeosols and polygenetic soils may furnish information from which it is possible to reconstruct past climatic and topographic conditions. It has been found that palaeosols are generally associated with buried landscapes, with exhumed portions of ancient landscapes and with relict features of previous geomorphic cycles. "Kankar pans" or calchi layers in the soil are interpreted as having been formed during periods of relative aridity. As a matter of fact, detailed studies of soil profiles and soil series along with geomorphic studies may give a fairly complete record of the climatic changes that have taken place in Rajasthan since the early quaternary time down to the present.

The study of the soils of Rajasthan has been undertaken from two points of view (i) as an indicator of past climatic history, and (ii) as a denominator of the morphogenetic regions of Rajasthan. The scheme of soil classification is based on soil taxonomy - a new comprehensive system by the soil Survey Organisation, Department of Agriculture, Jaipur (1975).

The soils of Rajasthan have been formed from the soil material derived from various rock formations under different sets of climatic conditions. The major portion of the soil mass in the state has been formed by alluvium and aeolians derived from different sources of parent material. Alluvium has been derived either from the country rocks of the surrounding areas or from the submountain regions of the Himalavas through fluvial agencies. Aeolian deposits of western Rajasthan have been transported either from coastal areas or from the local material. The alluvium is mostly of mixed minerology having been brought down by the rivers and tributaries from different sources of origin. Alluvium has been greatly modified by or intermixed with aeolian sands.

The comparative study of geology (Fig.1.1) and soil map (Fig.2.2) may give an idea about the relationship of parent material with soil types. Five factors of soil development namely, climate, vegetation, living organisms, relief and time have greatly influenced the parent material of the soils of Rajasthan. These factors simultaneously have acted to produce the soils and lead towards maturity.

CLIMATE

Climate is the most influential factor in determining soil formation both present and past. As has been discussed earlier the state presents a great variety of climatic environments. The western part of Rajasthan is characterized by low rainfall, high temperature and high wind speed: Conditions that are conducive to physical weathering. The soils are usually coarse textured with minimal profile development and in the absence of enough water for percolation through the soil there is less of chemical weathering. Edaphic conditions under low rainfall are not conducive to the addition of organic matter also the high prevailing temperatures inhibit the conservation of organic matter. The evapotranspiration rate is high hence in coarse textural soils evaporation losses are low as the surface soil acts as much owing to low capillarity. In medium and heavy textured surface soils, however, high temperatures cause excessive evaporation. accumulation of salts is also high in such soils. Constant removal and deposition of sand and silt under high velocity winds prevent horizonation of soil profiles. In general, however, the climate in this part of Rajasthan is not conducive to the development of soil profile except in isolated patches.

Rainfall is high in the south and south eastern part of Rajasthan. As a result there is more percolation of water through the soil and the processes of hydration, solution and decomposition are entranced. Chemical weathering is more pronounced in this region under hyperthermic condition. With high percolation rate the processes of elluviation and illuviation in the soil profile are more marked. Relatively higher rainfall also causes finer breakdown of minerals through chemical weathering resulting in medium to fine textured soils. As higher rainfall supports better vegetation and a higher microbial population, organic matter content in the soils is also relatively high. The movement of finer particles and lime to the lower layers has resulted of hard pans or kankar pans in many places.

FIG. 2.1

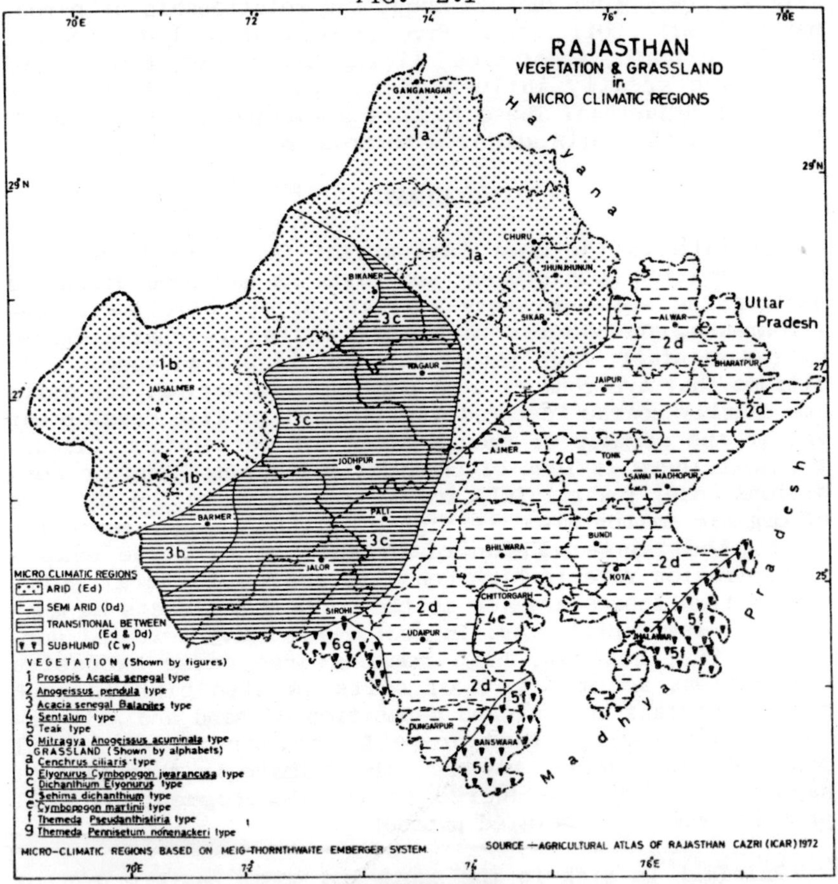

VEGETATION

Vegetation cover modifies climatic influences on the soils. In particular in the erosion of the surface soil by water and wind is prevented by plants. Also the layer of decaying organic matter on the soil surface traps the rainfall and reduces the direct effect of washing away of the soil by flowing water. Thick plant cover reduces the force of winds, thereby lessening wind erosion hazards and also regulating to a certain extent the evaporation rate of water from the soil surface. The spares and scattered vegetation in western Rajasthan does contribute towards the horizonation of soil profiles (Fig.2.1). On the other hand, in the eastern and south eastern parts of the state vegetative cover plays a significant role in reducing soil erosion in the addition of organic residues, and in the development of soil profiles.

LIVING ORGANISMS

The role of living organisms in soil formation has not received much attention in Rajasthan. Unquestionably the kind of climatic zone, the type of soil, the soil moisture content and the replenishment of organic matter do affect microbiological activity (Mathur, al. 1974-75).

TOPOGRAPHY

The sloping lands in the hilly regions facilitate the removal of products of weathering from the upper slopes to the lower pediment slopes. Much of the rain water is also lost through runoff and so these areas have less moisture percolation through the soils. This erosion and runoff result in shallow or truncated profiles of uplands. Washed material is deposited down below in the valleys and depressions where soils are usually deep and of finer textures. In the western part of Rajasthan, slopes have a very minor role to play in the formation and development of soils; the influence is observed in localized depressions and in interdunal flats. Water drained from surrounding areas collect in the lower depressions along with the finer soil particles making these soils in the depressions heavier in texture and such forms are called pan or khadins. On the side slopes of some of the stabilized dunes slight profile development with Cambic horizons can be observed.

The south and south-eastern part of the region east of the Aravallis are mostly hilly and undulating plateau and, therefore, varying degree of soil profile development may be seen. The soils at the foothills, in valleys and depressions, are deep, dark coloured and heavier in texture. They have moderate to well-developed 'B' horizons. Soils of the mountainous slopes and uplands are shallow, gravelly and light yellowish-brown to yellowish-brown in colour. They lack or have a very thin 'A' horizon.

TIME

The influence of time on the soils of western Rajasthan is very limited. Only time is more congenial for development of soils of eastern Rajasthan, where profile development is strong with well-defined horizons.

CLASSIFICATION OF SOIL OF RAJASTHAN

The first attempt to group the soils of the state was made in 1958 by compiling all the available information. This compilation was modified from time to time as knowledge about soils of the state developed. In delineating soil boundaries, soil colour, depth, parent material, soil texture, calcium carbonate content, accumulation of salts, etc. were considered. Although the system is useful in many ways it could not achieve the main purpose of soil classification. With the development of "Soil Taxonomy", the classification of soils on a sound scientific basis was established and the interpretation of soil survey data for planning land use was greatly simplified. Soil surveys in Rajasthan are gradually switching over to this new system of soil classification.

According to the old system of soil classification, soils are grouped into three orders-zonal, inter-zonal and azonal. The zonal order is divided into six suborders with one to many great groups in each suborder. The inter-zonal order is divided into three suborders with three great groups. These groups were further subdivided into various subgroups and further into families, series, types and phases. Roy and Sen (1968) have classified the soils of Rajasthan according to this system.

The schematic classification and grouping of Rajasthan soils into this classification is given in Table 2.1

According to the new classification there are six categories of classification, (i) Order (the broadest category), (ii) Suborder, (iii) Great group, (iv) Subgroup, (v) Family and (vi) Series (the most specific category). The orders proposed in this system are believed to cover all the soils of the world. With the exception of the order-entisol, which roughly corresponds to the azonal soils of the old system, the new orders show little resemblance to those formerly used.

TABLE 2.1 **The position of soils of Rajasthan in the whole system of soil classification**

Order	Suborder	Great Soil Group	Great group in Rajasthan
Zonal	a. Soils of the old zone	Tundra	
	b. Light coloured podzolised soils of timbered regions	Podzol soils, brown, podzolic soils, grey brown podzolic soils, red yellow podzolic soils, grey wooded soils	
	c. Soils of forested warm temperatre and tropical regions	A variety of latosols are recognized. They await detailed classification	
	d. Soils of forest grass land transition	Degraded chernozem soils, non-calcic brown or shantung brown soils	
	e. Dark coloured soils of semi-arid, subhumid and humid grass lands	Prairic soils, reddish prairic soils, thernozem soils, chestnut soils, reddish chestnut soils	
	f. Light coloured soils of arid regions	Brown soils, brown soils Serozem soils, red desertsoils	Brown soils saline phase Sierozem red desertic soils
Intra zonal soils	a. Hydromorophic soils of marshes, swamps, flats and seepage areas	Humic-grey soil. alpine meadow soils, bog soils, ground bog soils, planosols, ground water podzols, ground water latosols	-
	b. Halmorphic (saline and alkali) soils of imperfectly drained arid regions, litoral deposits	Solanchaks (saline soils), Solonetz soils (Alkali soils) Soloth soils	Saline soils of depression
	c. Calcic Imporphic soils	Brown forest soils Rendzina soils	Red loam
Azonal (no sub-orders) soil		Lithosols	Red and yellow soils, yellowish brown soils of foot hills.
		Regosols (Includes dry sands)	Desert soils, sand dunes
		Alluvial	Old alluvium, recent alluvium black soils.

Some soils of Rajasthan have been classified according to this new classification (Raychaudhari, 1963, FAO reports (1971), Mathur and Shankarnarayana, (1972); Shankarnarayana and Hirekarur,(1972). The soils of the state have been grouped under five orders, namely, Aridisols, Alfisols, Entisols, Inceptisols and Vertisols. Soils have further been subdivided at suborder and great group levels under these five orders (Fig. 2.2). The names of the soil orders occurring in the state and their subdivisions up to the great group level along with their equivalents in the old system of soil classification are given in Table 2.2.

Fig. 2.2

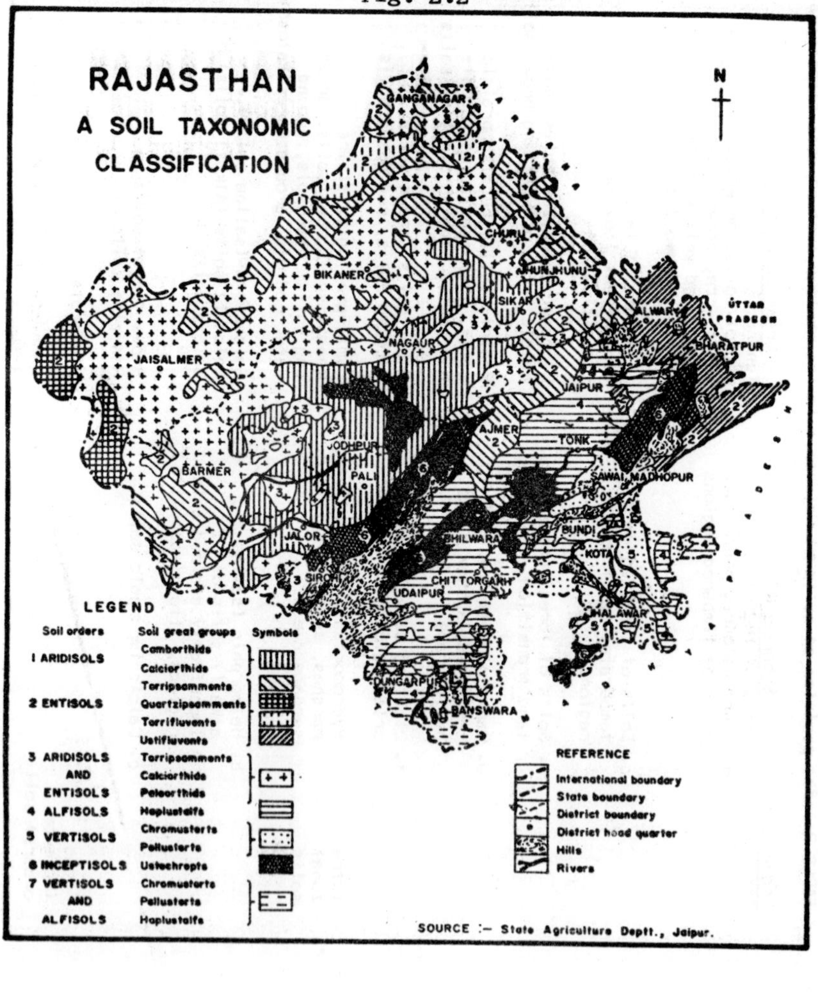

TABLE 2.2. **Approximate equivalents of old and new system of soils classification**

New comprehensive system			Approximate equivalent in the old system
Order	Suborder	Great group	
1. Aridisols	Orthids	a. Camborthids b. Calciorthids c. Salorthids d. Paleorthids	Seirozems, desert soils Saline soils of depression
2. Alfisols	Ustalfs	a. Haplustalfs	Red loam, black soils, brown soils, yellowish brown soils of foot hills, alluvial soils
3. Entisols	Psamments	a. Torripsamments b. Quartzipsamments	Desert soils and sand dunes
	Fluvents	c. Torrifluvents d. Ustifluvents	Alluvial soils (old & recent)
4. Inceptisols	Ochrepts	a. Ustochrepts	Brown soils, red and yellow soils of foot hills
5. Vertisols	Usterts	a. Chromusterts b. Pellusterts	Black soils

Source : G.M. Mathur ed al 1974-75 Monograph-2 Soil Survey Organization, Department of Agriculture, Rajasthan, Jaipur.

Aridisols

Aridisols occur extensively in the western half of Rajasthan divided by the Aravalli axis and some parts of Alwar, Jaipur and Ajmer districts which fall on the eastern half. Aridisols are associated with soils of the order entisols.

Aridisols are mineral soils mostly found in dry climates. Except where there is ground water or irrigation, the soil layers are dry throughout for most part of the year. Consequently, they have not been subjected to intensive leaching. The surface horizon or horizons are normally light in colour and are of a soft consistency when dry. If textures are fine, individual pads may be hard. They have one or more diagnostic horizons owing to the accumulation of salts, carbonates or silicate clays or of cementation by carbonates or silica. There may be alternation with the parent material without any significant accumulations. Aridisols have a low percentage of organic carbon.

Most of the surface is bare for most of the time or if gravel is present in the parent material, the surface has a desert pavement that is formed by deflation of the fine earth. The vegetation consists of xerophytic plants.

The major portion of aridisols in Rajasthan is made up of the suborder orthids which has one or more pedogenic horizons, but has no argillic or nitric horizon. Orthids commonly have horizons of accumulation of soluble salts and carbonates. They are salic, calcic, gypsic, petrocalcic, petrogypsic or cambic horizons and duripans, singly or in combinations. These horizons form the basis for defining the great groups of this suborder. Camborthids, calciorthids, salorthids and paleorthids are the great groups that commonly occur in Rajasthan. These are briefly discussed here.

Camborthids

Soils of this great group occur in parts of Jalore, Churu, Jodhpur, Pali, Barmer, Sikar and Jhunjhunu districts of Rajasthan and are associated with calciorthids. Without profile examination it is difficult to distinguish between these two great groups. Besides calciorthids, camborthids are associated at many places with other soils of the desert region.

Camborthids have a cambic horizon. They are brownish to reddish in colour and have relatively uniform textures with depth except if the soils are stratified floodplain soils. They have light coloured surface horizon over an altered horizon (cambic) that generally is red or brown or a horizon with well-developed soils structure. There is either no pronounced horizon of carbonate accumulation (a calcic horizon) nor has the soil any free carbonates thicker than 18 cm in the upper horizons. Camborthids are the youngest of the Aridisols, though they may be some what older if they had a few or no carbonates in the parent material and receive little calcareous dust. The cambic horizon is generally prismatic with the tops of the prisms only 2 cm to 5 cm below the surface. The prisms are weak and can easily be overlooked. Morphological description of a typical pedon of this group has been described here.

Horizon	Depth (cm)	Morphological description
A_0	0-18	Yellowish brown, loam, granular, many thin roots, no effervescence with dilute acid, well-drained
B_2	10-65	Weak red, loam, granular, slightly friable, many roots, no efferves-

		cence with dilute acid, well-drained
B_2^1	65-115	Light yellowish-brown, loam, coherent, slightly friable and gritty, very few thin roots, profuse effervescence with dilute acid, lime in fine form and in streaks, few concretion, weathered granite and calcigneisic, well drained
C_1	115-160	Yellowish-brown, loam, coherent, friable and gritty, profuse effervescence with dilute acid, lime in fine form and in streaks, many concretions 5-10 mm weathered rock material with lime

Source : Soil Survey Organization, Department of Agriculture, Jaipur.

At the subgroup level typic camborthids cover most of the area of this great group. At the family level camborthids of coarse loamy to fine loamy mixed, calcareous, hyperthermic classes have been identified.

Calciorthids

Calciorthids occur in Rajasthan in parts of Jalore, Jodhpur, Pali, Barmer, Nagaur, Sikar, Churu, Sriganganagar and Jhunjhunu districts where they are associated with camborthids and salorthids.

These soils have a calcic or gypsic horizon that has its upper boundary within one metre of the surface but have no petrocalcic horizon or duripan that has its upper boundary within one metre the surface. These soils also do not have a salic horizon above the calcic horizon, they may or may not have a cambic horizon. The lime may have come from the parent material or from atmospheric dust. The lime in the calcic horizon may be soft land powdery or hard as pendants from gravels and this is thought to be the earliest stage of the formation of the calciorthids. With continuing accumulation of lime, the pores are filled and the calcic horizon becomes plugged with lime and gets cemented. If still later stages, the percolating water is stopped by the calcic horizon and the lime is deposited in low places as laminae until the surface is smoothed. This is the petrocalcic horizon, recognized as the basis for the petrocalcic calciorthids. The limited rainfall is unable to remove the lime completely from the surface unless the textures are sandy.

The calciorthids recognized in this region are mostly typic. At the family level the calciorthids are identified as coarse loamy or fine loamy, mixed, hyperthemic classes. A typical pedon of this group is described here:

Horizon	Depth (cm)	Description
A_1	0–17	Dark yellowish-brown, fine sandy loam, single grain, friable common fine pores, common fine roots, calcareous clear smooth boundary, ($CaCO_3$: 5.4%)
C_2	17–42	Dark yellowish-brown, fine sandy loam, single grain, friable, common fne pores, common fine roots, calcareous, =clear smooth boundary, ($CaCO_3$: 9.8%)
C_3	42–69	Yellowish-brown, fine sandy loam, very weak fine angular blocky, friable common fine pores, common fine roots, few hard $CaCO_3$: 18.0%)
C_4Ca	69–114	Pale brown, fine sandy loam, very weak fine angular blocky, fine roots, frequent hard $CaCo_3$ concretions up to 60–80 mm calcareous, abrupt smooth boundary ($CaCo_3$: 10.0%)
C_5	114–141	Yellowish-brown, loamy fine sand, single grain, friable, common fine pores, common fine roots, very few hard $CaCo_3$ concretions upto 5 mm, calcareous, clear smooth boundary, ($CaCo_3$: 7.0%)
C_6	141–175	Yellowish-brown, fine sand, single grain, friable, many fine pores, few very fine roots, calcareous, ($CaCo_3$: 3.4%)

Salorthids

Soils of this great group of salorthids are found scattered in small patches in association with torrifluvents and calciorthids in the Ghaggar floodplain in Sriganganagar district. It is difficult to separate out the boundaries of these soils from the soils of other great groups. Salorthids also occur in small areas near the Sambhar, Pachpadra, Deedwana lakes, the little Rann of Kutch area in Jalore and in

Barmer district.

These are the very salty soils of wet places in the desert where capillary action and evaporation concentrate the salts into a salic horizon. They are primary soils of depressions where the groundwater saturates the soil at some time of the year in some subhorizons within one metre of the surface. The concentration of salts is usually within 75 cm and very often it is within a few centimetres of the surface. Salorthids have at least two percent salt in some subhorizons and the product of the thickness of horizon in centimetres and percentage of salt by weight is 60 or more.

While the salorthids may have water held at tensions of more than 15 bars in some surface horizons, in most soils the water is held at much lower tensions but the osmotic pressure owing to their salts content makes the soils physiologically dry.

A typical pedon of this great group is described here:

Location Rajasthan Canal Project topographic map No. 64, 100 metres north-east of stone No. 144E/3975. The profile is situated in a level irrigated field.

Horizon	Depth (cm)	Description
A_{11}	0-15	Brown silty clay loam, moderate medium subangular blocky, hard many very fine pores, very few very fine roots, common quartz (3 mm), calcareous, clear smooth boundary
A_{12}	15-45	Brown silty clay loam, moderate medium subangular blocky, firm moist, many fine pores, very few very fine roots, common fine yellowish brown mottlings, calcareous, clear smooth boundary
C_1sa	45-74	Brown silty clay, moderate coarse angular blocky, very firm moist, many very fine pores, many medium faint yellowish-brown mottlings, calcareous, clear smooth boundary
C_2	74-95	Brown silty clay, moderate coarse angular blocky, very firm moist, many very fine pores, many medium yellowish, dark brown, very dark

		brown, prominent mottlings, calcareous, abrupt smooth boundary
C_3	95-160	Brown, silty clay, moderate coarse angular blocky, very hard, dry, many very fine pores, many fine to medium prominent yellowish, dark brown, very dark, brown mottlings, calcareous gradual smooth boundary
C_4	160-171	Very dark greyish-brown fine sandy loam, coherent, hard dry, many very fine to fine pores, calcareous, clear smooth boundary
C_5	171-181	Dark greyish-brown, greyish-brown loamy fine sand, coherent, slightly hard, dry, many fine to fine pores, calcareous, clear smooth boundary
C_6	181-190	Yellowish-brown loamy fine sand, coherent slightly hard dry, many very fine to fine pores, common medium distinct mottlings, calcareous

Source : UNDP/FAO Technical Report 1.9 (1971).

Paleorthids:

Paleorthids occur in association with calciorthids and torripsamments in the western part of Rajasthan, they are scattered and cover proportionate a small area.

These soils have a petrocalcic horizon. Given a parent material that is rich in carbonates, or given a regular additions of carbonates in dust, the calcic horizon in time tends to become plugged with carbonates and cemented into a hard, massive continuous horizon which is referred to as a petrocalcic horizon. Ordinarily, a petrocalcic horizon has its upper boundary close to the surface and is quite thick. These soils generally occur on gentle slopes that have been stable for long periods. The soils are calcareous throughout and the colour ranges from reddish-brown to nearly white. These soils occur in small patches around Jaisalmer and in some parts of Barmer and Nagaur.

Alfisols

Alfisols occur in parts of Jaipur, Alwar, Bharatpur, Sawai Madhopur, Tonk, Bhilwara, Chittorgarh, Banswara, Udaipur, Dungarpur, Bundi, Kota and Jhalawar districts.

Alfisols are mature soils with medial to maximal profile development. These soils have an argillic horizon, moderate to high base saturation and water held at less than permanent wilting point (15 bars) during at least three months in the year when the soil is warm enough for plants to grow. Alfisols generally have a period when evapo-transpiration exceeds precipitation and one or more horizons drop well below field capacity, or reach wilting point. This is the normal moisture regime of soils with argillic horizons. Water movement through the solum is adequate to remove free carbonates from the fine earth in the epipedon and from most of the argillic horizons but inadequate to remove a substantial part of the exchangeable bases held by the soil. The rate of liberation of bases by weathering or addition of bases by wind or water, approaches the maximum capacity of leaching to remove bases the leaching of bases from the soil may occur almost every year or may be infrequent. In addition to these properties, alfisols may have fragipans, duripans, natric horizons, petrocalcic horizons, plinthite or other features that are used to define the various great groups.

Most of the area covered by alfisols in Rajasthan have only one suborder, ustalfs. The area in which this suborder occurs in the state has a warm rainy season, with the moisture moving through the soil to deeper layers only in occasional years. Haplustalfs form the great group of this suborder.

Haplustalfs:

The soils of haplustalfs occur in the eastern and south-eastern part of Rajasthan in Udaipur, Chittorgarh, Bhilwara, Ajmer and Jaipur districts.

These soils have a gradual or clear upper boundary of argillic horizon but no natric horizon they have no duripan with its upper boundary within one metre of the surface, and no petrocalcic horizon with its upper boundary within 1-5 m of the surface. Haplustalfs are reddish to brownish in colour.

The different haplustalfs recognized in the area typic, vertic, lithic haplustalfs. At the family level all the haplustalfs are coarse to fine loamy, mixed, hyperthemic, calcareous or non-calcareous. A typical pedon of this great group has been described here:

Horizon	Depth (cm)	Description
A_p	0-15	Yellowish-brown, loam, crumb, loose, many thin and thick roots, very mild effervescence with dilute acid, well-drained
B_2t	15-55	Dark yellowish-brown, loam, blocky, friable many thin roots, well-drained, few small medium and big white concretions, mild effervescence
C_{ica}	55-137	Light yellowish-brown, structureless, weakly cemented, very few thin roots, well-drained, lime in mixed form
C_{2ca}	137-180	Light yellowish-brown, loam, massive, friable, no roots, moderately drained, profuse effervescence, lime in powder form

Source : Soil Survey Organisation, Department of Agriculture, Jaipur

Entisols

Entisols are probably the dominant soils in the state. Psamments cover a large area in association with aridisols in western and central Rajasthan. Almost all the districts falling in the western half of the state are covered to a lesser or greater extent by these soils. Orthents cover a large area in Barmer and small areas in Jaisalmer and Bikaner districts. Fluvents are found in Sriganganagar, Alwar and Bharatpur districts.

Entisols lack well-developed horizons and have minimal profile development. Soils have poor development of horizons. Only fragments of which can be identified. Many soils ochric epipedon. The reasons for the lack of horizons vary. Some soils have been recently deposited and the time has been too short for horizons to be formed. Other soils are on steep, actively eroding slopes, or on floodplains that receive new deposits of alluvium at frequent intervals. Some Entisols, however, are very old and are mostly composed of quartz or other resistant minerals that do not alter to form horizons.

In Rajasthan, the important suborders of Entisols encountered are psamments, orthents and fluvents.

The psamments consist of two great groups, the Torripsamments and the quastzipsamments.

Torripsamments:

Torripsamments occur on sand dunes and in flood plains. The dunes cover up to 80 per cent of the area at many places in western Rajasthan. In this area torripsamments are found either as pure units or in association with calciorthids, camborthids and paleorthids. They also cover a part of the area extending over Jaipur, Alwar and Ajmer districts which fall on the east of the Aravallis.

These soils are characteristically of arid climate. They may have any mineralogy: quartz, mixed or even gypsum, and may have any colour.These soils have a low water holding capacity and when dry are devoid of vegetation and are subject to blowing or drifting. Vegetation consists of a few ephemeral plants or a partial cover of xerophytic plants.

These soils fall under the typic and ustic subgroups in some places. At the family level all the torripsamments are of the sandy or coarse loamy, mixed hyperthemic family.

A typical pedon is described here:

Location	Rajasthan canal project area topographic map P^4-10 near stone $P^4$239/5 (approximately 20°40' N and 71°15'E)

Horizon	Depth (cm)	Description
C_1	0–10	Light yellowish-brown, fine sand, non-coherent, loose, many fine pores, common very fine roots, slightly calcareous, gradual and smooth boundary
C_2	10–145	Pale brown, very friable loamy, sand, non-coherent, common very fine continuous pores, calcareous, few lime concretions, few very fine roots, abrupt and wavy boundary
C_3	145–170	Pale brown, loamy fine and non-coherent, friable, many fine pores, very few fine roots, contains 15-20% calcium carbonate concretions up to

		1 cm in diameter
C_4	170-230	(Augering) fine sand

Source : Rajasthan State Soil Survey Unit, Department of Agriculture, Bikaner.

Quartzipsamments

Soils occurring on the western border of the state covering small part of Jaisalmer and Barmer districts come under this great group of quartzipsamments. They occur in association with soils of other great groups especially torripsamments and other great groups of aridisols.

Quartzipsamments are freely drained quartz sands which may be white or stained with shades of brown, yellow or red. The important property of such a soil is that the first two metres or so is uncoated quartz sand. The sand fraction is to the extent of 95 per cent or more, composed of quartz, zircon, tourmaline, rutile or other normally insoluble crystalline minerals that do not weather to liberate iron or aluminum.

Two great group, torrifluvents and ustifluvents are found under the suborder - fluvents.

Torrifluvents:

Torrifluvents occur in association with calciorthids and salorthids in the Ghaggar floodplain area of Sriganganagar district. Stratification of the material is normal in these soils and organic carbon is irregularly distributed in the profile. These soils are very salty at many places.

Torrifluvents are the fluvents that have a torric or aridic moisture regime, they comprise recent water deposited sediments, primarily in floodplains, alluvial fans and deltas of rivers but not in back swamps where drainage is poor. These soils are mostly alkaline or calcareous and generally saline under natural conditions.

A typical pedon is described here :

Location - Near Rawatsar (Sriganganagar) about 75 m north of stone No. 153E/4038. Eastern block area of Rajasthan Canal Project

Horizon	Depth (cm)	Description

A_p	0–16	Light yellowish-brown, fine sandy, loam, weak fine granular, slightly hard when dry, friable when moist, many fine pores, abundant fine to very fine roots, very few lime nodules
C_1	16–47	Dark yellowish-brown, fine sandy loam, weak fine subangular blocky, friable moist, many fine pores, abundant fine to very fine roots, few lime nodules (3 mm), calcareous, gradual smooth boundary
C_2	47–75	Yellowish-brown, fine sandy loam, weak fine subangular blocky, friable moist, many fine pores, abundant fine to very roots common lime nodules (5 mm), calcareous, gradual smooth boundary
C_3	75–111	Yellowish-brown, moist, fine sandy loam, weak fine subangular blocky, friable moist, many fine pores, common lime nodules (7-10 mm) calcareous, gradual smooth boundary
C_4	111-152	Yellowish-brown, fine sandy loam, weak to moderate medium angular blocky, friable moist, many fine pores, few very fine roots, common to abundant lime nodules (10-15 mm), calcareous, abrupt smooth boundary.
IIC_5	152–172	Olive, silt loam, moderate medium angular blocky, firm moist, many fine pores, many yellow, distinct mottlings, calcareous, clear smooth boundary
IIC_6	172–200	Olive, silt loam, moderate medium angular blocky, firm moist, many fine pores, many yellow distinct mottlings, calcareous

Source: Rajasthan state Soil Survey Unit, Department of Agriculture, Bikaner and UNDP/FAO Tech. Report 1.9. (1971).

Ustifluvents:

The soils that belong to the ustifluvents great group occur on the north-eastern and eastern portion of Rajasthan covering parts of Alwar and almost the whole of Bharatpur district. These soils are subject to constant flooding during the rainy season. Ustifluvents are similar to torrifluvents described earlier except that they occur in a comparatively high rainfall areas and have an ustic moisture regime. These soils show irregular distribution of organic carbon, calcium carbonate, clay and silt in the profile.

Ustifluvents of Rajasthan at the subgroup level fall in the category of typic, vertic, aquic ustifluvents. A typical pedon of this great group has been described here:

Horizon	Depth (cm)	Description
A_p	0–25	Light yellowish-brown, sandy loam crumb, slightly hard, many thin and long roots well-drained
C_1	25–60	Brownish yellow, sandy loam, granular, slightly firm, many thin and long roots, well-drained
C_2	60–132	Yellowish-brown, sandy loam, single grain, slightly friable, many thin and long roots, well-drained
C_3	132–180	Yellowish-brown, sandy loam, single grain, friable, many thin and long roots, very mild and localized effervescence few small rounded kankars, well-drained

Source : Soil Survey Organisation, Department of Agriculture, Jaipur.

Inceptisols

Inceptisols occupy a major area in Rajasthan along the foothills of the Aravallis in the districts of Sirohi, Pali, Udaipur, Bhilwara, Chittorgarh and in some parts of the alluvial plains in Jaipur, Alwar, Sawai Madhopur and Jhalawar.

Inceptisols have altered horizons that have lost some bases or iron and alluvium. These soils lack argillic spodic, natric, gypsic or petrogypsic horizons. The general sequence of horizons is an ochric epipedon over a cambic horizon with or without a duripan and fragipan.

In Rajasthan, as stated above, these soils occur dominantly along the foothills of the Aravallis and the Vindhyans. Most of the soils have an ochric epipedon and occur in an ustic moisture regime. Such soils are classified under the great group ustochrepts.

Ustochrepts:

Ustochrepts are the reddish and brownish soils of the semi-arid regions in Rajasthan and mostly occur along the foothill slopes. They are generally calcareous and shallow to deep. These soils cover extensive areas in Bhilwara, Udaipur, north-western Sawai Madhopur and in parts of district Sirohi, Pali, Jaipur and Chittorgarh. The soils have an ochric epipedon over a cambic horizon.

The common subgroups are typic, lithic, paralithic, vertic and fluventic ustochrepts.

Vertisols

Vertisols cover a large area in Jhalawar, Kota and Bundi districts and a smaller area in Sawai Madhopur, Bharatpur, Dungarpur, Banswara and Chittorgarh districts.

Vertisols are clayey soils that have deep wide cracks at some time of the year. They have many common properties associated with their fine texture. Besides the high shrink-swell potential, they are compact and very slowly permeable in the lower layers. They are very sticky and plastic when wet and have high water holding capacity but low available water. The base exchange capacity is moderately high to high. The changes in the soil moisture seems to be essential to the genesis of vertisols. They may form in alluvium or in plains where temperatures are usually high and where there is a marked hot dry season during the year. The average high temperatures the average medium rainfall, the nature of the clay minerals (expending type), the changing moisture conditions caused by seasonable fluctuations and temporary water stagnation in smooth relief cause deep cracking and churning of soils, the promotion of slicken sides, and in some cases, self, Mulching. These open cracks may be one centimetre or more wide at a depth of 50 cm or more. Normally the cracks are open to the surface, but if the soils are under cultivation, the cracks may open only up to the base of the plough layer.

Owing to the churning of soil, horizon differentiation is usually not well-defined. Vertisols may or may not have any diagnostic horizon. Nevertheless, variations in moisture regimes are wide, as are also the associated colours. The soils have variable base saturation, content of carbonates and deep depth to lithic or paralithic contact.

Under the order vertisols is the suborder usterts whose associated great groups viz. chromusterts and pallusterts are described here.

Chromusterts

Chromusterts cover a large area in association with pellusterts in parts of Bundi, Kota and Jhalawar districts in the south-eastern part of Rajasthan, and to a lesser extent in Chittorgarh, Dungarp r, Banswara and Bharatpur districts. Soils are classified as deep and medium black soils in the earlier soil classification.

In chromusterts, the creacks are open for 90 cumulative days or more in most years but are closed for 60 days consecutive days or more in most years at a time when the soil temperature is continuously above 8°c.

Chromusterts are divided into the subgroups entic, paleustollic and typic chromusterts. The chromusterts in the state belong to clayey to fine clayey, calcareous or non-calcareous, mixed or montmorillonitic, hyperthemic families.

A pedon description of typic chromusterts is given here:

Location - 2 km east of Malpura (Panchayat Samiti, Ladpura), Kota. The profile was located in a cultivated field.

Horizon	Depth (cm)	Description
A_p	0-10	Dark-brown clay, dark-brown moist, medium moderate crumb, hard, friable, sticky, abundant very fine roots, common fine and very fine roots, common fine and very fine continuous inped vertical interstitial open pores, non-effervescent, diffuse boundary, PH 8.9 (10 cm thick) (49.4% clay)
	10-56	Dark-brown moist, clay, medium moderate subangular blocky, friable, sticky, abundant very fine roots, many fine continuous inped vertical interstitial open pores, non-effervescence common slicken sides,

		diffuse boundary, PH 8.6 (46 cm thick) (50.4% clay)
C_2	56-115	Dark-brown moist clay, medium moderate subangular blocky, firm, very sticky, abundant very fine roots, many fine continuous oblique inped tubular open pores, common slicken sides, non-effervescent, diffuse boundary, PH 8.4 (59 cm thick), (51.4% clay)
C_3	115-140	Dark-brown moist, clay, medium coares subangular blocky; very sticky, few very fine roots, many fine continuous, oblique inped tubular dendriticpores, very slightly effervescent, PH 8.4 (25 cm thick), (51.4% clay)

Source : Department of Agriculture, Soil Survey Unit, Kota (1973).

Pellusterts

These pellusterts occur in association with chromusterts over a large area covering Bundi, Kota, Jhalawar, Chittorgarh, Dungarpur, Banswara and Bharatpur districts.

These soils are dominantly grey to black in all subhorizons throughout the upper 30 cm in more than half of each pedon. Surface horizons may range from black to grey. Pellusterts occur in level plains and depressions. These soils crack and the cracks remain open for 90 cumulative days or more in most years but are closed for 60 consecutive days or more in most years at a time when the soil temperature is continuously above 8°c at 50 cm depth.

A pedon description of typic pellusterts is given here:

Location- Kotra (Jhalawar district and tehsil Jhalrapatan)

Horizon	Depth (cm)	Description
	0-	Dark grey, clay, blocky, slightly hard, many thin roots, mild effervescent, well-drained, many small white concretions.
	22-50	Very dark greyish-brown, clay, blocky, firm, many thin roots, mild

	effervescent, well-drained, many small white concretions
50-107	Very dark-brown clay, blocky, firm, many thin roots, mild effervescence, moderately drained, many small and few medium white concretions. very few rusty mottlings
107-157	Very dark-brown, clay, blocky, firm, mind effervescence, moderately drained many small and few medium white concretions, few rusty brown mottlings.
157-180	Very dark-grey, clay, massive plastic, mild effervescence, imperfectly drained, few small yellow concretions, many rusty-brown mottlings.

Source : Mehta et al... (1962).

THREE

Present Climate

Climatic factors which control all dynamic processes, take precendence over all structural components. Every climatic zone has its own specific complex of dynamic processes in terms of present day activity. This means that the significance of the structural features of the earth's crust changes in relation to current dynamic processes. In different climatic zones, for example, the morphological hardness of rocks differs. This is also the case in Rajasthan. It has been observed that the exogenic processes operating today in the state are in accordance with the present climate. The action of wind is particularly predominant in arid Rajasthan while wind plays a minor role in the south-eastern part of the state. The fluvial process produces a sequence of landform assemblages in eastern and southern Rajasthan, but has little effect in the western region. If one travels from the eastern part of the state towards the west, a network of gullies on the foothills on established dunes and along the stream courses can be observed but towards the west of the Aravallis this feature is completely absent and is replaced by the features produced by the work of strong wind erosion. Similarly, physical weathering is most dominant in western Rajasthan while it is moderate in semi-arid eastern Rajasthan and minimum in the southern and eastern parts of the state. Chemical weathering is negligible in the arid parts of Rajasthan while physico-chemical and bio-chemical processes are predominant in the south and south-eastern regions.

Although Rajasthan exhibits several geomorphic forms such as pediments, playas, fossilized dunes, older river cources, older floodplains and terraces that seem to be abnormal features in the present day climate. But these features are the products of the earlier climate of the Pleistocene period. Therefore, to appreciate the past climates of

Rajasthan it is necessary to analyse present day climatic conditions.

Rajasthan lies within a region which according to the classification of Koppen (1931), Thornthwaite (1931) and Meigs (1953) may be described as having arid (Ed), semi-arid (Dd), transitional between Ed and Dd, and subhumid (CW) zones (Fig.2.1). The area under study receives limited rainfall. The rainfall variablility in high and evaporation losses are very high. Almost 90 per cent of the rainfall occurs between the months of June and September. During the remaining months of the year rainfall is light and sporadic and several consecutive rainless months and perhaps years are not uncommon in the arid part of the state. Annual and daily ranges of temperature are high. Marked seasonal contrasts in humidity, cloud, wind speed and velocity and radiation are significant.

METEOROLOGICAL CONDITIONS

Rajasthan occupies the west north-west portion of India and hence the pattern of atmospheric circulation with slight variations is the same as in north-west India. The principal seasons can be divided into

- Winter (January and February).
- Pre-Monsoon (hot weather—March, April and May).
- South-west monsoon (rainy season—June, July, August and September).
- Post-monsoon (retreating monsoon—October, November and December).

Post-Monsoon and Winter

The normal date of the withdrawal of the monsoon is between 15 September and 30 September when an anticyclone starts forming over the state. This anticyclone is on the periphery of the Great Siberian high and, therefore, temperate continental air prevails over the region. The monsoon period of a shallow low that so far prevailed over Rajasthan is slowly transformed to a high. The pressure gradient round the Siberian high is strong but it gradually weakens down towards the south. As a result of the high pressure prevailing over Pakistan, Rajasthan and adjoining areas, the upper winds at all levels become westerly or north-westerlies. These north-westerly winds swing with the high and appear as north-easterlies i.e. south of Rajasthan. Being a part of the temperate continental air mass, the wind is practically dry with low humidity and so the sky is cloudless and morning temperatures are low.

A jet stream make its appearance in late October at about 200 mbs (12 km) and as the season advances it comes as low as 300 mbs.

The western disturbances start making their appearance from late November. These western disturbances are the cause of the winter rainfall over Rajasthan. These induced lows have no frontal features and the normal tracks are in the ENE direction. These disturbances entrap the same polar continental air mass which causes the cold wave. In the month of november there is an overlapping of the western disturbances affecting western Rajasthan. These disturbances increase from two to three in December but in some years there may be none at all. The movement of these lows is very slow— five degrees in a day, sometimes even less. Faster movements occur under the influence of a mid or upper tropospheric trough. The rainfall is sharp and of a short duration, sometimes accompanied by hail owing to the low freezing level. Heavy rains occur only when an occluded front is formed. Inversions are a common feature during this period hence mist prevails in the morning. Fog occurs in the wake of the western disturbance after rain and rapid clearance of clouds. In the winter months, namely, January and February, the Siberian high strengthens itself and therefore, greatly influences the wind pattern of Rajasthan. By now the anticyclone establishes itself over Rajasthan and the temperate continental air mass, which has now become cooler and drier, prevails over the region. The winds are strong westerly and north-westerly and temperatures fall sharply during the night and at many places reach zero degree celsius. The frequencies of the western disturbances increase from January to February, on an average rising to four. After the passing of the western disturbances, owing to low night temperatures frosting occurs. Otherwise these winter months are characterized by clear skies and low temperatures.

Pre-Monsoon

The weakening of the intense winter Asiatic anti-cyclone commences in March and continues progressively throughout the season. With the progressive northward march of the sun the equatorial trough rapidly shifts northward. The jet stream, which is a permanent feature, gradually shifts northward and migrates to a high level (200 mbs). Owing to the increased heating of the land the isobars have a flat pressure distribution. In April the land low begins to establish itself. A weak cell makes its appearance over west Rajasthan and the adjoining areas of Pakistan. Upper winds upto two kilometres are controlled by the surface pattern of the isobars. The wind blows from west to east and is colder and drier than that below. A temperate continental (Tc) air mass prevails over Rajasthan.

The western disturbances continue to affect Rajasthan as the air is now drier. During these months dust storms occur and are most frequent in April and half of May, after which the incidence sharply drops. Upper air inversions occur in May owing to which the air is dust laden and on many days the sky is hazy.

Monsoon

In the month of June the thermal low over Rajasthan strengthens itself and also the associated trough extends itself over the Gangetic Plain. The jet stream which had already started migrating northward from the month of April has by now moved north of 30° north where its velocity is considerably diminished. The stage is now set for the flow of the south-west monsoon. The low which is prominent in June, July and August weakens in September. The normal time for the monsoon to break over eastern Rajasthan is by the end of june and by the first week of July the whole of Rajasthan is under the influence of the equatorial maritime airmass. This airmass recedes after 15 September. The intensity of the low is responsible for drawing in the airmass and because of this it gives a pulsating character to the monsoon flow. Though Rajasthan gets the Arabian sea branch of the south-western monsoon earlier the rainfall is scanty except for some heavy rain in the western part of the state. Eighty per cent of the annual rainfall of Rajasthan is caused by the monsoon lows which form over the head of Bay of Bengal, travel along the trough and on reaching Rajasthan either as lows or depression, unburden their moisture. The average number of these lows is about one per week in the month of July and less in September. The track of these lows is northerly in the beginning of the season and takes a southerly course in the month of September when southern Rajasthan gets heavy rainfall. It has been observed that owing to the passage of these monsoon lows between 17°N and 20°N in September. Jodhpur and adjoining areas get heavy showers.

Because of the difference in air temperatures the frequency of thunderstorms is greater in the early part of the monsoon. Thunderstorms also occur when there is a new surge of the monsoon.

RAINFALL CHARACTERISTICS

The rainfall in Rajasthan is extremely seasonal, 90 per cent falling between July and September, which is also the season of high evaporation. The rainfall is highly variable both in space and time. Some stations may have no rainfall for an entire year. In other years much more than the normal annual total may fall in a sudden downpour as happened in Jaipur on 19 July, 1981 with 353.6 mm in just 36 hours and also in Jodhpur on 15 July, 1979 with 400 mm in just 24 hours.

Rainfall has an important bearing on geomorphic processes such as chemical weathering, fluvial erosion, accelerated soil erosion etc. Catastrophic rainfall especially in arid and semi-arid areas brings about considerable changes on the surface form than continuous rain in subhumid and humid area. For the purpose of rainfall analysis (mean annual) 50 years averages have been studied. Besides this, 40 years averages have also been analysed (1939-79). Table 3.1 shows the mean annual rainfall average of 50 years. It is evident from the table that the highest rainfall has been recorded at Mount Abu (1630.1 mm), Digod (1075.9 mm), Chipabarod (1044.9 mm), Rajsamand (1053.7 mm), Kishangarh (1017.90 mm), Chittorgarh (1008.8 mm) and Danpur (1006.50 mm). All these stations fall within Banswara, Udaipur, Sirohi, Chittorgarh and Kota which are south-eastern districts of Rajasthan. This area gets its rainfall from both the Arabian sea branch of the south-west monsoon and also the Bay of Bengal branch. This area is characterized by the high Aravalli ranges and other Vindhyan hills hence, owing to orographic effects rain bearing monsoon winds tend to give more rain to this region. The mean rainfall decreases going from the south-east to north-west. The areas that lie east of the Aravalli mountain also get higher rainfall as compared to stations west of the Aravallis. The pattern of rainfall completely changes west of Aravallis (fig. 3.1). Minimum rainfall is recorded at Ramgarh (136.4 mm), Jaisalmer (178.5 mm) and Fatahgarh at the extreme western arid part of the state. The rainfall in Ganganagar district (extreme northern part) at Karanpur (190.8 mm), Anupgarh (191.4 mm) and Padampur (201.2 mm) is also very low. The rainfall in this district increases in the northern part, the rainfall at Bhadra is 384.5 mm.

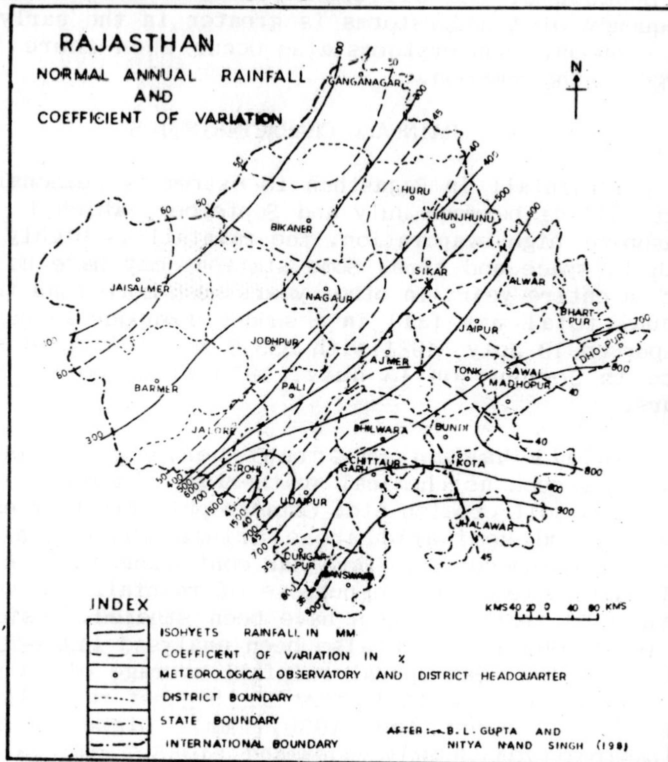

FIG. 3.1

Source : Gupta and Singh (1981)

On the basis of the amount of rainfall received the state can be divided into five rainfall regions. These are:

1. Arid Region - 100-300 mm or less rainfall.
2. Transitional between Arid and Semi-arid - 300-500 mm rain.
3. Semi-arid - 450-600 mm rainfall.
4. Subhumid - 500-750 mm rainfall.
5. Humid - 750 and above up to 1650 mm.

Monthly Distribution of Annual Rainfall

The monthly contribution to annual rainfall has been computed for different stations using 40 years data (Gupta and Nityanand, 1981). It is apparent from Tables 3.2 and 3.3

that the distribution of normal annual rainfall during different months is in good agreement. An interesting feature is that the percentage contribution of the SW monsoon to annual rainfall is about 92 per cent in eastern Rajasthan and 86 per cent in western Rajasthan. However, on the whole the state gets 67 per cent of normal rainfall up to 31st May, 89.9 per cent during the south-west monsoon period, and only 3.4 per cent during the remaining period.

Rainfall Intensity

Rainfall intensity also has a bearing on denudational processes. The annual rainfall intensity of 12 representative stations has been calculated. Table 3.4 giving rainfall intensity indicates that rainfall intensity decreases from the south-east to the north-west. Mount Abu which is in the extreme south has an intensity of 32.77 mm, Udaipur north-east of Mount Abu has an intensity of 18.64 mm, Kota and Jhalawar which are in the extreme south-east have intensities of 21.03 mm and 21.99 mm, respectively. The rainfall intensities of Jaipur and Ajmer which represent semi-arid climatic conditions are 17.32 mm and 17.69 mm, respectively. The lowest rainfall intensity has been recorded in Ganganagar (15.19 mm) which represents the extreme northern portion of the arid region of Rajasthan. Even in the arid region there is an increasing trend of rainfall intensity from north-east to south-west. Barmer which is in extreme south-west of the desert region has intensity of 18.14 mm. From this analysis it is evident that there is a positive correlation between mean annual rainfall and mean annual rainfall intensity, both the elements show a decreasing trend from the south-east to north-west.

It has also been observed that rainfall intensities are higher in the monsoonal season. A week prior to the monsoon a dry spell prevails, owing to which in many parts of Rajasthan soils are pulverized and loose cohesion. It has been pointed out by several workers that if abrupt rainfall of high intensity occurs on such a non-coherent soils, the soil may be beaten into a pasty semifluid mass before runoff begins, and, therefore, the soil surface becomes protected by surface detention (Horton, 1945). Sometimes, rainfall of unusual intensity causes cloudburst floods which are characterized by the carrying off of large volumes of solid and surprisingly little water as runoff.

TABLE 3.1 **Mean Annual Rainfall of 50 Years (in mm)**

Stations	
1. Ajmer District	
Ajmer	520.0
Sawar	531.9
Kekri	607.1
Pisagan	398.5
Goela	451.7
Beawar	468.5
Jawaja	556.9
Tatgarh	592.8
2. Alwar District	
Alwar	639.7
Kishangarh	606.3
Mundawar	572.0
Laxmangarh	570.1
Tijara	590.2
Nimrana	555.9
Govindgarh	586.6
Kotkasim	749.9
3. Banswara District	
Banswara	929.2
Garhi	856.1
Kushalgarh	1017.9
Danpur	1006.5
Shergarh	859.2
Khandu	840.0
Arthuna	886.8
Lutharia	866.2
Jagpura	882.9
Sallopat	947.7
Bhuagra	1010.5
Sajjangarh	915.2
4. Barmer District	
Barmer (a) (MO)	314.0
Sheo 207.6	
Siwana	344.4
Panchpadra	270.1
Balotra	278.5
Chortan	238.5
5. Bharatpur District	
Bharatpur	673.5
Kanwan	646.5
Nadbai	618.5
Bayana	644.4
Dholpur	722.8
Bari 721.9	
Rajakhera	691.6
Basin	631.3
Sepao	702.9
Sri Mathura	661.8
6. Bhilwara District	
Bhilwara	748.5
Shahpura	657.0
Jahajpur	835.3
Sahada	555.6
7. Bikaner District	
Bikaner	306.1
Lunkaransar	233.3
Magra	227.0
Palana	286.6
Surpura	265.8
8. Bundi District	
Bundi	758.6
Hindoli	754.0
K. Patan	779.4
9. Chittorgarh District	
Chittorgarh	1008.8
Kapasan	781.6
Pratapgarh	908.9
Nimbahera	709.5
10. Churu District	
Churu	367.6
Ratangarh	353.9
Sujangarh	372.3
Sardarsahar	280.5
Rajgarh	342.4
Taranagar	301.8
Dungargarh	261.2
11. Dungarpur District	
Dungarpur	732.6
Sagwana	700.9
Dhambola	825.3
12. Ganaganagar District	
Ganganagar	303.3
Karanpur	190.8
Padampur	201.2
Raisinghnagar	206.6
Anupgarh	191.4
Suratgarh	227.1
Hanumangarh	288.5
Nohar	307.8
Bhadra	384.5
13. Jaipur District	
Jaipur (a)	597.9

Chaksu	536.9
Amber	636.8
Jamua Ramgarh	576.7
Bairath	569.0
Kotputli	516.1
Dausa	554.9
Lalsot	633.3
Dudu (Mojamabad)	481.9
Sanganer	557.7
Baswa	574.6
Naraina	468.0
Chomu	522.3
Paota	504.1
Sambhar	494.4
Samod	534.2

14. Jaisalmer District

Jaisalmer	178.5
Ramgarh	136.4
Fatehgarh	182.6

15. Jalore District

Jalore	363.0
Samchor	380.5
Jaswantpura	471.5
Binmal	470.2

16. Jhalawar District

Jhalawar	1089.3
Deeg	941.6
Pirawa	938.8
Bakani	900.2
Aklera	985.7
Khanpur	1128.4

17. Jhunjhunu District

Jhunjhunu	387.4
Chirawa	408.9
Khetri	560.6

18. Jodhpur District

Jodhpur	366.0
Phalodi	235.8
Bilara	434.4
Shergarh	264.8

19. Kota District

Kota	785.6
Mangral	834.0
Sangod	703.6
Indergarh	879.9
Antah	819.4
Atru	994.3
Baran	888.2
Itawah	739.4
Shahbad	864.3
Kishanganj	953.1
Chhipa Barod	1044.7
Digod	1075.9

20. Nagaur District

Nagaur	309.9
Didwana	356.8
Parbatsar	389.3
Merta city	418.8
Nawa	468.8

21. Pali District

Pali	411.0
Jaitaran	378.2
Dasim	626.1
Bali	564.3
Sojat	472.1

22. Sawai Madhopur District

Sawai Madhopur	887.6
Khanda	718.3
Gangapur	656.8
Hindaun	660.4
Toda Bhim	589.4
Mahuwa	579.4
Karoli	708.6
Sapotra	741.2

23. Sikar District

Sikar	468.6
Nim ka Thana	504.7
Danta Ramgarh	426.0
Sri-Madhopur	489.7

24. Sirohi District

Mount Abu	1630.1
Sirohi	574.2
Sheoganj	493.7
Abu Road	847.4

25. Tonk District

Tonk	667.2
Malpura	513.6
Niwai	602.1
Todaraisingh	611.6
Aligarh	651.3
Deoli	719.1

26. Udaipur District

Udaipur	638.1
Sarara	702.7
Bhim	570.5
Khewara	672.6
Rajsamand	1053.7

Source of Data : Indian Meteorological Department, New Delhi.

TABLE 3.2 **Mean Monthly Rainfall, 40 Years Average–1939–79 in (in mm)**

Station/Month	J	F	M	A	M	J	JL	SA	S	O	N	D	Total
Ajmer	7.8	6.7	5.4	3.2	14.3	51.9	181.8	169.1	75.2	13.0	4.0	4.3	542.5
Alwar	12.3	10.8	8.5	4.7	17.0	58.4	193.8	210.2	134.7	15.9	2.5	6.6	675.4
Anupgarh	4.2	4.4	7.3	3.8	6.1	17.6	55.9	64.9	26.6	3.1	0.6	5.2	199.7
Banswara	3.1	2.7	2.2	0.6	6.3	100.3	326.5	311.9	175.8	25.2	7.8	2.5	964.9
Barmer	3.2	3.3	4.4	1.9	7.5	28.3	81.0	114.3	35.0	2.8	3.0	0.9	285.6
Bharatpur	10.7	13.6	8.4	4.2	11.0	44.8	196.4	224.1	112.6	16.8	2.3	4.0	648.9
Bhilwara	5.7	3.2	4.3	1.0	3.7	55.4	228.8	232.8	117.7	18.5	6.4	3.0	680.5
Bikaner	5.3	7.1	6.4	4.7	12.7	28.0	86.0	86.9	45.0	6.3	1.8	3.9	294.1
Chittorgarh	5.4	1.8	4.9	1.2	10.7	64.3	262.3	323.0	151.2	23.4	5.2	1.5	854.9
Churu	8.9	6.8	5.2	4.7	11.7	38.1	117.8	109.1	50.2	7.7	1.8	4.4	366.4
Dholpur	12.9	8.0	7.0	4.0	9.1	52.7	221.0	256.9	722.3	22.8	4.1	5.0	725.8
Durgapura	2.7	1.8	2.1	1.4	9.4	93.4	272.1	212.2	128.1	17.4	8.8	1.2	750.6
Ganganagar	7.7	7.8	10.8	5.1	10.8	26.6	71.1	73.7	34.2	1.3	1.5	5.6	256.2
Jaipur	10.3	10.0	6.7	4.8	14.8	56.3	192.6	215.1	83.9	17.3	3.0	6.4	621.2
Jaisalmer	2.1	5.0	3.2	2.2	7.0	14.0	58.6	69.1	22.2	1.5	1.4	2.1	188.4
Jalore	2.5	4.3	3.3	1.1	9.9	32.3	121.0	139.7	158.5	8.0	2.2	0.9	383.7
Jhalawar	12.9	2.8	5.1	1.8	5.5	101.6	325.7	321.2	161.5	18.6	6.4	5.7	968.8
Jodhpur	4.0	5.0	4.1	2.7	10.6	32.3	113.4	128.9	57.4	7.7	2.7	1.5	370.3
Kota	6.5	4.4	3.2	3.2	10.3	70.2	280.3	249.0	120.4	14.2	6.4	4.2	772.3
Mount Abu	4.5	4.8	2.7	2.4	15.2	103.0	594.6	617.6	235.0	18.6	8.3	1.9	1608.6
Nagaur	4.3	7.3	4.6	3.0	13.2	31.8	102.8	113.5	38.4	5.2	1.6	3.5	329.2
Phalodi	3.3	4.6	3.5	3.2	10.0	24.8	75.4	85.0	29.3	3.1	1.2	1.9	245.3
Sawai Madhopur	7.9	5.9	5.6	1.6	7.6	70.1	320.6	324.0	121.2	11.4	3.0	4.4	883.3
Sikar	8.7	8.0	6.6	3.2	16.1	44.5	157.0	152.9	56.8	5.1	3.3	4.6	467.5
Tonk	7.5	5.7	4.5	2.1	10.4	65.2	238.8	221.5	94.2	14.9	5.0	3.7	673.6
Udaipur	6.3	3.4	5.8	2.8	17.9	79.8	202.6	187.4	107.8	16.3	6.6	2.3	639.0

Source : Gupta and Nitya Nand (1981).

TABLE 3.3 **Cumulative Total of Normal Monthly Rainfall (in percentage)**

Station/Month	J	F	M	A	M	J	JL	SA	S	O	N	D
Ajmer	1.4	2.6	3.6	4.1	6.7	16.9	49.5	81.5	95.8	98.4	99.1	100
Alwar	1.8	3.4	4.7	5.4	7.9	16.6	44.8	76.5	96.3	98.8	99.1	100
Anupgarh	2.1	4.3	8.0	9.9	13.0	21.8	49.8	82.3	95.6	97.2	97.5	100
Banswara	0.3	0.6	0.8	0.9	1.5	11.7	45.3	77.7	96.4	99.1	99.7	100
Barmer	1.2	2.3	3.9	4.5	6.8	16.4	44.5	85.1	97.8	98.8	99.7	100
Bharatpur	0.6	3.6	4.9	5.6	7.3	14.3	44.3	79.5	96.3	99.0	99.4	100
Bhilwara	0.8	1.1	1.7	1.9	2.4	10.3	42.9	78.0	96.2	99.0	99.5	100
Bikaner	1.8	4.1	6.2	7.8	12.1	21.3	50.0	80.0	95.7	97.9	98.5	100
Chittorgarh	0.6	0.7	1.3	1.4	2.6	10.1	39.9	77.7	96.4	99.3	99.7	100
Churu	2.6	4.3	5.7	6.9	10.1	20.5	51.4	82.1	94.0	98.2	98.7	100
Dholpur	1.7	2.7	3.7	4.2	5.4	12.8	43.1	78.6	95.6	98.9	99.4	100
Durgapura	0.4	0.6	0.9	1.1	2.3	14.7	51.3	79.0	96.5	98.9	99.8	100
Ganganagar	2.9	3.8	10.2	11.9	16.1	25.7	52.8	83.6	96.6	97.2	97.7	100
Jaipur	1.6	3.1	4.2	4.9	7.2	15.8	46.5	81.6	95.5	98.5	99.0	100
Jaisalmer	1.2	3.8	5.6	6.8	10.6	17.8	48.3	85.7	97.8	98.5	99.2	100
Jalore	0.7	1.7	2.6	2.9	5.1	13.5	44.4	81.4	97.3	99.4	99.8	100
Jhalawar	1.4	1.6	2.1	2.3	2.8	13.1	46.0	79.4	96.7	98.7	99.4	100
Jodhpur	1.1	2.4	3.6	4.3	7.4	15.5	45.4	80.5	96.8	99.0	99.7	100
Kota	0.9	1.5	1.9	2.1	3.4	11.9	48.7	80.5	96.8	98.8	99.5	100
Mount Abu	0.3	0.6	0.8	1.0	2.0	8.5	45.0	83.5	98.5	99.6	99.9	100
Nagaur	1.1	3.4	4.8	5.7	9.7	19.5	49.5	84.3	97.1	98.6	99.1	100
Phalodi	1.3	3.0	4.5	5.7	9.7	19.5	49.8	85.1	97.4	98.7	99.2	100
Sawai Madhopur	0.9	1.4	2.0	2.2	3.1	11.0	47.6	83.9	97.7	99.1	99.5	100
Sikar	1.8	3.5	5.0	5.7	9.2	18.0	51.2	84.8	97.0	98.4	99.1	100
Tonk	1.1	1.8	2.5	2.8	4.4	14.0	48.9	82.2	96.5	98.8	99.5	100
Udaipur	0.9	1.4	2.3	2.8	5.5	17.9	49.1	79.0	96.5	99.4	97.7	100
Average	1.3	2.5	3.8	4.4	6.7	15.8	47.3	81.3	96.6	98.7	99.2	100

TABLE 3.4

Rainfall Intensity (in mm)

Climatic Region	Station Intensity	Annaul Rainfall
Arid region	1. Ganganagar	15.19
	2. Bikaner	15.55
	3. Phalodi	16.35
	4. Jodhpur	18.10
	5. Barmer	18.14
Transitional zone (AD-SA)	6. Sikar	15.69
Semi-arid region	7. Jaipur	17.32
	8. Ajmer	17.69
Subhumid region	9. Mount Abu	32.77
	10. Udaipur	18.64
Humid region	11. Kota	21.03
	12. Jhalawar	21.99

Catastrophic rain, that is, rains of unusual intensity which occur during the monsoon period, effect the hill slopes and uplands and also the valley slopes. During the period of catastrophic rains, linear erosion on slopes operates together with gravitational processes. Rains of high intensity cause acceleration of the processes of weathering and of denudation in deforested areas. In such regions most of the surfaces are bare with scanty bushes, hence very heavy rainfall causes considerable changes in the landscape. The rainfall data (Table 3.5) from 15 July 1981 to 25 July 1981 illustrate how catastrophic rains affect the geomorphic processes. The table indicates that maximum rainfall during this period was received in Jaipur, Tonk and Ajmer districts while station of western Rajasthan received lesser amounts of lain. The intensity of rainfall at Jaipur on 17 July 1981, and 18 July 1981 was of a high order (Table 3.6). The intensities have been calculated for every 18 minutes and at times it found to be as high as 22.0 mm. In the span of one hour there was continuous rainfall amounting to 24.0 mm. On 18 July the intensity was as high as 50.0 mm in one hour, which is extremely high when compared with the mean annual intensity of rainfall of 17.32 mm. Because of this catastrophic rainfall of higher intensity considerable destruction along the stream courses occurred. The dynamic equilibrium of the ecosystem was upset, stream courses were rejuvenated, new gullies were formed, land-slides took place in the eastern hill zones, a considerable amount of sand covered the southern and eastern parts of the township, and a number of railway and road bridges were breached.

TABLE 3.5 **Rainfall Data of Rajasthan : From 15 July, 1981 to 25 July, 1981** (Rainfall in mm).

Station	15	16	17	18	19	20	21	22	23	24	25
Jaipur A/P	79.4	00.6	7.2	162.5	328.0	134.8	14.7	17.8	53.2	13.0	1.0
Jaipur city	35.9	-	4.2	235.8	353.6	228.6	14.2				
Kota A/P	32.2	05.8	5.4	-	4.8	22.5	63.1	58.1	4.3	4.8	-
Tonk	23.8	00.0	19.6	56.2	134.2	-	5.4	-	-	-	1.3
Sawai Modhopur	35.8	00.6	4.8	57.4	68.0	61.6	9.6	46.8	3.2	15.8	0.0
Bharatpur	00.0	02.5	7.7	-	2.7	5.4	8.5	1.4	0.4	0.0	6.0
Alwar	2.2	35.6	8.6	21.8	47.2	1.8	-	0.0	13.6	5.0	35.5
Ajmer	0.0	0.0	38.4	157.5	24.8	-	0.4	4.2	7.5	12.0	0.8
Banswara	23.0	.0	6.0	14.5	15.0	0.4	-	0.0	-	21.0	4.0
Bhilwara	10.0	-	8.3	3.6	-	Trace	44.0	68.4	0.6	1.4	4.2
Chittorgarh	3.0	0.0	4.7	0.0	0.2	-	20.0	10.0	34.2	17.0	13.8
Dungarpur	2.6	12.0	5.4	-	0.0	0.0	-	0.0	0.0	17.0	2.2
Jhalawar	Flood	-	-	-	-	-	-	-	-	-	-
Jhunjhunu	7.0	-	6.0	15.0	-	-	0.0	1.5	7.0	-	-
Sikar	61.3	0.0	70.1	60.10	67.1	18.0	0.0	39.3	110.0	0.0	0.0
Mount Abu	4.0	1.3	25.2	25.2	3.5	1.8	0.3	4.0	60.7	13.7	27.2
Udaipur	0.0	2.0	6.0	0.0	-	0.0	0.0	1.8	10.2	23.6	1.8
Jodhpur A/P	-	-	-	-	-	-	-	30.8	Trace	4.2	0.6
Jodhpur city	0.0	Trace	2.3	0.0	0.0	0.0	0.0	-	-	-	-
Jalore	0.0	0.0	2.9	0.0	0.0	0.0	-	8.9	1.6	7.2	4.4
Nagaur	0.0	-	65.1	3.4	-	0.0	-	-	5.4	4.0	0.0
Churu		28.6	0.0	73.5	11.10	0.0	Trace	0.7	20.5	Trace	Trace
Pali(Er. road)	1.0	1.0	1.0	-	1.0	0.0	0.0	14.4	4.0	0.4	0.4
Bikaner	10.0	0.4	127.2	-	-	-	0.0	7.3	2.8	0.8	0.0
Sriganganagar	6.7	26.0	14.4	0.0	0.0	0.0	0.0	-	0.0	0.0	36.9
Pilani	66.4	16.2	3.8	4.4	-	-	0.0	-	-	-	-
Phalodi	0.0	0.0	3.4	0.0	0.0	0.0	0.0				
Deogarh	0.0	10.0	7.2	-	-	-	0.0				
Siwana	-	0.0	2.6	0.0	-	0.0	0.0				
Sojat		0.0	0.0	13.6	4.6	0.0	0.0				
Pokaran	0.0	0.0	3.0	0.0	0.0	0.0	0.0				
Sirohi	0.0	0.0	5.0	0.0	-	0.0	0.0				
Lunkaransar	2.0	4.0	52.0	2.0	-	-	7.0				
Suratgarh	-	-	6.0	0.0	60.0	0.0	2.0				
Dholpur	31.7	0.0	-	18.0	5.2	-	1.0				
Rawat Bhatta	3.5	0.0	1.4	0.0	4.5	1.2	44.0				
Jaisalmer	0.0	Trace	0.2	0.0	0.0	0.0	0.0				

Source : Indian Meteorological Department Jaipur

TABLE 3.6 **Rainfall Intensity at Jaipur from 17 July, 1981 to 18 July, 1981**

17 July, 1981 Rainfall (mm)

0845	0900	0915	0930	0945	1000	1015	1030	1045	1100
00	00	Tr	Tr	0.5	0.25	0.25	0.0	0.0	1.1
1115	1130	1145	1200	1215	1230	1245	1300	1315	1330
0.0	0.0	0.0	0.0	0.0	0.0	0.0	0.0	0.0	0.1
1345	1400	1415	1430	1445	1500	1515	1530	1545	1600
0.0	0.0	0.0	0.0	0.0	1.25	1.0	0.0	0.1	0.5
1615	1630	1645	1700	1715	1730	1745	1800	1815	1830
1.0	2.5	1.6	2.0	1.0	2.0	1.0	9.2	0.8	2.8
1845	1900	1915	1930	1945	2000	2015	2030	2045	2100
5.0	1.5	1.0	1.2	0.8	1.5	1.0	1.2	1.7	0.6
2115	2130	2145	2200	2215	2230	2245	2300	2315	2330
0.6	0.5	0.6	0.6	0.4	0.1	0.2	0.0	Tr	0.1
2345	2400	0015	0030	0045	0100	0115	0130	0145	0200
0.2	0.3	0.0	0.0	0.0	0.1	2.0	0.4	.0.1	1.3
0215	0230	0245	0300	0315	0330	0345	0400	0415	0430
0.2	0.1	1.8	1.3	2.0	4.5	6.2	8.0	4.3	2.2
0445	0500	0515	0530	0545	0600	0615	0630	0645	0700
1.8	2.0	0.7	2.5	4.6	10.9	7.2	3.8	2.5	1.7
0715	0730	0745	0800	0815	0830				
2.4	4.5	3.0	5.5	22.0	4.9	Total	162.5		

18 July, 1981 Rainfall (mm)

0830	0845	0800	0915	0930	0945	1000	1015	1030	1045
18.0	22.0	10.5	4.0	4.5	1.0	0.5	1.0	5.0	7.0
1100	1115	1130	1145	1200	1215	1230	1245	1300	1315
3.0	2.5	10.0	1.5	2.0	1.5	3.2	1.3	2.0	2.0
1330	1345	1400	1415	1430	1445	1500	1515	1530	1545
1.0	1.5	3.00.4	0.0	3.0	2.0	1.0	1.8	0.5	
1600	1615	1630	1645	1700	1715	1730	1745	1800	1815
0.6	1.5	0.5	0.5	0.5	0.7	0.7	0.5	0.2	0.8
1830	1845	1900	1915	1930	1945	2000	2015	2030	2045
0.7	0.8	1.8	0.8	0.6	1.0	2.5	3.0	2.0	2.0
2100	2115	2130	2145	2200	2215	2230	2245	2300	2315
2.0	3.0	5.0	7.5	1.6	1.4	9.0	9.5	0.9	0.0
2330	2345	2400	0015	0030	0045	0100	0015	0130	0145
0.0	0.0	0.2	0.6	0.8	0.6	0.6	0.0	0.0	0.2
0200	0215	0230	0245	0300	0315	0330	0345	0400	0415
0.3	0.7	1.0	1.0	1.0	1.0	1.0	1.0	1.0	0.7
0430	0445	0500	0515	0530	0545	0600	0615	0630	0645
0.7	3.0	3.7	4.8	2.0	3.5	6.0	4.5	14.5	8.5
0700	0715	0730	0745	0800	0815	0830	0845		
13.0	13.0	11.0	4.0	14.0	11.0	11.0	6.0	Total	324.2

Rainfall Variability

The coefficient of rainfall variability is a measure of the variation and uncertainty of rainfall. The greater the coefficient of variability, the greater is the variability of rainfall. The coefficient of rainfall variability in this study has been based on the data of 26 stations of Rajasthan from the years 1939 to 1979. The rainfall coefficient of variability, both mean and monthly, of 75 years mean has also been calculated. the mean of 75 years is slightly less than the average of 40 years. Tables 3.7 and 3.8 show that the coefficient of variability decreases from north-west to south-east. As a matter of fact there is a negative correlation between mean annual rainfall and coefficient of variability. As has been pointed out earlier, rainfall in Rajasthan decreases from the south-east to the north-west which the coefficient of variability increases from the south-east to the north-west. From this analysis of mean annual rainfall and mean annual coefficient of variability it can be inferred that rainfall show a decreasing trend in the arid parts of Rajasthan, at; the same time there is a greater possibility of uncertainty of rainfall.

The mean monthly coefficient of variability shows slightly different results as it increases from the south-east to the north-west (Fig. 3.1). The monthly coefficient of variability of the arid region at Ganganagar has been computed as 90.409 per cent, and increasing from the north-east to the south-west, it is maximum at Barmer (160.56 per cent). Mount Abu which is in the extreme southern part of Rajasthan has a monthly value of coefficient of variability of 168.51 per cent.

Rainfall Fluctuations

Rainfall data pertaining to the period 1891 to 1981, that is, 91 years, are represented in Figures 3.2 and 3.2a. The figure reveals that Rajasthan is a land of rainfall fluctuations and variability. It is interesting to note that in western Rajasthan almost nil rainfall was recorded at Jaisalmer in 1899, about 88 per cent less than the normal rainfall, whereas the maximum rainfall of the same station is 226 mm in 1917, that is, 75 times more than that of a dry year.

Similarly, in eastern Rajasthan Sawai Madhopur received the highest amount of rainfall 701 mm in the year 1945 against the average rainfall of 348 mm.

TABLE 3.7 **Coefficient Variability of Rainfall (40 Years Average 1939–79) (as Percentage)**

Station/Month	J	F	M	A	M	J	JL	SA	S	O	N	D
Ajmer	193.3	173.1	194.4	281.2	139.2	92.6	67.1	72.6	110.2	249.3	222.5	258.1
Alwar	139.8	127.8	174.1	227.7	135.9	90.7	49.5	57.7	85.0	244.0	272.0	215.2
Anupgarh	159.5	179.5	230.1	207.9	170.5	138.1	98.5	92.0	146.6	354.8	483.3	309.6
Banswara	200.0	266.7	322.7	450.0	225.4	85.9	59.0	60.4	96.5	230.2	235.9	428.0
Barmer	331.2	230.3	325.0	257.9	197.3	159.7	87.6	111.9	131.1	260.7	343.3	355.6
Bharatpur	124.3	154.4	201.0	250.0	182.7	102.9	54.4	63.6	97.0	491.7	269.6	245.5
Bhilwara	245.5	237.5	337.5	340.0	229.7	89.5	62.2	61.2	99.4	321.1	323.4	320.0
Bikaner	139.6	200.0	206.2	183.0	163.0	102.9	86.2	87.5	133.3	288.9	327.8	218.0
Chittorgarh	251.8	216.7	304.1	333.3	142.1	89.0	52.2	54.1	100.5	200.4	330.8	306.8
Churu	194.5	164.7	184.9	195.7	126.5	110.5	76.7	81.5	115.9	283.1	255.6	247.7
Dholpur	123.3	171.2	230.0	192.5	157.1	103.0	59.3	54.7	78.5	195.6	361.1	218.0
Durgapura	296.3	272.2	507.5	328.6	241.5	102.1	64.4	63.2	48.2	264.4	300.0	341.7
Ganganagar	129.9	143.6	155.6	172.5	138.0	144.0	93.0	89.3	157.9	223.1	253.3	162.5
Jaipur	139.8	166.0	174.6	206.2	129.0	113.3	56.4	62.8	95.8	212.7	213.3	203.1
Jaisalmer	186.2	190.3	234.6	249.6	183.8	189.1	88.2	112.7	158.9	257.1	344.2	293.5
Jalore	256.0	251.2	403.0	463.6	233.3	118.6	80.2	94.3	139.2	303.7	318.2	488.9
Jhalawar	141.9	185.7	243.1	238.9	263.8	81.7	59.0	55.0	93.9	233.9	218.7	235.1
Jodhpur	250.0	194.0	322.0	214.8	162.3	120.4	77.7	85.6	123.3	300.0	296.3	280.0
Kota	152.3	240.9	209.4	271.9	157.3	88.0	53.8	55.7	94.8	218.3	293.7	238.1
Mount Abu	200.0	243.7	274.1	266.7	230.3	90.1	71.6	76.3	127.0	295.2	384.3	294.7
Nagaur	204.6	223.3	234.8	230.0	128.8	119.5	72.4	98.9	133.3	226.9	293.7	231.5
Phalodi	206.1	176.1	255.7	265.6	115.9	129.8	88.9	92.2	149.8	348.4	333.3	294.7
Swai Madhopur	154.4	233.9	233.9	338.5	185.5	147.9	70.6	61.2	86.0	230.7	386.7	277.3
Sikar	151.2	170.0	192.4	200.0	131.1	89.9	74.6	74.0	103.9	259.0	242.4	234.8
Tonk	173.3	205.3	204.4	338.1	220.2	109.6	63.4	64.0	91.7	216.7	248.0	216.2
Udaipur	201.6	197.1	417.2	285.7	198.3	81.4	49.7	62.1	99.3	216.6	292.4	365.5

Source : Gupta and Nitya Nand (1981)

TABLE 3.8 **Coefficient of Variability, Rajashtan**

Based on Data of 75 years (1891-1965)

Climatic Region	Station	Coefficient Variability (%) Annual	Monthly
Arid Region	Jaisalmer	66.63	-
	Jodhpur	51.81	150.75
	Bikaner	93.37	132.98
	Barmer	58.88	160.56
	Ganganagar	-	90.40
	Phalodi	-	145.24
Transitional zone (AD-SA)	Jalore	53.98	-
	Pali	47.48	-
	Nagaur	49.25	-
	Sikar	38.31	133.25
	Jhunjhunu	37.51	-
Semi-arid region	Tonk	44.41	-
	Jaipur	42.73	142.64
	Ajmer	39.27	133.12
	Alwar	34.97	-
	Bharatpur	33.52	-
	Sawai Madhopur	46.31	-
Subhumid region	Sirohi	44.90	-
	Udaipur	35.65	136.80
	Mount Abu	-	168.51
Humid region	Banswara	33.53	-
	Kota	42.72	149.78
	Jhalawar	-	146.26

Source : Indian Meterological Department, New Delhi.

The overall study of the graphs (Figs.3.2 and 3.2a) indicates that during this period of ninty-one years there were periods of high rainfall and very low rainfall. However, large-scale climate change can not be inferred from it, small-scale climatic ameliorations may be deduced.

MOISTURE REGIME

Moisture regime plays a significant role in Rajasthan in defining vegetation types and also intensity of weathering processes. For the analysis of moisture regimes, the data compiled by Subramaniam and Prasad Rao (1981)[1] have been utilized. The analysis was carried out using the data of 14 stations of Rajasthan. Using Thornthwaite's formula the authors have computed moisture indices both on an annual as well as seasonal basis (south-west monsoon—June-September, post-monsoon October-november, cold weather December-February, pre-monsoon March-May.). The results are presented in Table 3.9.

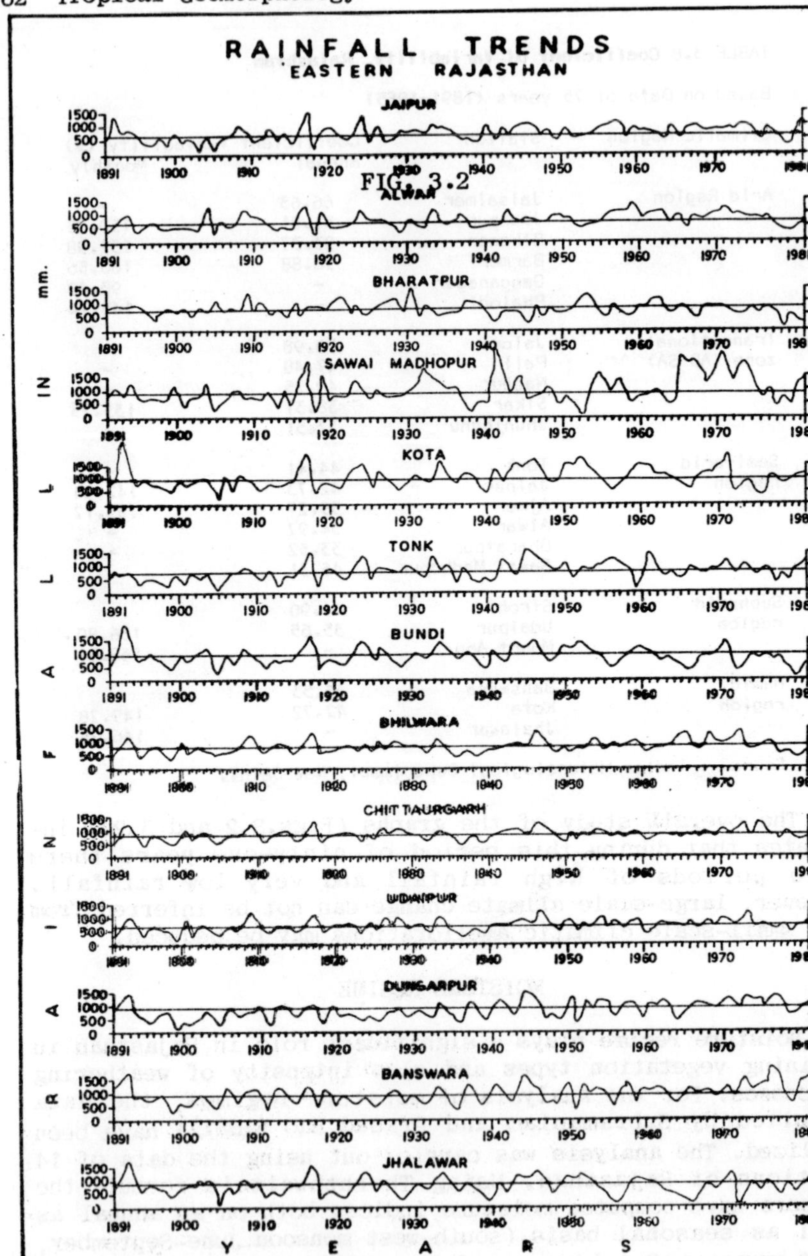

DATA SOURCE :— I.M.D

FIGURE 3.2

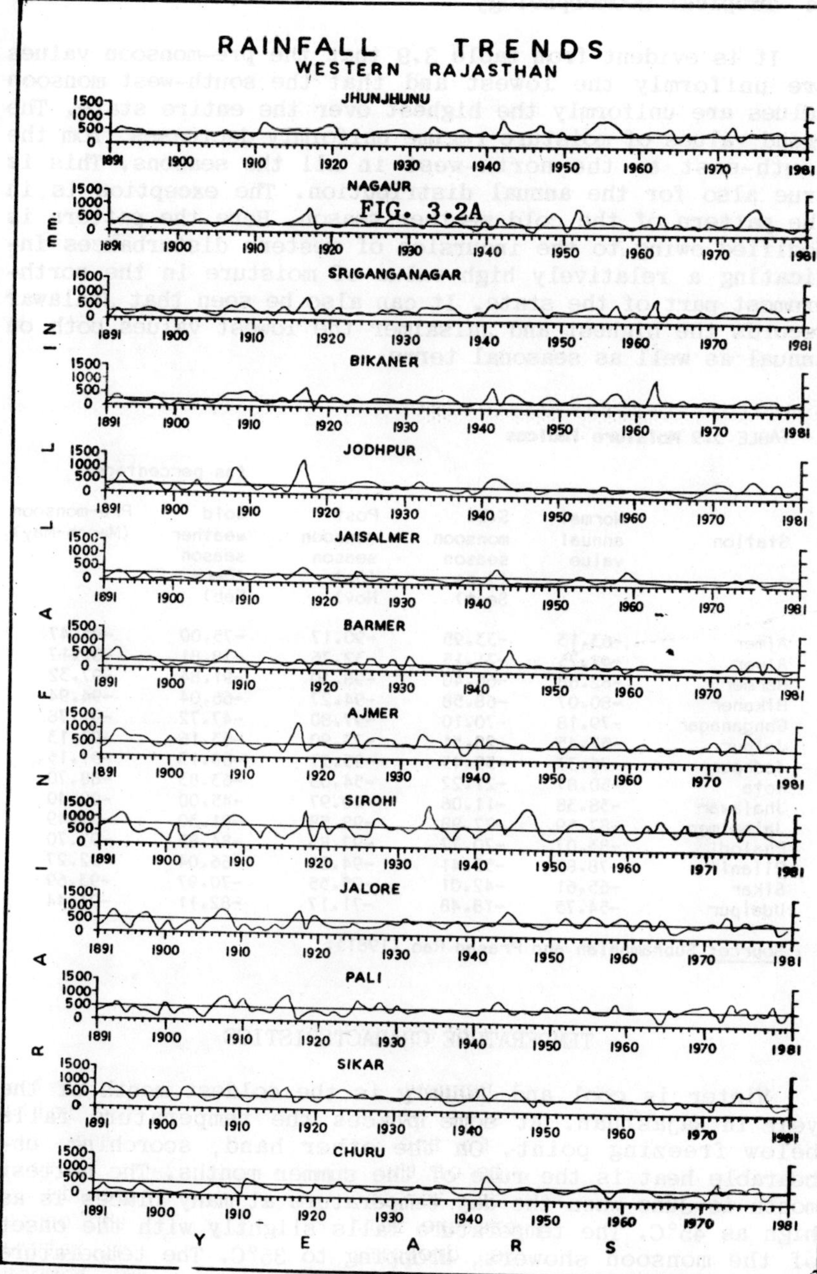

FIGURE 3.2A

DATA SOURCE :— I.M.D.

It is evident from table 3.9 that the pre-monsoon values are uniformly the lowest and that the south-west monsoon values are uniformly the highest over the entire state. The trend values of moisture regime uniformly decreases from the south-east to the north-west in all the seasons. This is true also for the annual distribution. The exception is in the pattern of the cold weather season. Here the pattern is modified owing to the incursion of western disturbances indicating a relatively high value of moisture in the northernmost part of the state. It can also be seen that Jhalawar records the highest and Jaisalmer the lowest values both on annual as well as seasonal terms.

TABLE 3.9 **Moisture Indices**

(as percentage)

Station	Normal annual value	S.W. monsoon season (June-Sept)	Post-monsoon season (Oct-Nov)	Cold weather season (Dec-Feb)	Pre-monsoon (March-May)
Ajmer	-63.13	-33.95	-90.17	-75.00	-95.47
Alwar	-42.75	-21.15	-32.76	-38.81	-82.17
Barmer	-82.68	-66.40	-98.30	-91.84	-97.32
Bikaner	-80.07	-68.58	-94.27	-66.04	-94.94
Ganganagar	-79.18	-70.10	-97.80	-47.72	-91.78
Jaipur	-57.45	-28.51	-77.90	-63.16	-93.13
Jodhpur	-76.33	-55.45	-95.37	-89.13	-97.15
Kota	-50.87	-21.22	-54.03	-63.83	-88.70
Jhalawar	-38.38	-11.06	-20.97	-45.00	-90.40
Jaisalmer	-87.59	-77.98	-98.58	-91.30	-97.49
Phalodi	-83.07	-70.72	-97.83	-84.85	-97.70
Pilani	-78.83	-53.41	-94.29	-66.04	-92.27
Sikar	-65.61	-42.01	-95.55	-70.97	-93.69
Udaipur	-54.75	-18.48	-71.17	-82.11	-93.44

Source: Subramanian and Prasad Rao (1981).

TEMPERATURE CHARACTERISTICS

Winter is cool and January is the coldest month of the year in Rajasthan. At some places the temperature falls below freezing point. On the other hand, scorching, unbearable heat is the rule of the summer months. The hottest month is June when the day temperature at many places is as high as 45°C. The temperature falls slightly with the onset of the monsoon showers, dropping to 35°C. The temperature steadily decreases from mid-October to reach its lowest point in the second half of december or first half of January.

The analysis of temperature conditions given nere is based on the average data of 12 stations from 1931 to 1960. This data (Table 3.10) has been represented in the form of temperature charts in which monthly maximum, monthly minimum, and monthly average temperatures have been depicted. It is evident from the temperature charts (Fig.3.3) of various stations that the trends of temperature rise and fall is more or less the same in all the stations. The stations of the arid region record maximum monthly and daily, and minimum monthly and daily temperatures. The stations falling in the humid region as represented by Mount Abu, Jhalawar and Kota have slightly different trends. In the case of Mount Abu it is evident that there is less variation in all the values of temperatures. The temperature never exceeds 35°C in May and never falls below 5°C in the month of January. The mean monthly temperature in January at Mount Abu is 14°C while at Jhalawar it is 10°C and at Kota 18°C. The maximum temperature in the three stations observed in the month of May.

Range of Temperature

The range of temperature both daily and monthly has an important bearing on the processes of mechanical weathering. The daily and monthly range of temperature have been calculated. The mean monthly temperature shows a decreasing trend from the south south-east to north-west. The station of the semi-arid region are represented by Jaipur and Ajmer: they have a monthly range of temperature of 24.1°C at Jaipur and 41.8°C at Ajmer (Fig.3.4). The monthly range of temperature is lowest in Jaipur. The range of temperature at Mount Abu is 31.4°C which is higher than that of Jaipur. Ganganagar has the highest mean monthly range of temperature, 46.97°C. The monthly range of temperature decreases in the north-west (Bikaner 46.1°C, Phalodi 43.90°C) and again increases in the south-west (Barmer 46.3°C). The study of the data of mean daily range of temperature also shows an increasing trend from the south-east to the north-west. Mount Abu has the lowest value of mean daily range of temperature (8.4°C). The highest mean daily range of temperature of Ganganagar is 15.9°C. It is interesting to note that the daily range of temperature is highest in the month of November in all the stations. The monthly range of temperature is highest in the months of February and March in all the stations of Rajasthan. From this analysis of the range of temperatures it can be inferred that the maximum physical weathering processes in the form of thermal expansion occur in the months of

TABLE 3.10 Temperature Data

Temperature in degree °C

Months	Daily Maxi-mum	Daily Mini-mum	Highest in the Month	Lowest in the Month	Daily range of te-mpera-ture	Monthly range of tem-tempera-ture	Mean monthly tempera-ture
STATION-JAIPUR							
January	22.0	8.3	26.8	2.7	13.7	21.1	14.5
February	25.4	10.7	31.2	4.8	14.7	26.4	18.5
March	30.9	15.5	36.7	9.3	15.4	27.4	23.0
April	36.5	21.0	41.2	15.5	15.5	25.9	28.2
May	40.6	25.8	44.5	20.8	14.8	23.7	32.6
June	39.2	27.3	43.4	23.0	11.9	20.4	33.2
July	34.1	26.6	39.5	22.9	8.5	16.6	31.2
August	31.9	24.3	36.1	22.1	7.6	14.0	29.1
September	33.2	23.0	36.5	19.9	10.2	16.6	28.2
October	33.2	18.3	36.1	13.8	14.9	22.3	24.9
November	29.0	12.0	32.7	7.8	17.0	24.9	20.2
December	24.4	9.1	28.3	4.2	15.3	24.1	16.2
STATION-AJMER							
January	22.2	7.3	26.8	3.3	14.9	23.5	15.0
February	25.3	9.9	30.7	3.8	15.4	20.9	17.2
March	30.7	15.7	36.0	9.6	15.0	26.4	22.8
April	35.9	21.9	40.1	15.5	14.0	24.6	27.8
May	39.5	27.3	42.9	21.3	12.2	21.6	32.1
June	38.1	27.7	41.6	22.2	10.4	19.4	31.9
July	33.2	25.6	38.2	22.7	7.6	15.5	30.4
August	30.9	24.3	35.0	22.4	6.6	12.6	28.7
September	32.1	23.7	35.5	20.2	8.4	15.3	27.8
October	32.9	17.8	35.3	12.5	15.1	22.8	23.9
November	28.9	10.9	32.0	6.8	18.0	25.2	19.4
December	24.4	7.7	28.1	3.3	16.7	24.8	15.7
STATION-PHALODI							
January	22.7	6.6	28.3	2.5	16.1	25.8	15.4
February	26.3	9.9	32.6	4.4	16.4	28.2	18.5
March	32.1	15.9	37.8	9.7	16.2	28.1	23.7
April	37.6	21.8	42.4	16.0	15.8	26.4	29.2
May	41.6	26.6	45.6	22.4	15.0	23.2	34.0
June	40.8	28.2	44.5	23.9	12.8	20.6	34.2
July	36.9	27.1	41.8	24.2	9.8	17.6	33.0
August	34.5	25.8	39.5	23.4	8.7	16.1	31.4
September	35.5	24.6	39.0	22.2	10.9	16.8	30.6
October	35.2	19.7	38.5	13.7	15.5	24.8	26.1
November	30.3	13.1	34.3	8.7	17.2	25.6	21.5
December	24.7	8.0	29.3	4.2	16.7	25.1	16.7
STATION-JODHPUR							
January	24.6	9.5	29.6	4.5	15.1	25.1	17.0
February	27.9	12.0	33.3	6.3	15.9	27.0	19.8
March	33.3	17.1	38.8	11.2	16.2	27.6	25.0
April	38.3	22.4	42.8	16.9	15.9	25.9	29.8
May	41.6	27.3	45.5	22.5	14.3	23.0	34.0
June	40.1	28.5	43.6	24.3	11.6	19.3	33.9
July	35.7	26.8	40.8	24.0	8.9	16.8	32.4

August	33.2	25.2	38.0	23.1	8.0	14.9	30.5
September	34.7	24.1	38.7	21.9	10.6	16.8	30.3
October	35.7	19.6	38.6	15.5	16.1	23.1	27.0
November	31.4	13.9	34.9	10.0	17.5	24.9	22.4
December	26.7	10.7	30.8	6.1	16.0	24.7	18.4
STATION-JHALAWAR							
January	25.1	9.4	30.3	4.4	15.7	25.9	17.3
February	28.4	11.4	34.0	5.8	17.0	28.2	19.9
March	33.9	16.4	38.6	10.8	17.5	27.8	24.7
April	38.6	22.0	42.5	16.9	16.6	25.6	29.7
May	42.0	27.3	44.7	22.8	14.7	21.9	33.7
June	39.1	27.5	43.5	23.1	11.6	20.4	33.3
July	32.3	24.9	38.1	22.5	7.4	15.6	30.3
August	30.6	24.1	34.8	22.1	6.5	12.7	28.4
September	31.6	23.2	35.0	21.0	8.7	14.0	28.0
October	33.5	18.3	35.9	13.6	15.2	22.3	24.7
November	29.8	12.2	33.2	8.0	17.6	25.2	20.6
December	26.5	9.6	30.2	5.3	16.9	24.9	17.7
STATION-UDAIPUR							
January	24.2	7.7	28.5	3.1	16.4	25.4	15.8
February	27.6	9.7	33.0	3.8	17.9	29.2	18.4
March	32.3	15.1	36.4	9.8	17.2	26.6	23.1
April	36.0	20.2	34.9	14.1	15.8	25.8	27.0
May	38.6	24.9	42.0	20.3	13.7	21.7	31.1
June	35.9	25.3	39.9	21.3	10.6	18.6	30.6
July	30.7	23.9	35.1	21.7	6.8	13.4	28.4
August	29.3	22.9	32.9	20.7	6.4	12.2	26.8
September	30.9	22.1	34.3	18.3	8.8	16.0	26.3
October	32.0	18.9	34.4	11.9	13.1	22.5	23.1
November	29.1	11.0	31.8	[illegible]	18.1	24.7	19.4
December	26.3	8.3	29.2	4.4	18.0	24.8	16.8
STATION-SRIGANGANAGAR							
January	20.5	7.7	25.2	0.6	15.8	24.6	12.9
February	24.1	7.5	27.6	2.1	16.6	25.5	21.0
March	22.6	11.1	35.8	6.3	18.5	29.5	27.3
April	36.3	18.2	42.1	12.5	18.1	29.6	32.1
May	41.5	24.0	46.1	18.2	17.5	27.9	34.2
June	42.1	28.0	46.2	22.3	14.1	23.9	33.7
July	38.8	28.1	43.5	21.0	10.7	19.5	32.2
August	37.3	26.9	40.6	23.3	10.4	16.8	30.0
September	36.6	24.1	40.2	19.9	12.5	20.63	24.9
October	35.0	17.0	38.2	11.7	18.0	26.5	19.5
November	29.9	8.9	33.6	5.5	20.5	28.1	14.9
December	23.3	5.5	27.8	2.0	17.8	25.8	14.8
STATION-BIKANER							
January	22.3	5.0	27.9	0.8	17.3	27.1	14.3
February	26.1	8.2	32.2	3.4	17.9	28.8	17.8
March	31.8	14.4	37.7	8.3	17.4	29.4	23.0
April	37.6	20.8	42.3	14.3	16.8	28.0	28.3
May	42.0	27.2	45.8	21.9	14.8	23.9	33.8
June	41.7	29.3	45.1	24.0	12.4	21.1	34.5
July	38.9	28.0	40.0	24.4	10.9	16.4	32.6
August	36.1	26.6	38.0	23.9	9.5	14.1	30.6
September	36.6	25.2	37.9	22.1	11.4	15.8	30.0
October	35.3	18.6	38.5	13.2	16.7	25.3	25.8
November	30.3	10.0	34.4	6.2	20.3	28.2	20.3

December	24.7	5.6	29.8	2.0	19.1	27.8	15.9

STATION-MOUNT ABU

January	19.3	9.3	23.2	4.8	10.0	18.4	14.0
February	21.2	11.5	25.7	5.9	9.7	19.8	15.6
March	25.2	15.9	30.0	10.2	9.4	19.8	20.1
April	29.4	20.0	33.2	15.0	9.4	18.2	24.1
May	31.5	22.3	35.1	17.7	9.2	17.4	26.4
June	29.1	20.5	32.9	17.9	8.6	15.0	25.4
July	24.3	19.3	28.9	17.7	5.0	11.2	23.3
August	22.5	18.3	26.6	16.8	4.2	9.8	21.7
September	24.1	18.4	28.2	16.3	5.7	11.9	27.2
October	26.6	17.4	29.4	14.2	9.2	15.2	21.8
November	24.1	13.5	26.9	10.2	10.6	16.7	18.5
December	21.2	11.2	25.0	6.5	10.0	18.5	15.7

STATION-KOTA

January	24.5	10.6	30.1	6.5	13.9	23.6	18.3
February	28.5	13.1	34.2	8.5	15.4	25.9	21.2
March	34.1	18.5	39.2	13.4	15.6	25.8	26.3
April	39.0	24.4	43.1	19.1	14.6	24.0	31.3
May	42.6	29.7	45.9	24.6	12.9	21.3	35.2
June	40.3	29.5	44.7	24.1	10.8	20.6	34.4
July	33.3	26.4	39.5	23.3	6.9	16.2	31.4
August	31.7	25.4	35.8	23.1	6.3	12.7	29.4
September	33.1	24.7	36.3	22.3	8.4	14.0	29.3
October	34.5	21.0	37.0	16.7	13.5	20.3	26.8
November	30.8	14.8	34.1	11.0	16.0	23.1	22.5
December	226.7	11.3	30.6	7.9	15.4	22.7	19.2

STATION-BARMER

January	24.7	10.2	29.7	5.5	14.5	24.2	17.6
February	28.6	13.3	34.0	7.6	15.3	26.4	20.8
March	33.9	18.7	39.1	12.4	15.2	26.7	25.7
April	38.9	24.1	32.3	18.0	14.8	25.3	30.6
May	41.9	26.8	46.0	22.2	15.1	23.8	34.1
June	40.3	27.3	44.0	23.5	13.0	20.5	33.7
July	36.1	20.3	41.0	23.5	9.8	17.5	32.2
August	33.8	25.1	38.2	22.7	8.7	15.5	30.4
September	35.3	24.5	39.4	22.0	10.8	17.4	30.7
October	36.00	21.7	38.9	16.7	14.3	22.2	27.8
November	31.5	15.9	35.1	11.3	15.6	23.8	23.2
December	27.0	12.1	31.4	7.5	14.9	23.9	19.4

STATION-SIKAR

January	22.0	5.8	27.3	-0.5	16.2	27.8	13.9
February	26.2	8.2	32.2	-1.9	18.0	30.3	17.0
March	31.5	14.5	36.1	6.7	17.0	29.4	21.4
April	36.4	19.0	40.7	11.8	17.4	28.9	26.2
May	39.7	24.3	43.3	18.7	15.4	24.6	31.0
June	39.1	27.5	42.2	21.6	11.6	20.9	32.0
July	34.7	26.0	39.5	22.5	8.7	17.0	31.0
August	33.1	24.7	37.6	22.1	8.4	15.5	29.8
September	33.7	23.1	36.7	19.6	10.6	17.1	28.1
October	33.1	17.1	36.2	10.9	16.0	25.3	23.5
November	29.1	10.4	32.3	4.4	18.7	27.9	18.3
December	24.8	6.7	28.8	1.2	18.1	27.6	15.0

Source : Indian Meteorological Department, New Delhi.
Note : Based on observations during 1931-60.

November, February and March. The range of temperature is higher in western Rajasthan than in the eastern part of the state hence the physical weathering processes should also be greater in western Rajasthan.

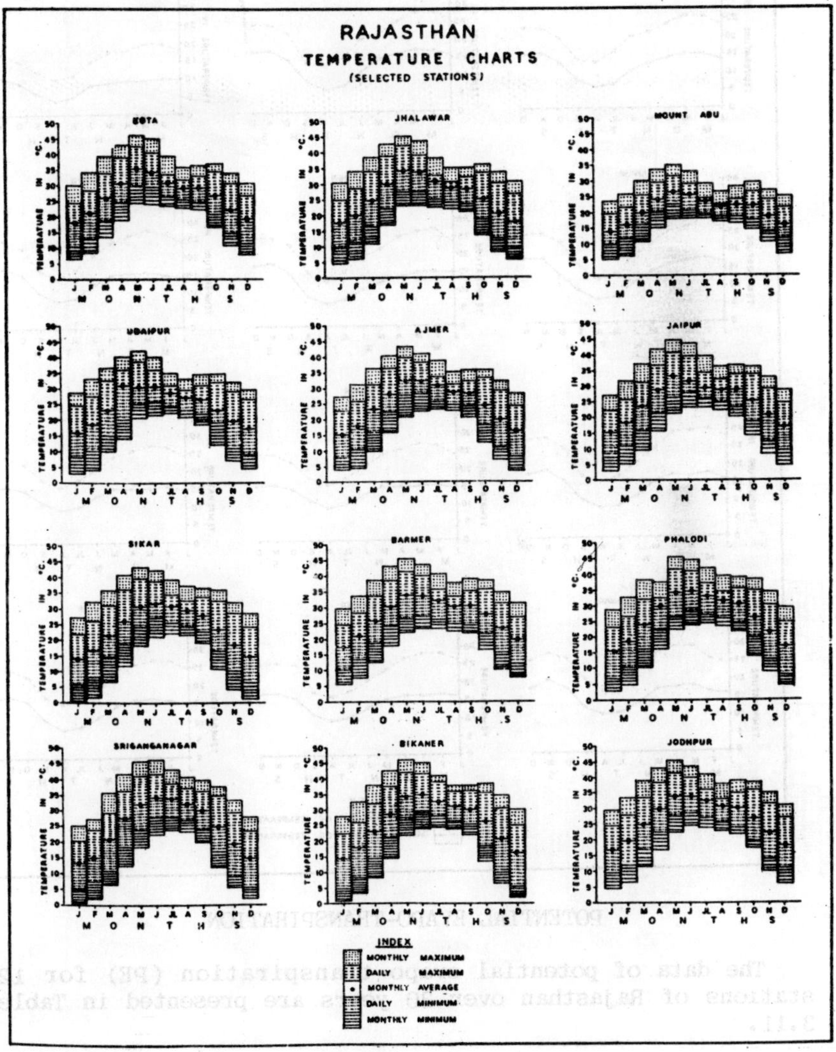

FIGURE 3.3

FIGURE 3.4

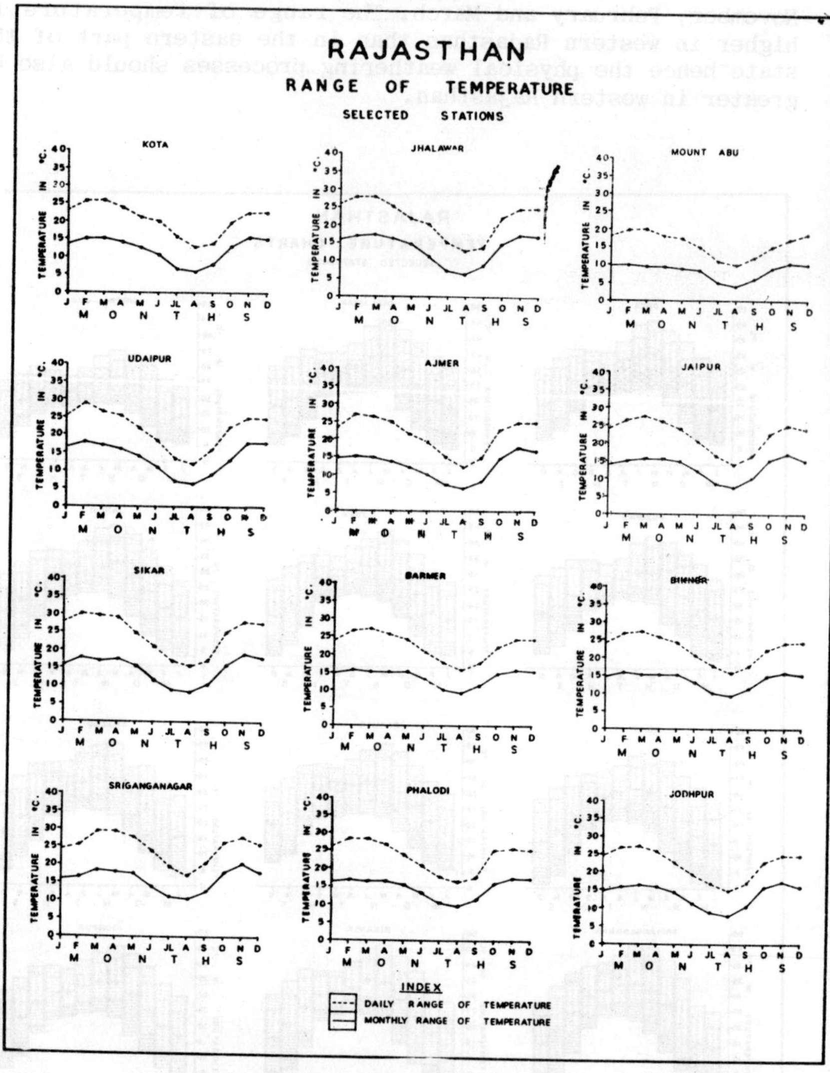

POTENTIAL EVAPO-TRANSPIRATION

The data of potential evapo-transpiration (PE) for 12 stations of Rajasthan over 30 years are presented in Table 3.11.

TABLE 3.11 Potential Evapotranspiration (PE) over Rajasthan

(In mm)

Station	Jan	Feb	Mar	Apr	May	Jun	July	Aug	Sept	Oct	Nov	Dec	Total
Rajasthan West													
Ganganagar	43.0	65.0	115.3	164.6	217.3	247.9	236.3	193.7	161.4	115.6	60.7	40.4	1662.2
Bikaner	53.3	75.3	131.4	172.4	236.8	258.3	228.2	196.6	177.7	124.7	68.3	48.8	1772.4
Phalodi	72.1	87.9	159.2	209.8	293.7	295.5	225.1	185.5	186.0	154.9	104.4	66.4	2041.0
Jodhpur	83.7	101.3	156.3	192.3	264.7	260.3	184.0	145.2	156.1	134.3	87.7	76.5	1843.0
Jaisalmer	70.6	91.6	153.2	203.5	281.0	317.4	247.6	210.7	192.0	147.0	83.9	64.1	2063.2
Barmer	81.0	100.5	159.7	206.7	259.4	239.4	178.1	152.3	163.6	147.1	92.2	77.1	1857.7
Rajasthan East													
Sikar	51.9	73.8	126.8	166.6	215.1	214.1	155.9	133.0	134.4	113.9	68.4	48.7	1503.2
Jaipur	75.6	98.2	152.7	196.4	257.4	242.6	161.7	126.4	144.8	137.2	84.7	67.0	1745.2
Ajmer	55.2	76.5	130.0	174.5	243.0	229.3	155.8	128.3	142.1	113.9	66.4	50.6	1566.3
Kota	61.6	80.3	131.4	166.9	225.0	210.3	138.6	120.6	133.3	122.4	75.4	58.3	1524.7
Udaipur	59.3	78.5	125.3	158.9	204.4	178.7	119.5	102.3	115.3	112.3	70.9	55.2	1381.2
Jhalawar	66.9	86.9	136.4	171.8	229.9	213.5	140.8	125.8	129.6	118.2	75.0	60.7	1555.9

Source : Indian Meteorological Depeartment, New Delhi.

TABLE 3.12 **Mean Wind Speed in Rajasthan (1951-60)**

Climatic	Station	Months												Mean Wind Speed
		Jan	Feb	Mar	Apr	May	Jun	Jul	Aug	Sep	Oct	Nov	Dec	
Humdi region	Kota	2.2	2.3	3.9	5.2	8.2	9.1	8.2	6.9	5.5	3.0	1.7	1.7	4.8
	Jhalawar	3.6	4.1	4.9	6.0	9.0	11.5	9.6	7.8	5.8	2.9	2.1	2.6	5.8
Subhumid Region	Mount Abu	5.0	6.0	7.3	8.3	10.4	11.9	11.6	11.1	7.6	4.4	3.6	3.7	7.6
	Udaipur	2.2	2.6	3.6	4.6	6.3	7.4	6.3	4.7	3.7	2.3	1.5	1.5	3.9
Semi-arid region	Jaipur	11.0	11.5	12.0	13.1	16.5	17.8	16.2	12.4	12.3	10.0	7.6	8.1	12.4
	Ajmer	3.4	4.4	6.3	8.4	13.1	14.6	11.6	9.5	8.4	4.0	2.3	2.5	7.4
Transitional zone	Sikar	6.1	6.2	7.9	8.0	9.4	11.6	9.2	7.8	7.3	5.1	4.3	4.9	7.3
Arid region	Ganganagar	4.0	5.0	6.4	6.8	8.0	10.7	13.3	8.0	6.2	4.7	3.3	3.4	6.7
	Bikaner	4.6	5.1	6.5	7.3	10.1	13.3	12.8	11.0	9.4	5.3	3.5	3.7	7.7
	Phalodi	10.0	8.8	12.9	14.1	20.7	25.6	23.6	19.4	16.6	11.6	11.8	8.3	15.3
	Jodhpur	8.9	8.8	9.8	10.2	15.0	18.5	16.6	12.9	10.6	6.6	5.8	7.3	10.9
	Barmer	7.6	7.5	9.1	10.6	12.9	14.2	12.4	10.6	9.6	7.2	5.3	6.4	9.5
	Jaisalmer	8.6	8.2	10.9	12.7	18.3	27.2	24.8	21.7	16.1	8.5	5.5	6.5	14.1

Source : Hydrology Department, Jaipur, Government of Rajasthan.

The rate of PE increases from the south-east to the north-west. The minimum PE is recorded at Udaipur (1381.2 mm) and the maximum at Jaisalmer (2063.2 mm). In western Rajasthan there is an increasing trend of PE from the north-east to the south-west except at Barmer where it is 1857.7 mm. In eastern Rajasthan , Jaipur has the highest rate of PE (1745.2 mm). When we study the mean monthly data of PE it has been observed that the maximum rate of PE occurs in the months of May, June and July. Jaisalmer and Phalodi in arid Rajasthan have the highest rates of PE, 281.0 mm and 293.7 mm, respectively, in the month of May. Similarly, June is the month in which highest rates of PE are recorded at Jaisalmer and Phalodi (317.4 and 295.5 mm respectively). The stations of eastern Rajasthan have a lesser rate of PE as compared to western part in the month of May. This decreasing trend of PE is because of the earlier withdrawal of the monsoon in eastern Rajasthan.

WIND VELOCITY, SPEED AND DIRECTION

Table 3.12 indicates that highest wind velocity tends to occur in early and middle summer when the south-westerlies are blowing. The highest wind velocity is recorded in the month of June and ranges from 9.1 km/hour at Jaisalmer. The minimum wind speed 1.5 km/hour is recorded at Udaipur and 11.8 km/hour at Phalodi in the month of November. The mean wind velocity is highest at Phalodi (15.3 km/hour) and minimum at Udaipur (3.9 km/hour).

The dominant characteristic of the winds in Rajasthan is their tendency towards an almost complete reversal between the winter north-eastern monsoon and the summer south-westerly monsoon. This is illustrated in Fig. 3.5. It is also evident that the mean wind direction at Bikaner, Phalodi, Jodhpur, Barmer, Udaipur and Mount Abu is from the south-west. At Ganganagar and Jaipur on most days the wind blows from the north-westerly, however, wind directions in the state from the south-west and the north-east.

It would be in the fitness of things that the wind velocity and direction should have an effect on the intensity of abrasion by wind and the formation of dunes in the desert of Rajasthan. It has been shown that the orientation of the dunes in Barmer, Bikaner, Churu and Jhunjhunu is related to the direction of the wind. The ventifects in Jaisalmer area are the product of winds of high velocity.

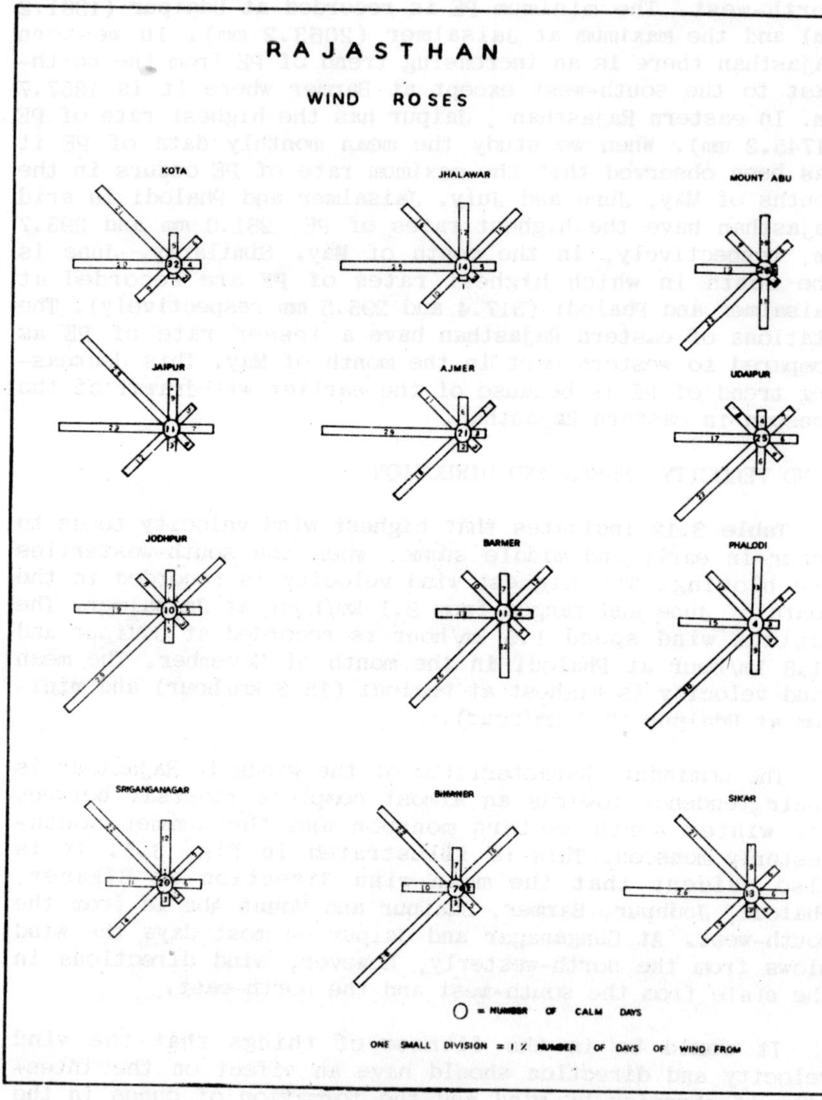

FIGURE 3.5

NOTE

Thornthwaite Moisture Index is: $IM = \frac{WS-WD}{WN}$ expressed as percentages, where WS = Annual Water Surplus, WD = Annual Water Deficit, and WN = Annual Water Need.

FOUR

Palaeoclimate and Desertification

Rajasthan bears incomparable geomorphological, archaeological, palynological and pedological evidences of climate change. The landforms created by various geomorphic processes under different climatic phases have been used to construct the chronological sequences of the climatic changes in different parts of the world during the Quaternary period. In Rajasthan, especially in the desert region, several, climato-morphogenetic studies have been conducted by Singh, (1971), Singh, et al. (1974), Singh (1977), Vishnu Mittre (1976), Ghose (1964, 1965), Singh and Ghose (1977). Ghose et al. (1977), Roy and Pandey (1971), Abhichandani and Roy (1966) and other workers from Central Arid Zone Research Institute, Jodhpur: Stein (1942), Raikes (1968), Sood et. al. (1978), and Yash Pal et. al. (1980). Allchin, Goudie and Hegde (1978), Meher Homiji (1973), Bryson and Barries (1967), Wadia (1960), Ahmad (1969), Verstappan (1970), Agrawal, et. al. (1978, 1979, 1980), Mishra et. al. Rajaguru (1975, 1980), Whilhelmy (1969). In the present chapter, however, the geomorphological, archaeological, palynological characteristics of Rajasthan have been critically examined, supported by detailed field investigations and recent reporting of bauxite and laterite deposits on the Deccan trap in south-eastern Rajasthan, to reconstruct the chronological framework of the past climatic changes during the Quaternary, and even the Tertiary Period.

Various evidences of palaeoclimatic changes in Rajasthan are being discussed under the following heads:

1. Geomorphic Evidences, which include:
    a. paleo-drainage channels;
    b. kankar pan-calcret as an indicator of climatic changes;
    c. burried pediments;
    d. sand-dunes-fossil dunes and reddened dune sands;

e. inter-dunal flats (pans);
f. boulders and gravels; and
g. laterite and bauxite.

2. Archaeological evidences.
3. Palaeo-botanical evidences.

GEOMORPHIC EVIDENCES

Paleo-drainage Channels

The fluvial landform features such as paleo-channels, vast older alluvial graded plains, rocky pediments and paleo-flood plains and terraces created by the well-integrated drainage system of the rivers like Luni, Sukri-Jawai, Ghaggar and Yamuna, the buried channels of inland drainage of Kantli, Sota, Dohan, Barah, Banganga and the Paleo-flood plains of the Chambal system provide evidence of a prolonged humid phase in the Pleistocene period. Several workers, namely, Ghose (1980), Singh and Ghose (1977) of the Central Arid Zone Research Institute have identified the prior drainage channels of the Luni river system on aerial photographs (Fig. 4.1). Associated with these old courses are lines and clusters of saline depressions such as Pachbhadra and Thob (west of Jodhpur), Sambhar lake, Kuchaman and Didwana and certain ranns near Pokaran (Lawar-ka-Rann, Dediaka Rann, Jhalaria-ka-Rann). The old river courses as studied by scientists of the Central Arid Zone Research Institute adequately account for the origin of salts in the pans. However, this idea has been refuted by Sood and Pathan (1982) of the Space Application Centre, Ahmedabad who on the basis of landsat imageries has pointed out that the alignment of salt lakes are guided by major lineaments and not the courses of prior channels. Further, on the basis of landsat imageries, Yeshpal, et al. (1980) have studied the paleo, environmental changes of the lost Saraswati river (Fig. 4.2). They have pointed out that during the period 4-5 millennia BP, north-western Rajasthan was much greener with the Saraswati flowing through it. Some of the present rivers joined together to form the Saraswati which probably discharged into the Rann of Kutch, the Nara without joining the Indus. Again the Sutlej once flowed into Ghaggar. The paleo-yamuna changed its course at least three times before joining the Ganga through its present course. Yashpal and coworkers have attributed paleo-environmental changes in the north-western subcontinent to climatic, tectonic and anthropo-genetic factors. Tectonic, factors according to them, assumed overwhelming importance as the

Saraswati was structurally controlled by the "Enelchelon Fault"

The study of landsat imageries, especially in the north-eastern and eastern part of Rajasthan has been substantiated by field studies by the present author who has identified paleo-flood plains and the migration of the river Banas at Uncha (District Chittor) and Berach and the inland drainage system of north eastern Rajasthan including the streams, Kantly, Sabi, Dohan, Barah and Banganga. These inland stream courses formerly were a part of the integrated system of the Ghaggar-Yamuna system but owing to the migration of the Yamuna towards the east and of the Ghaggar towards the north-west, these streams were left as beheaded streams (Sharma, 1983). The paleochannels of these streams have been identified on landsat imageries and coarse fluvial sand underlain by gravel have been observed in the field. The field observations near Kolayat in the Bikaner area also indicate the existence of prior stream courses through this area as evidenced by fluvial sand and gravels near the Kolayat-Bikaner road. Paleo-flood plains have been identified on the Karauli plateau of the tributaries of the Ghambir river. Near Uncha a wide floodplain containing big boulders and cobbles mixed with red sand is widespread on the tertiary planation surface. The existence of a gravel ridge in Nagaur district 70-75 km west of the main Aravalli range is also a fossil feature which indicates strong fluvial conditions suggesting that the sediment was carried by braided river channel from the Aravallis in the Quaternary period (Allchin goudie and Hegde 1978).

Sen (1971) has also concluded that the Rajasthan desert is true desert owing its origin to the same climatic oscillations during the Pleistocene era which resulted in the formation of middle and low latitude deserts. The desiccation of the Saraswati or the Ghaggar had nothing to do with the origin of this desert. A subsidence in the Ganges delta aided by a later uplift during sub-recent times in the Aravalli-Delhi axis led to the gradual rise of the Eocene sea floor of the Indian desert. This resulted in the shifting of the Yamuna, which was hitherto a tributary of the Saraswati of the Indus system to the Ganges system. The increased aridity, owing to the rise of the Himalaya and lowering of the Aravallis probably caused the shifting and desiccation of the rivers such as the Saraswati that made the desert gradually uninhabitable and made the people of the Saraswati and the Ghaggar valleys as nomadic. Uneconomic exploitation of resources and deforestation contributed

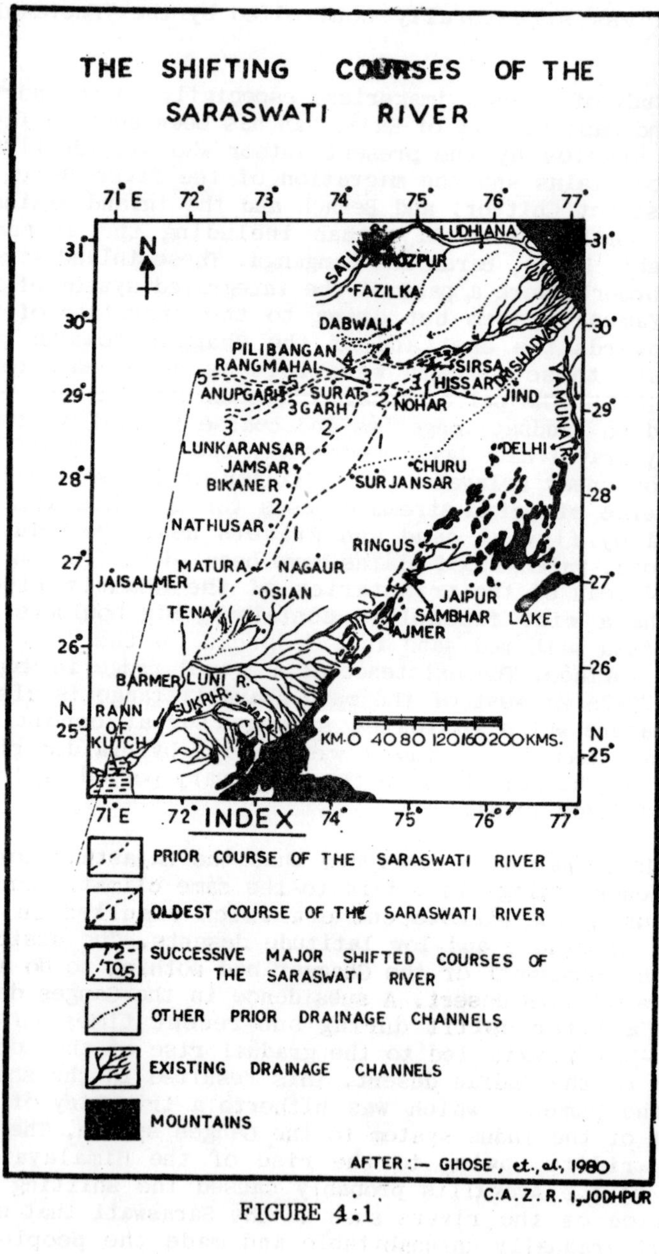
THE SHIFTING COURSES OF THE SARASWATI RIVER
LUDHIANA
FIROZPUR
SATLUJ
FAZILKA
DABWALI
PILIBANGAN
RANGMAHAL
SIRSA
HISSAR
JIND
ANUPGARH
SURAT GARH
NOHAR
YAMUNA R.
LUNKARANSAR
DELHI
JAMSAR
CHURU
BIKANER
SURJANSAR
NATHUSAR
RINGUS
MATURA
NAGAUR
JAISALMER
OSIAN
TENA
JAIPUR
SAMBHAR LAKE
AJMER
BARMER
LUNI R.
SUKRI R.
RANN OF KUTCH
KM.0 40 80 120 160 200 KMS.
INDEX
PRIOR COURSE OF THE SARASWATI RIVER
OLDEST COURSE OF THE SARASWATI RIVER
SUCCESSIVE MAJOR SHIFTED COURSES OF THE SARASWATI RIVER
OTHER PRIOR DRAINAGE CHANNELS
EXISTING DRAINAGE CHANNELS
MOUNTAINS
AFTER :— GHOSE, et., al., 1980
C.A.Z.R. I, JODHPUR

FIGURE 4.1

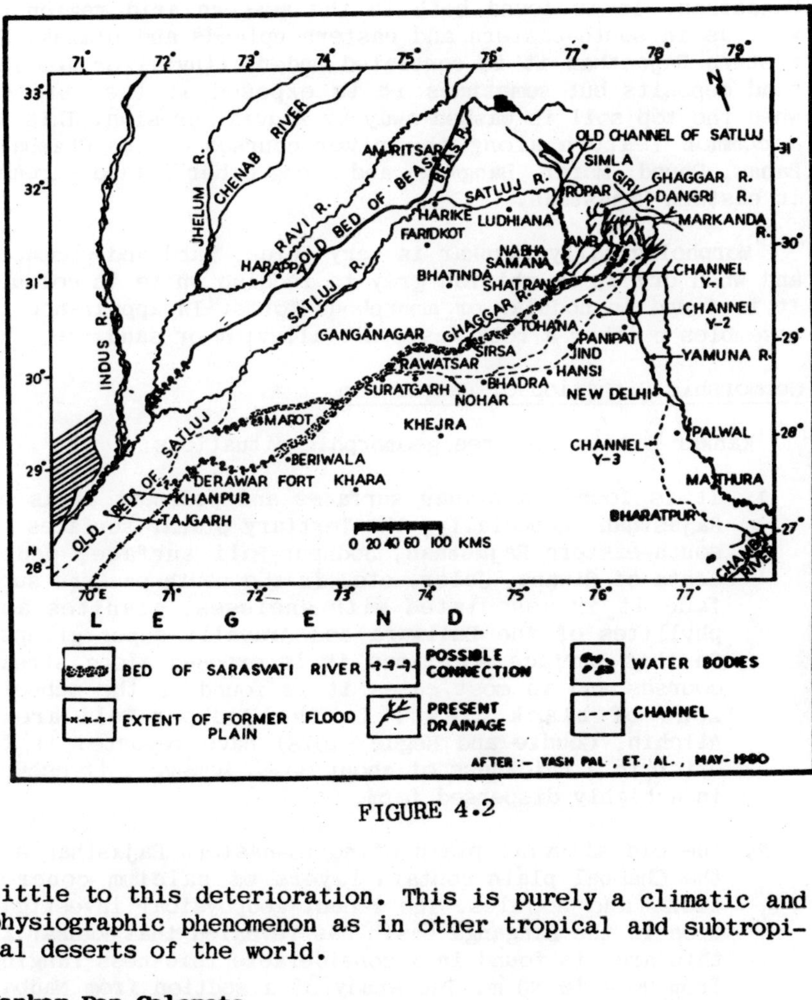

FIGURE 4.2

little to this deterioration. This is purely a climatic and physiographic phenomenon as in other tropical and subtropical deserts of the world.

Kankar Pan-Calcrete

Kankar ($CaCO_3$) or calcrete is quite ubiquitous in Rajasthan. It has been used to indicate climatic conditions though there is considerable divergence of views about its exact significance. Generally, kankar is taken as an indicator of aridity, but Allchin and Goudie (1971) have interpreted the kankarized sand dunes to mark the onset of humid conditions.

The kankar, known as <u>rotha</u> in the local language is the most common and widespread of all the quaternary products of Rajasthan. It is found both in the western arid region as well as in south-eastern and eastern uplands and plains. In western Rajasthan it is concealed under alluvial or aeolian sand deposits but sometimes it is exposed at the surface when the top soil is washed away by fluvial erosion. This is a common feature along the river courses of the Chambal, Banas, Dhund, Morel, Banganga and along other inland streams in eastern Rajasthan.

Morphologically, kankar is very thick, hard and compact, and when dry it is whitish grey to reddish white in colour. It is found in nodular or amorphous forms. In appearance it resembles neither a rock waste nor alluvium or sand.

Geomorphic Situation of Kankar Pan

Kankar exists in three geomorphic situations:

1. It is found on planar surfaces and pediment areas of Rajasthan, especially, the Tertiary planar surfaces of south-eastern Rajasthan, Jodhpur-Pali surface, pediments of Siwana, Jalor, etc. In the south-eastern surface it is associated with gneisses, granites and phyllites of the Bhilwara and Aravalli super groups. On this surface sometimes it is exposed along stream courses and in most cases it is found in the subsoil zone of black clays. In the Jodhpur-Pali area, Allchin, Goudie and Hegde (1978) have reported it to exist in a thickness of about 30 m. However, it occurs in a highly dispersed form.

2. The old alluvial plain of north-eastern Rajasthan and the Chambal plain contain layers of calcium concretions and nodules. The recent geophysical investigation in the Banganga basin has revealed that kankar in this area is found in a considerable thickness ranging from 20 m to 60 m. The study of a section from Nadbai to Chaksana reveals that it occurs below clay and sand layers at a depth of 20 m and continues up to 60 m (Fig.4.3). It is found admixed with clay and silt. In a section from Kumbher to Uchain the thickness is from 20 m to 40 m (Fig 4.4). In a section Fig. 4.5 across the Banganga it is found exposed on the surface, below it is found in pockets. The thickness here ranges from 20 m to 30 m. The thickness of the kankar layer is comparatively less in a section from Uchain to Kumher. However, it is not less than 10 m. It is interesting

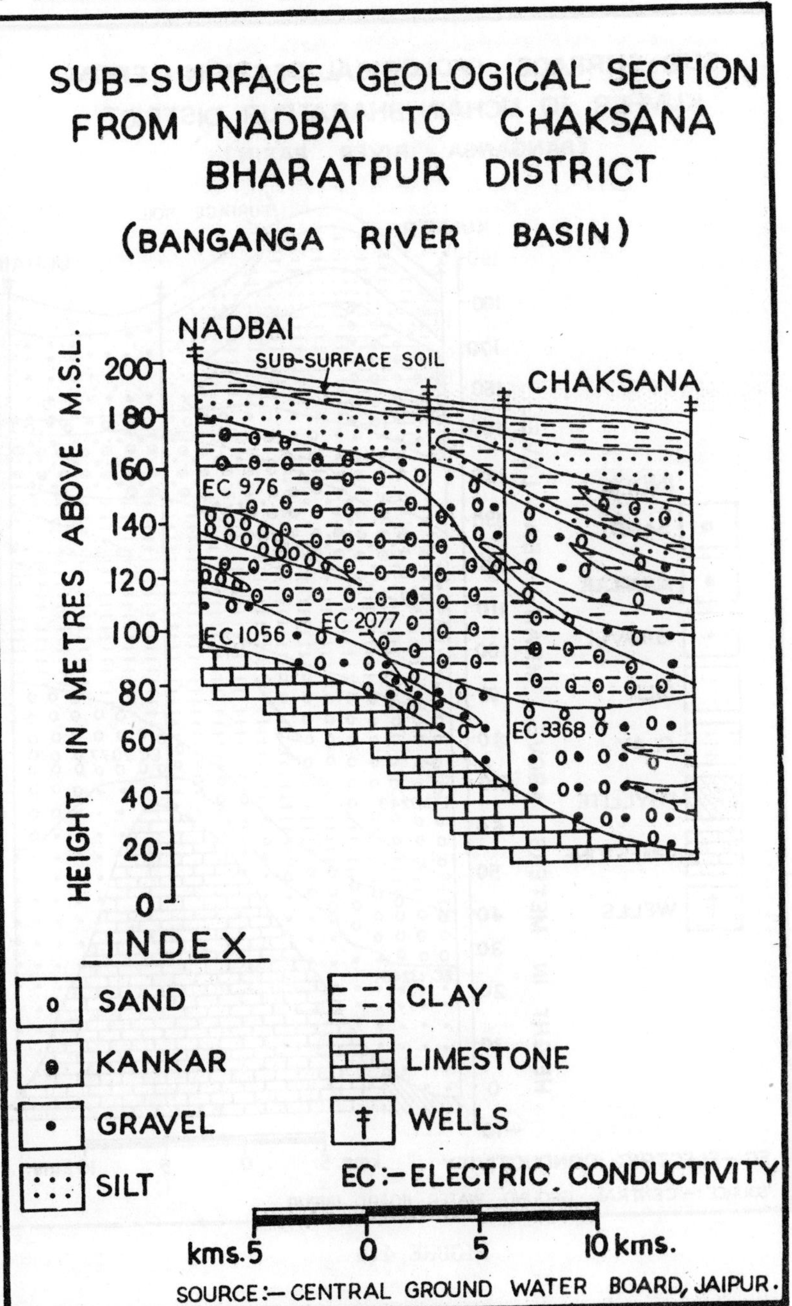
SUB-SURFACE GEOLOGICAL SECTION
FROM NADBAI TO CHAKSANA
BHARATPUR DISTRICT
(BANGANGA RIVER BASIN)
NADBAI
SUB-SURFACE SOIL
CHAKSANA
HEIGHT IN METRES ABOVE M.S.L.
200
180
160
140
120
100
80
60
40
20
0
EC 976
EC 1056
EC 2077
EC 3368
INDEX
SAND
KANKAR
GRAVEL
SILT
CLAY
LIMESTONE
WELLS
EC :- ELECTRIC CONDUCTIVITY
kms.5
0
5
10 kms.
SOURCE :- CENTRAL GROUND WATER BOARD, JAIPUR.

FIGURE 4.3

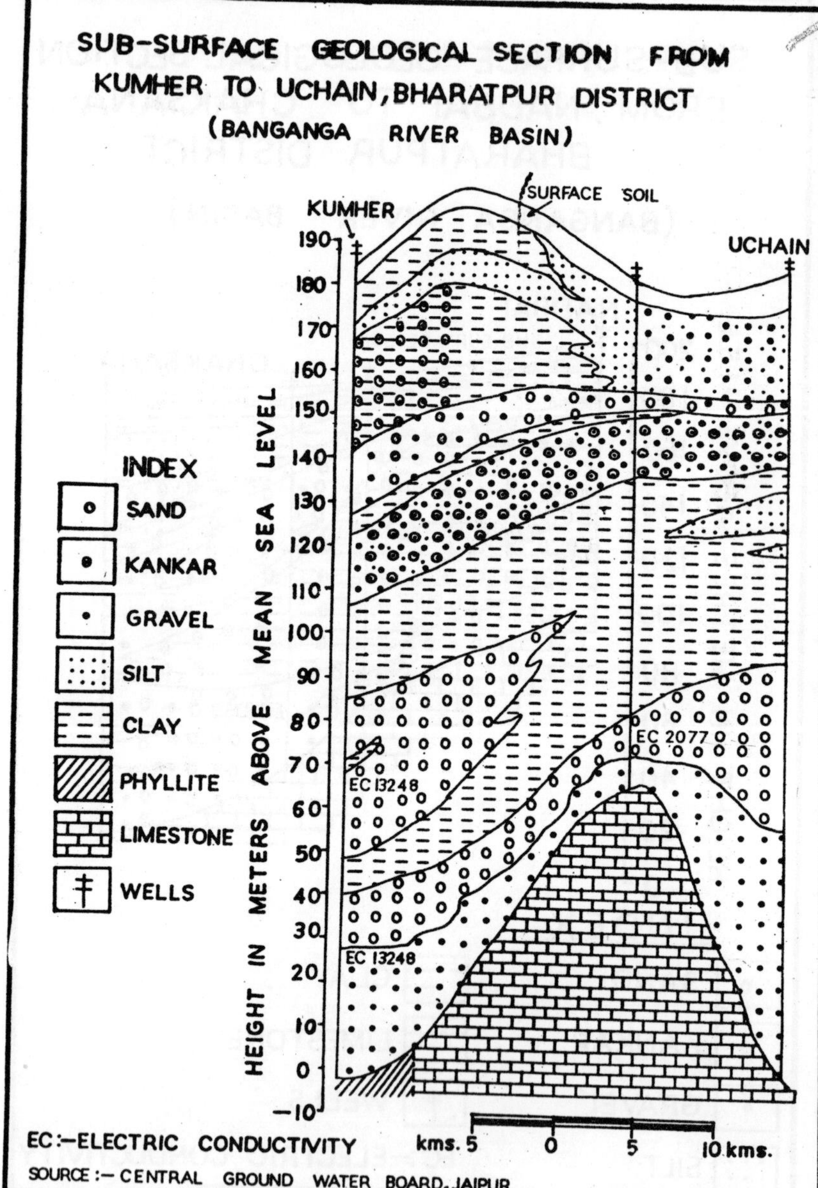
SUB-SURFACE GEOLOGICAL SECTION FROM
KUMHER TO UCHAIN, BHARATPUR DISTRICT
(BANGANGA RIVER BASIN)
SURFACE SOIL
KUMHER
UCHAIN
HEIGHT IN METERS ABOVE MEAN SEA LEVEL
190
180
170
160
150
140
130
120
110
100
90
80
70
60
50
40
30
20
10
0
−10
EC 2077
EC 13248
EC 13248
INDEX
SAND
KANKAR
GRAVEL
SILT
CLAY
PHYLLITE
LIMESTONE
WELLS
EC:—ELECTRIC CONDUCTIVITY
kms. 5
0
5
10 kms.
SOURCE:— CENTRAL GROUND WATER BOARD, JAIPUR

FIGURE 4.4

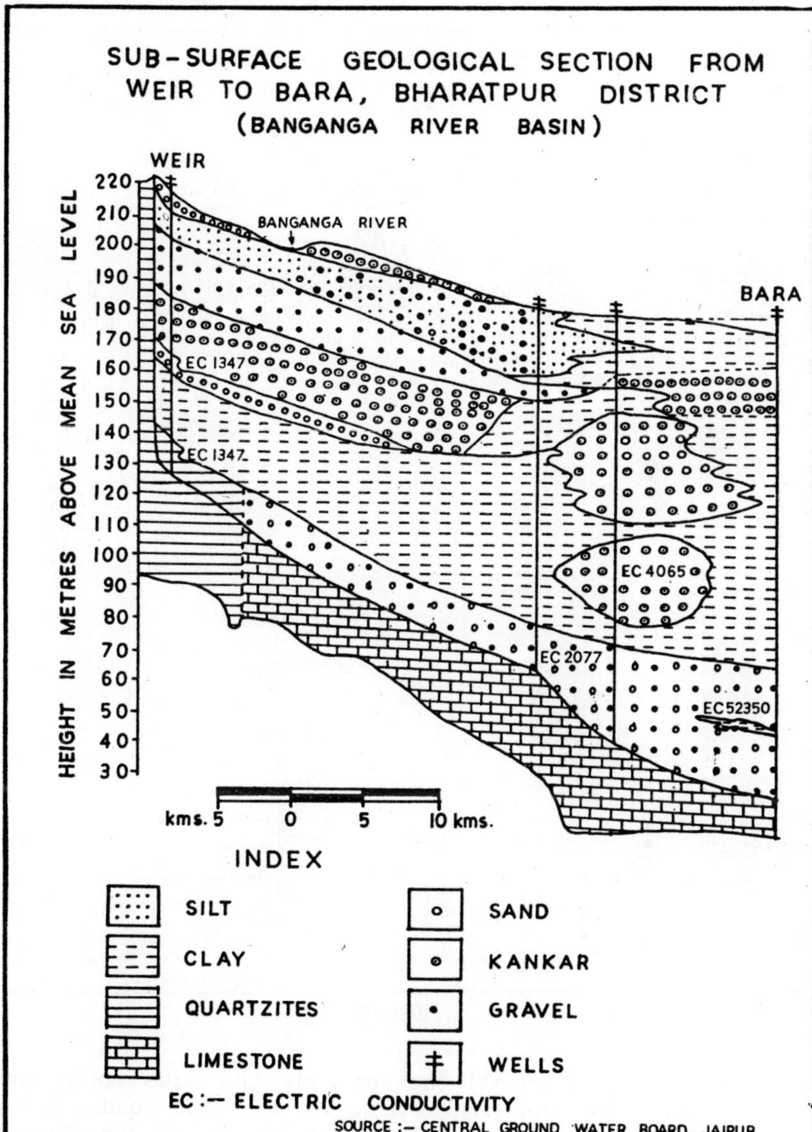
SUB-SURFACE GEOLOGICAL SECTION FROM WEIR TO BARA, BHARATPUR DISTRICT
(BANGANGA RIVER BASIN)
WEIR
BANGANGA RIVER
BARA
HEIGHT IN METRES ABOVE MEAN SEA LEVEL
220
210
200
190
180
170
160
150
140
130
120
110
100
90
80
70
60
50
40
30
EC 1347
EC 1347
EC 2077
EC 4065
EC52350
kms. 5 0 5 10 kms.
INDEX
SILT
CLAY
QUARTZITES
LIMESTONE
SAND
KANKAR
GRAVEL
WELLS
EC :— ELECTRIC CONDUCTIVITY
SOURCE :— CENTRAL GROUND WATER BOARD, JAIPUR

FIGURE 4.5

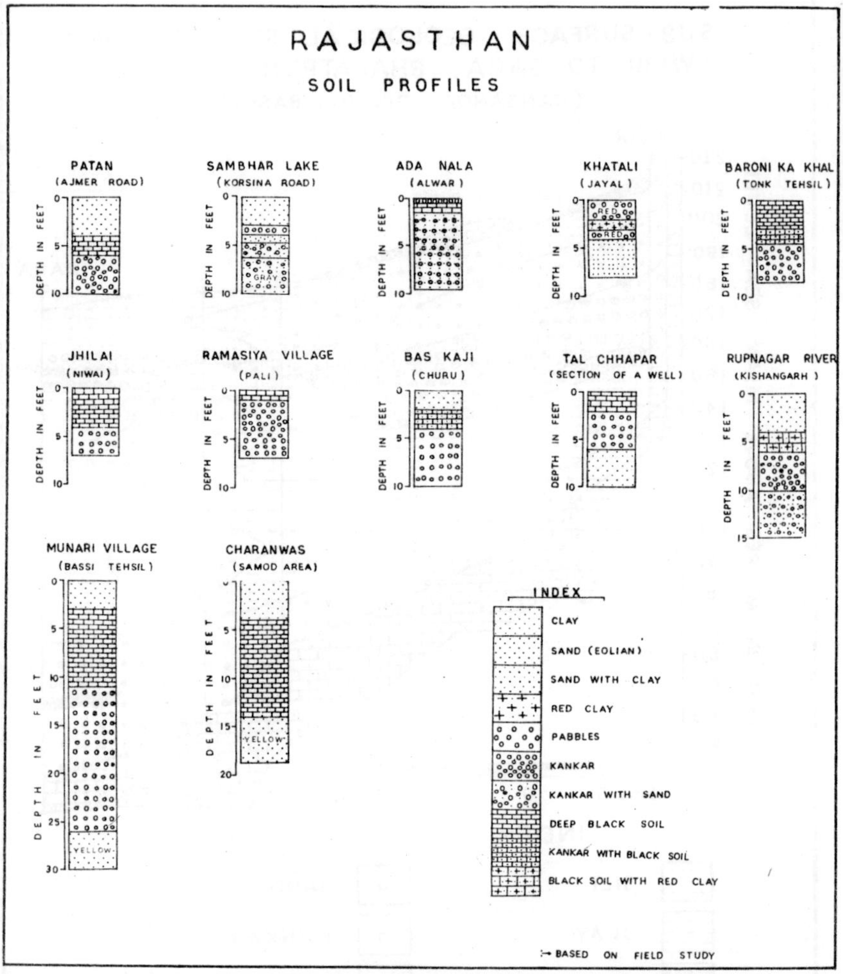

FIGURE 4.6

to note that in all the sections the base rocks are limestones and phyllites. The electric conductivity generally decreases from the upper to the lower layers.

The study of a number of other profiles from upland areas and nalas also give a clear picture of the thickness of the kankar pan (Figs.4.5 and 4.6). In a section at Munari Well (Bassi Tahsil), the thickness is about 6-7 m. The stream cuttings also give a fairly good picture of thickness. The average thickness is found to be about 2-3 m (Fig.4.6).

3. Kankar is also found in association with fossil dunes. It is widespread in western Rajasthan. The obstacle dunes at Jaipur, Banada, Jhir, Achroal, Lalsot, Newai, Bhairana (Bichun) and ChandSen in eastern Rajasthan are sites of fossil dunes and relict sands which contain kankar tubes. The interdunal flats in Bikaner, Churu, Ganganagar, Jaisalmer, and Jhunjhunu districts are also characterized by kankar pans in a dispersed form.

An analysis of kankar samples has been carried out. The selection of samples was made with a view to represent different geomorphic situations. The data has been tabulated in Table 4.1.

TABLE 4.1. **Geomorphic Situation of Kankar Pan**

Name of site	Geomorphic situation	$CaCo_3$%
Kota	Chambal river bank	16.151
Kota	Chambal river bank	12.730
Jaipur	Fossile dune at Moti Dungari	21.380
Ramasiya village	Interdunal flat western Rajasthan	13.414
Pinapiya village	Interdunal flat western Rajasthan	18.742
Dungagarh village	Interdunal flat western Rajasthan	19.035
Roisabsar village	Interdunal flat western Rajasthan	12.144
Malsar village	Interdunal flat western Rajasthan	15.418
Thar series	Level plain (Rajasthan Canal)	16.4
Sakhi series	Level plain (Rajasthan canal	9.2
Masitawali series	Flood plain (North-western Rajasthan)	19.2
Kavani series	Level plain (Rajasthan Canal)	17.8
Nachana series	Undulating plain (Western Rajasthan)	10.8
Badrasar series	Level plain (Rajasthan Canal)	17.2
Shobhesar series	Undulating surface(Rajasthan Canal)	11.6
Rupangarh	South-eastern upland (Kishangarh)	19.279

Table 4.1 indicates that the highest percentage of $CaCo_3$ is found in the Jaipur dune (fossil) at Jaipur. The percentage of $CaCo_3$ is also quite high in a sample of the planar surface of south-eastern Rajasthan. In the interdunal flats the percentages range from 12.144 to 19.035.

C^{14} dates of kankar:

The problem of carbon 14 dating of some kankar samples of Rajasthan was taken up by Dhir (1976) and by Agrawal and his team (1976-77) at the Physical Research Laboratory, Ahmedabad. In the Central Luni Basin, the Calchi deposits (kankar pan) within the alluvial material has been dated by the C^{14} method by Dhir as 28000 years to 30,000 years and the earliest age for the alluvial material is perhaps 40,000 years. Agrawal and Roy (1976) tried to date various fractions of kankar with radioactive carbon but found the range of > 40,000 BP infinite for C^{14} dating. However, using younger kankar samples they were able to estimate their age as summarized in Table 4.2.

Origin of kankars:

There is a little doubt that the kankar of Rajasthan is a product of epigenic changes of physico-chemical reactions. There are several models for the formation of kankar. La Touche (1902) believed that the carbonate pan has developed from the forminifera brought by storms from the Kutch region. Auden (1952) pointed out that kankar had no observed source of lime other than the narrow outcrops of the Vindhyans. Krishnan (1952) attributed the cause for such development to marine transgression up to the Mesozoic period. Ray-Choudhri and Sen (1952) are of the opinion that the kankar has developed owing to the weathering of minerals in the soil under humid conditions related to the earlier

TABLE 4.2. **Radiocarbon dating of calchi samples from Rajasthan**
Agrawal et al. (1977)

PRL No.	Site	Age B.P. (1/2 = 5730 years)	
358	Nalisar lake	13120 (Ist fraction)	+ 290 - 280
358	Nalisar lake	16030 (2nd fraction)	+ 380 - 370
359	Nalisar lake	16010 (Ist fraction)	+ 610 - 570
359	Nalisar lake	20580 (2nd fraction)	+ 940 - 850
359	Nalisar lake	22980 (3rd fraction)	+ 720 - 660
373	Gudlai nadi	12070 (Total)	+ 460 - 440
374	Gudlai nadi	15400 (Ist fraction)	+ 230 - 220
374	Gudlai nadi	23570 (2nd fraction)	+1090 - 960
374	Gudlai nadi	18840 (3rd fraction)	+ 600 - 560
375	Gudlai nadi	27220 (total)	+3280 -2340
376	Gudlai nadi	>40,000 (total)	
379	Chirai	23000 (total)	+ /10 - 650
378	Chirai	22180 (total)	+ 660 - 610
337	Chirai	28630	+2760 -2070

phase of the desert. Taylor et. al. (1955) suggested four factors responsible for the development of kankar, (i) rainfall less than 200 mm with net excessive evaporation, (ii) relatively stable or slowly degrading land surface, (iii) deep weathering of parent rocks, and (iv) effect of rainwater containing carbonic acid on weathered rocks. Roy, Chatterjee and Pandey (1969) are of the opinion that the carbonate pan has developed owing to the katamorphic processes in the weathered zones of regional geological formations, and the salts of which the pan is composed have developed from the rocks, owing to physiochemical reactions. The fact that kankar pan occurs on the old mid-Miocene surfaces (Bhilwara, Shahpura, Ajmer, Tonk, Deoli, Malpura areas) piedmont zones of western Rajasthan and in the older aggraded alluvial plains together with its great thickness (30-40 m) indicates that this process has been active for a very long time in the geological history of the state rather than for the last few thousand years.

The regions in which kankar pan develop are mostly covered by granitoid, malani rhyollits, gneisses and phyllitic rocks. These rocks are mainly composed of stable basic silicates. Under metamorphism some of the weakly held base metals are released when reacted upon by carbonic acid contained in rain water. Infiltrated rain water charged with carbonic acid starts chemical reactions in the weathered zone. When the carbonic acid becomes stronger than silicic acid, under the katomorphic conditions as stated by Roy et. al. (1969), new salts are formed, predominantly calcium carbonate. The oxidation and leaching processes owing to the intensified action of oxygen and carbon dioxide result in rock solution (crest) in the zone of weathering. These processes saturate the subsurface zone with calcium and magnesium bicarbonates, ferric oxide, sulphates and chlorides of the same metals, and with salts of sodium and potassium. The calcium and magnesium carbonates precipitate from the solution. In the deeper zones sulphates and chlorides also precipitated along with carbonates This process leads to the cementation of the subsurface formations by the filling of the interstices and crevices, and the coating of the rock surface by these salts.

In south-eastern Rajasthan and in the piedmont zones around isolated monadnocks, and inselbergs and in western Rajasthan layers of calcium carbonate are found above the parent rock-gneiss and granites. These rock fragments after mechanical weathering have been washed away from the higher to the lower slopes and have been deposited in bands. Later on, these fragments have been cemented with lime. The lime nodules of the pan found in the alluvial plains of the Banganga, Chambal, Gambhir, Morel and Banas rivers invariably consist of fragments of rocks of the surrounding areas.

It may thus be concluded that in the formation of kankar the parent rock is first mechanically disintegrated, and then the primary and secondary products that result from the chemical reactions, as stated above either remain in situ or are transported elsewhere. Such processes are very slow; however, they do take place under normal climatic conditions.

Kankar Pan and Climatic inferences:

Few attempts have been made so far to assess the paleoclimatic significance of the kankar pan. However, several studies by Allchin, goudie and Hegde (1978), Agrawal et. al. (1976, 1977, 1980), Roy, Chatterjee and Pandey (1969), Khandelwal (1975), Singh (1977) have demonstrated that the development of kankar is very much related to an arid climate with rainfall varying from 100 mm to 500 mm, high temperatures and prolonged droughts. Agrawal et. al. (1977) on the basis of stable isotope analysis, chemical analysis, semientological analysis and SEM (Scanning Electron Microscopy) data concluded that there is a clear sequence of wet and dry phases in the sediments of their study area. Different sites of the Rajasthan desert show that kankar is associated with a wet phase and pure sand with a dry phase. Allchin and Goudie (1971) have interpreted the presence of kankar to mark the onset of humid conditions. The palaeobotanical studies done by Jacob, Sastry and Sastry (1952), Singh (1967) and Khandelwal (1975), showing the occurrence of gestropods associated with kankar indicate that kankar was deposited under fresh water conditions. Voute (1964) has concluded that the crust of the Mediterranea area represents a deposit of the inter-fluvial periods of the Quaternary era. Verstappen et. al. (1969) stated that the formation of the kankar concretions was most likely to be associated with past climatic changes, alternating between less arid and more arid conditions. Blagoves chenskiy (1968) reports that in the Thar desert kankar deposits date from the very end of the Tertiary period but this idea has been discarded on the grounds that a kankar pan is generally associated with the older alluvium. In Rajasthan since it is found in older surfaces as well as in older alluvium. Therefore it may be inferred that the kankar formation was started after the Tertiary period and the process continues down to date.

Buried Pediments

The systematic geomorphological mapping of western Rajasthan by Surender Singh of the Central Arid Zone Research Institute, Jodhpur has revealed that there are numerous patches of buried pediments which can very well be picked up on aerial photographs. A few patches of pediplain and pediments exist in the Udaipur-Chittorgarh area and on Karauli plateau. These pediments now buried with aeolian and fluvial sediments, are fossil features and indicate the pleistocene humid climatic phase.

Fossil Dunes_Reddened Sand

The existence of stabilized and fossilized old dunes is an indication of former aeolian activity in the Quaternary period. The dissection of these forms provides further evidence of the dominance of fluvial activity over aeolian processes in the subsequent periods. A large number of relict dunes or fossil ergs have been identified in various parts of the world which give an indication of the palaeo climatic changes that have taken place. In North America Price (1958) and Hacks (1941) have reported in Texas and Arizona, and Smith (1965) has various phases of former aeolian activity in the sand hills of Nebraska. Tricart (1974) has used landsat imageries to identify fossil dunes in Arizona and north-east Brazil. Grove (1958) has studied the subsaharan zone of Africa. Grove and Warren (1968) in Sudan, Grove (1969) in the Kalahari, Flint and Bond (1968) in Rhodesia and Jennings (1969), Mabbutt (1971) and Bowler (1970) in Australia have studied fossil dune features and shown their significance as climatic indicators. Sarnthein (1972) has also identified some relict dunes in the persian Gulf.

In Rajasthan and adjoining areas an attempt has also been made to identify fossil dunes by Allchin, Goudie and Hegde (1978), Singh and Ghose (1977). The identification of fossil dunes in India was first made at Pavagarh hill, where large widespread sand accumulation was found against an outlier of the Deccan trap escarpment near Baroda in Gujarat (Allchin and Goudie, 1971).The extent of fossil dune features has also been mapped (Fig. 4.7). The author has identified the fossil sanddunes and sand fields beyond the limit demarcated by Allchin and Goudie. The eastern limit of the fossil dunes extends far up to Gangapur-Lalsot and up to Tonk-Newai in the south-east.

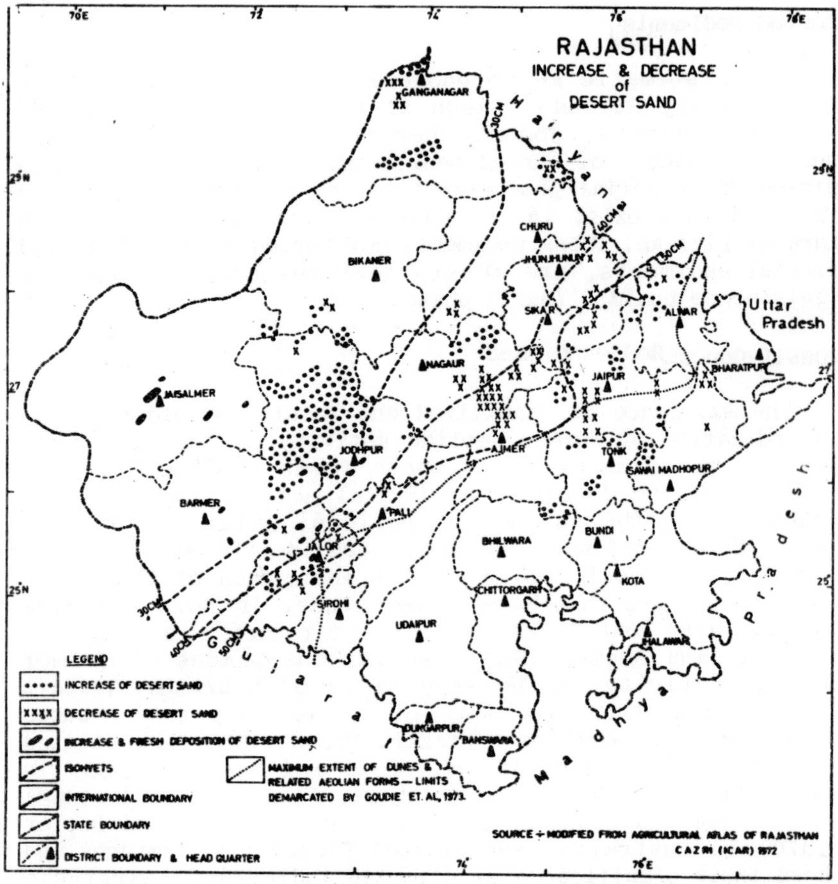

FIGURE 4.7

Fossil barchan dunes have been noticed on the bank of the Sabi river. They appear to be the remnants of earlier active barchans which have subsequently been stabilized. The stabilization and change in geometry of these dunes could have occurred because of high moisture content and growth of sparse vegetative cover. The fossil dunes present on the banks of the Sabi do not retain their characteristic shape owing to subsequent alteration by natural and human agencies. Their existing shape exhibits long parallel sand ridges with cultivated intervening portions. In only a very few cases has the crest been preserved. The possible wind direction responsible for the formation of these fossil dunes is the same as the one prevailing today. They are well ditected on photographs, by their medium grey tones, coarse texture, sparse vegetation and their characteristics topographic expression.

A description of some of the fossil dunes is given here:

Bhairana obstacle dunes (Bichun Hills)

Fossil sanddunes are also in existence at Bhairana north-west of the Bichun hill. These are obstacle dunes and are characterized by the red phase of their aeolian weathered sand (Fig.4.8). The dunes have now been dissected into deep gullies, hence their shape is not very well-preserved. Near the hill range, new aeolian features are also developing on the crest of the older dunes. This new

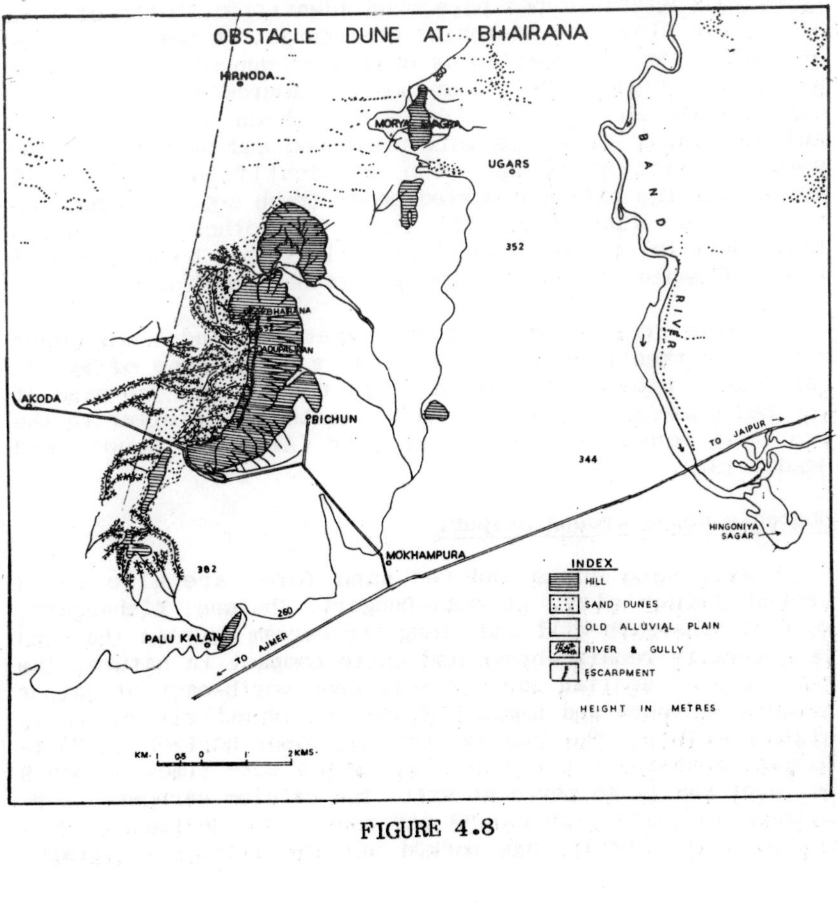

FIGURE 4.8

sand is of great thickness. The distinction between the old dune and new aeolian feature can be made on the basis of the compactness of the sand grains, colour tones and clay content. New sand is pale and yellow in colour and very loose.

Newai obstacle dunes:

The obstacle stabilized dunes at Newai are true fossil dunes. Southward from Sanganer much of the terrain is made of older aggraded alluvial plain and near (Fig. 4.9) Newai the surface is covered with sand, thus forming some striking obstacle dunes.

Newai lies 66 km. south of Jaipur on the Jaipur-Kota sub-highway. The flat topped monoclinal hill of Alwar quartzite dominates the landscape. West of this hill, two types of obstacle dunes have been identified. North of this hill there flows the perennial Kosi Ganga. On both the sides of this river lie continuous chains of dunes of both older and newer origin. The river has its source from these old dunes. Although the fossil dunes have been dissected into gullies their crest is well preserved and over the latter some microlith sites can also be identified. The fossil dunes near the hill are buried under fresh aeolian sand. Between these dunes and hill there is another stream which flows parallel to the monoclinal hill. It has been dammed at the confluence of the Kosi below Jalandhar Nath cave.

In this area there are three types of sand : an upper zone of currently reactivated sand, a lower zone of kankar sand, and dissected knolls of compact red sand. The sand of the red phase has also been deeply weathered, similar to the red sand found at Banada and Jhir (Allchin, Goudie and Hegde, 1978.

Obstacle dunes around Jaipur.

Fossil sand dunes and red sand forms are also found around Jaipur city. At Moti-Dungari, Jhalana, Kishangarh, west of Nahargarh hill and along the stream courses the sand is generally reddish-brown and quite compact in nature. The red phase of aeolian sand is well seen north-east of Jaipur between Sahipura and Ramgarh along the Dhund river. It is highly stable. The reddish sand of lower horizon at Moti-Dongari contains 5 per cent clay, which some times become 8 per cent and 12.40 per cent silt. The calcium carbonate percentage is quite high (21.38 per cent). The Geological Survey of India (1979), has worked out the litho-stratigraphy of the Quaternary deposits around Jaipur (Table 4.3.) :

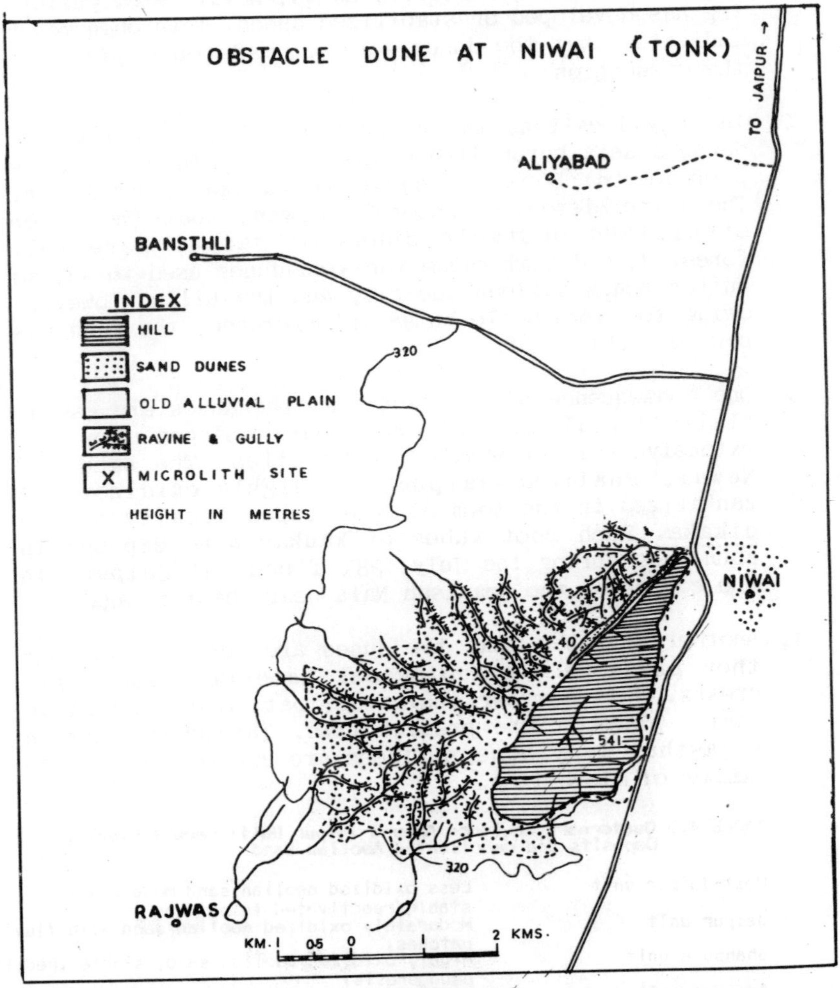

FIGURE 4.9

Evidences of relict nature of fossil dunes:

There is a great deal of evidence to the relict nature of the sand in Rajasthan. Some of the evidence is discussed here:

1. It has been observed that most of the dunes are extensively moulded by fluvial action. In the foothills of

Jaipur, Bhairana, Archroal, Samod, Newai, etc. gulleying has developed on stabilized dunes. This phenomenon indicates that the dunes were in existence prior to the dissection.

2. In many locations the natural vegetation on dunes is dense acacia scrub with grasslands; this is clearly seen at the Forest reserve at Pushkar near Ajmer. There are historical records showing that the older stabilized obstacle dunes of Jaipur were well forested, and that these forested dunes used to act as buffer zones between the city and the hills. However, owing to large scale human interference, the sand has now been shaved off.

3. One consequence of the fact that the dunes are essentially fossil is that they have developed upon them extensive soil or weathering profiles. The sands at Newai, Bhairana-Jaipur are highly oxidized and kankarized in the form of root tubes locally called dikaka. Such root tubes of kankar were exposed in abundance during the July 1981 floods at Jaipur. In the course of the Amanisah Nala near Shastri Nagar.

4. Another evidence that the dunes are very old is that they preserve archaeological material over their crests. This further indicates that no deposition is taking place at the present time. Throughout eastern Rajasthan many dune surfaces are covered by a large number of microliths.

TABLE 4.3 **Quaternary Sequence Around Jaipur Unit: Recent Fluvial Deposits and reactivated Aeolian Sand**

Post-Jaipur unit	Less oxidized aeolian sand moderately stable/reactivated in patches
Jaipur unit	Moderately oxidized aeolian sand with fluvial patches.
Shahpura unit	Highly oxidized aeolian sand, stable (pedalfar pedogenesis)
Archrol unit	Light grey and yellowish aeolian sand, stable (pedalfar pedogenesis)
Kalikharal unit	Highly oxidized fine sand with kanker and considerable clay fractions/aeolian /fluvial origi yet indeterminate
Pre-Kali Kharai unit	Fluvial sands, silts clay with boulder bed at the base

Pre-Cambrian schists and gneisses (in boulders)

Source: C. Madhav, et. al. (1979).

Interdunal Flats (Pans)

During the course of geomorphological investigations especially in the north-western parts of Rajasthan, it has been found that a morphological study of inter-dunal flats gives a good indication of paleo-climatic changes. The fluctuations in climate have left imprints in the sequence of deposits. The inter-dunal flats break the monotony of this otherwise complex dunal country. Most of the inter-dunal flats are small, dry or ephemeral lakes, contained in isolated sub-circular to sub-elliptical flats or depressions formed mainly of clay and silt rimmed by stabilized, coalesced, parabolic, longitudinal and transverse dunes in the north-western part around Taranagar, Rajgarh, Nohar, Churu, Bikaner, Nagaur, Jodhpur and Barmer.

In the districts of Bikaner, Churu, Nagaur and Jhunjhunu, most of the villages are named as sar, meaning a small lake or reservoir (Figs. 4.10 and 4.11). These reservoirs provide suitable sites for collecting surface runoff, and they generally occur in the area of interior drainage. They are temporary gathering ground for rainwater. The stormwater, after flowing for a short distance gets lost in these tracts.

Morphology of interdunal flats

For the study of interdunal flats or pans eight pans have been studied. The mechanical analysis of sediments have also been carried out. The study reveals that these are sub circular to sub-eliptical in shape. They are isolated, well-defined and enclosed. The majority of flats are enclosed by dunes of both stabilized and newer ones. These dunes range in height from 3 m to 20 m . Their width ranges from a few metres to one kilometer. The floor of these flats contain small basins, widely known as pans,(in local language sar), the word is sometimes used also to mean a shallow tectonic basin. The basin ranges in diameter from a few meters to 120-150 m. In ground plan they are of many shapes. Some are sub-circular to ovate. Some are irregular. The surface slope of these flats is less than 1°.

Pan deposits:

The texture of the surface sediments varies from loamy sand to sandy loan and clay loam in pockets (Figs.4.12 and 4.13), the clay being pure white. These are underlain by invariable thickness of kankar pan, occuring 30 cm to 180 cm

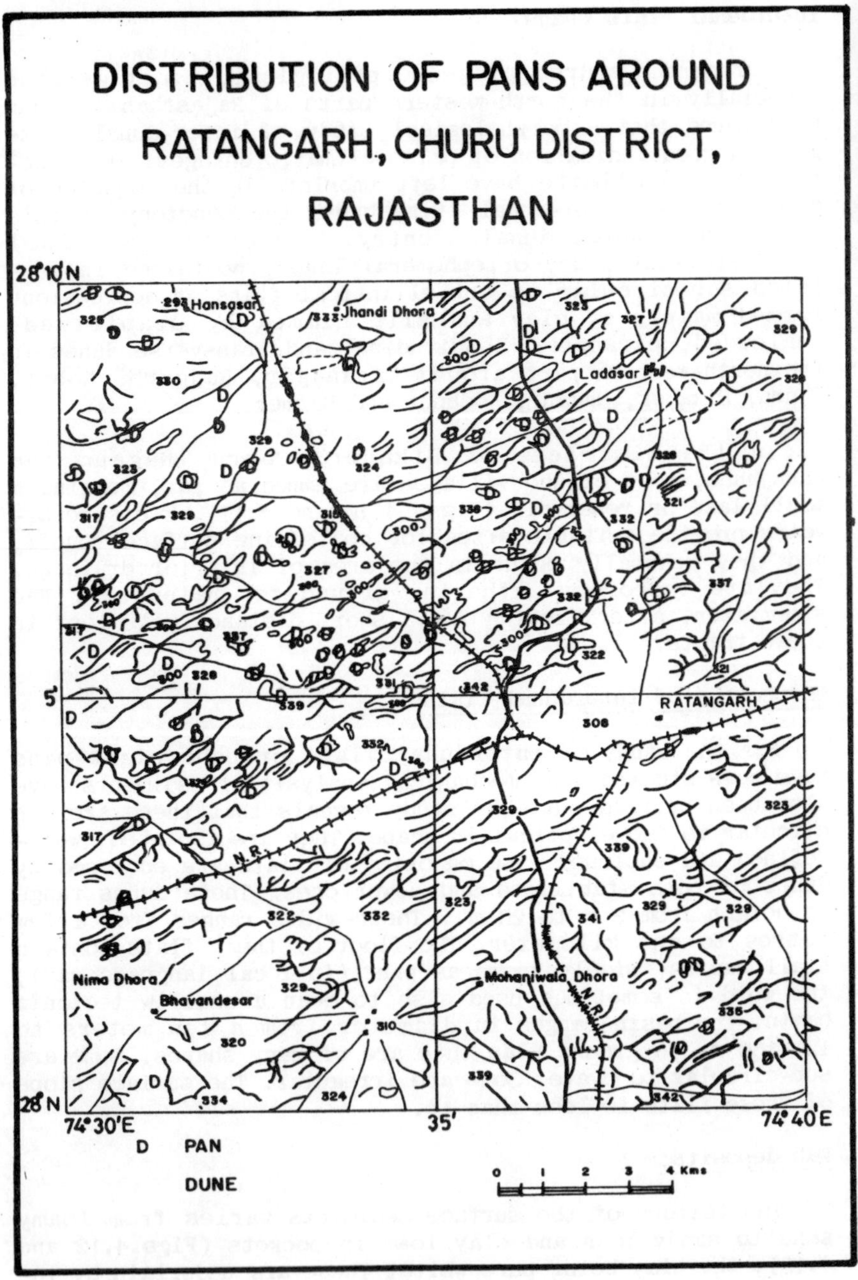
DISTRIBUTION OF PANS AROUND RATANGARH, CHURU DISTRICT, RAJASTHAN
28°10'N
Hansasar
Jhandi Dhora
Ladasar
RATANGARH
N.R.
Nima Dhora
Bhavandesar
Mohaniwala Dhora
5'
28°N
74°30'E
35'
74°40'E
D PAN
DUNE
0 1 2 3 4 Kms

FIGURE 4.10

below the surface sediments. A study of the profiles of all the sections reveals alternate layers of fresh sand, kankar, buried sand, clay and deep black soil indicating fluctuations of climatic conditions.

DISTRIBUTION OF PANS AROUND RATAN-NAGAR, CHURU DISTRICT, RAJASTHAN

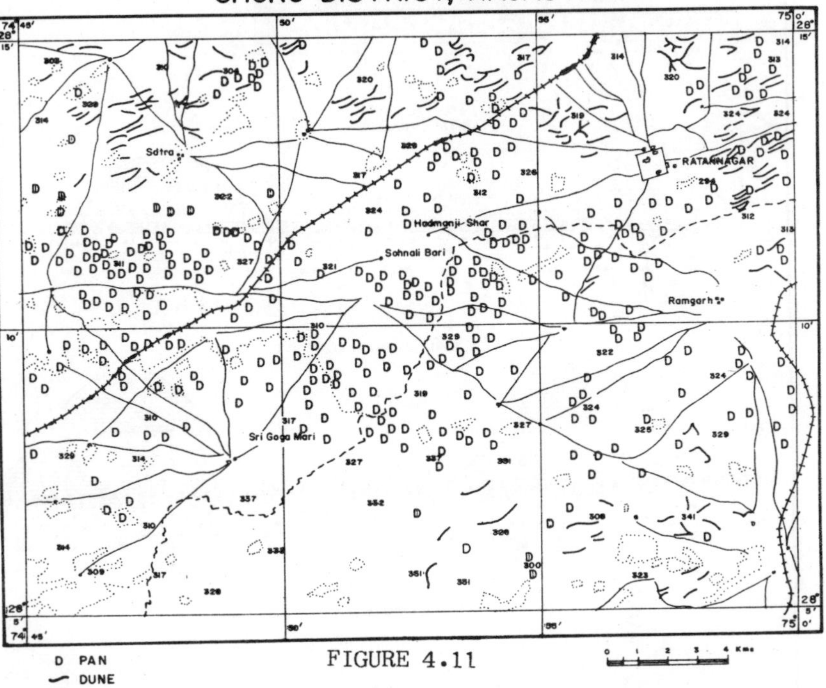

FIGURE 4.11

Interdunal flats of similar form have been identified in a number of present day arid and semi-arid areas, including South Africa (Du. Toit, 1908, 1926); South West Africa ((Jaeger, 1939); the High Plains of texas, USA (Gilbert, 1895, Evans and Meade, 1945, Reeves, 1960), the Simpson Desert Australia (Twidale, 1972); and Central Asia (Reeves, 1966). In these areas enclosed depressions with a characteristic downwind elongation are believed to have occured as a result of deflation of unconsolidated, often sandy, surface sediments by strong uni-directional winds. Passarge (1904) described the pans in the Ghana area and suggested the probable role of the game animals in their formation. However, Wayland (1953), disagreeing, with this hypothesis discounted the influence of game animals. Boocock and Van

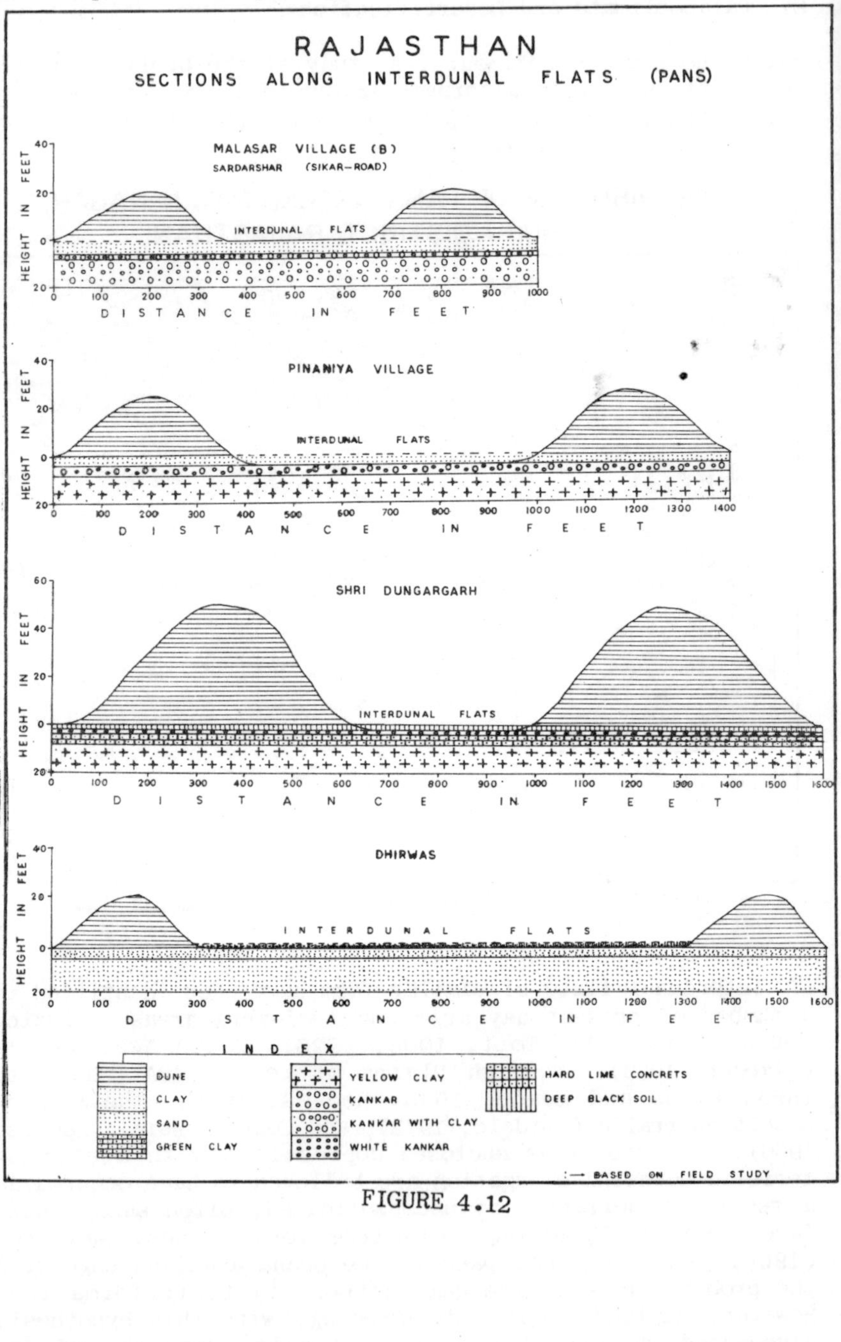

RAJASTHAN
SECTIONS ALONG INTERDUNAL FLATS (PANS)
MALASAR VILLAGE (B)
SARDARSHAR (SIKAR-ROAD)
INTERDUNAL FLATS
HEIGHT IN FEET
DISTANCE IN FEET
PINANIYA VILLAGE
INTERDUNAL FLATS
HEIGHT IN FEET
DISTANCE IN FEET
SHRI DUNGARGARH
INTERDUNAL FLATS
HEIGHT IN FEET
DISTANCE IN FEET
DHIRWAS
INTERDUNAL FLATS
HEIGHT IN FEET
DISTANCE IN FEET
INDEX
DUNE
CLAY
SAND
GREEN CLAY
YELLOW CLAY
KANKAR
KANKAR WITH CLAY
WHITE KANKAR
HARD LIME CONCRETS
DEEP BLACK SOIL
:— BASED ON FIELD STUDY

FIGURE 4.12

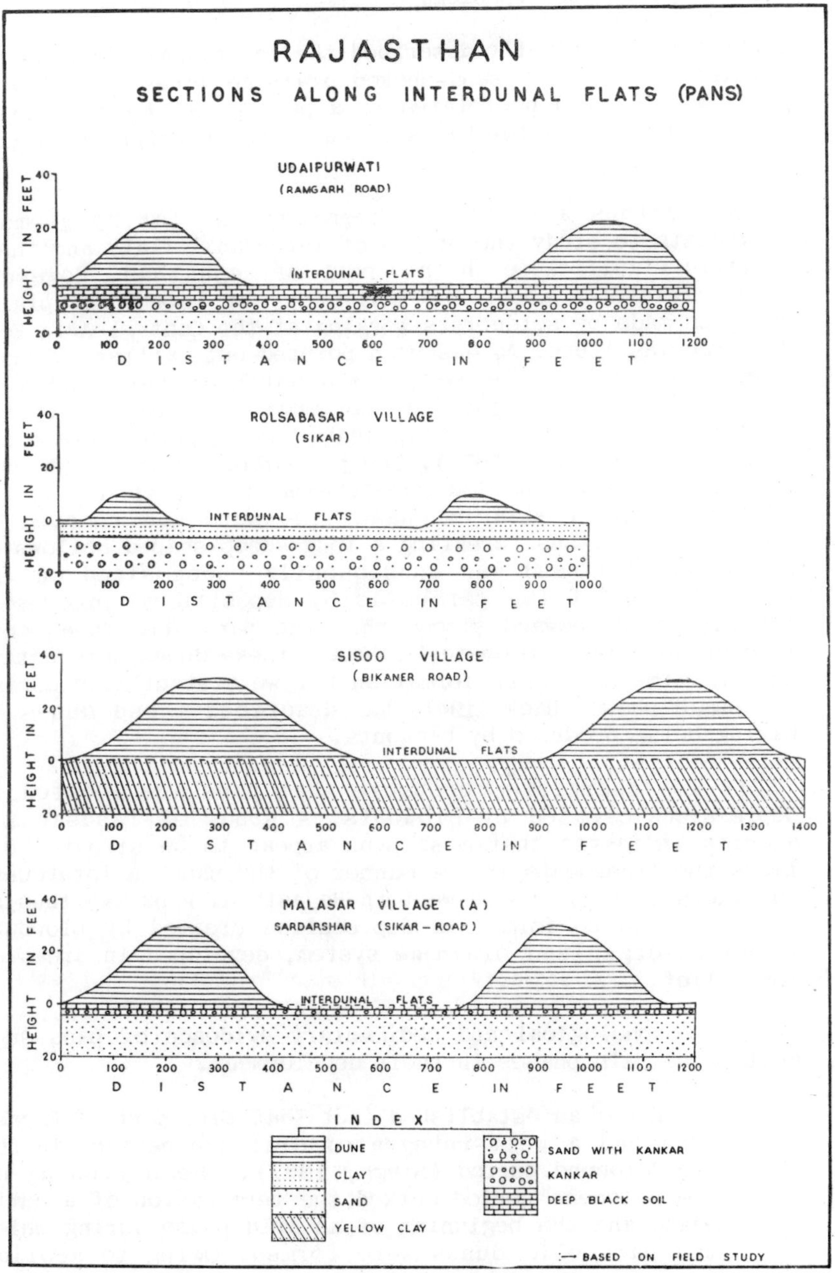
RAJASTHAN
SECTIONS ALONG INTERDUNAL FLATS (PANS)
UDAIPURWATI
(RAMGARH ROAD)
INTERDUNAL FLATS
HEIGHT IN FEET
DISTANCE IN FEET
ROLSABASAR VILLAGE
(SIKAR)
INTERDUNAL FLATS
SISOO VILLAGE
(BIKANER ROAD)
INTERDUNAL FLATS
MALASAR VILLAGE (A)
SARDARSHAR (SIKAR – ROAD)
INTERDUNAL FLATS
INDEX
DUNE
CLAY
SAND
YELLOW CLAY
SAND WITH KANKAR
KANKAR
DEEP BLACK SOIL
:— BASED ON FIELD STUDY

FIGURE 4.13

Straten (1962) briefly described the interdunal flats as the remnants of ancient sand-choked drainage lines, and lay in belts which could be related to a pattern of drainage lines and divides. Lancaster has attributed their origin to deflation.

No serious attempt has been made so far by geomorphologists to study the origin of interdunal flats and their climatic significance in this part of Rajasthan. However, casual reference about the morphology of interdunal plains has been made by scientists working at the Central Arid Zone Research Institute. As has been pointed out earlier the concentration of flats is more in the north-western parts of the arid zone of Rajasthan, an area characterized by the presence of maximum parabolic dunes (Verstappan, Saxena and Singh, 1976), Singh (1977). The parabolic dunes are an outcome of the erosional and depositional processes during the prolonged operated arid phase. The sand was removed by erosional processes, creating a long scoop-shaped hollows or parabolas. The sand was subsequently redeposited on the leeward side of the parabolas by depositional processes, forming steep leeward slopes thus the parabolic dunes come into being. Like the obstacle dunes, these dunes were active at the time of their formation but were stabilized during the humid phase. Hack (1941) has described these dunes as having being produced by blowouts.

It is, therefore, probable that pans are located in depressions created by blowouts. A study of landsat imageries indicates that most pans appear to be situated between the lineaments in the manner of the pans in interdunal corridors found in north-western Rajasthan. Pans may thus be considered to be formed in depressions created by blowouts in an un-organized drainage system, developed in areas of low relief.

The morphological and sedimentary evidence of pans suggest three main phases in their development.

1. It is now an established fact that this part of Rajasthan had a well-integrated drainage network in the pre-Holocene period (Singh, 1977). The beginning of the Holocene period marked the termination of a humid phase and the beginning of an arid phase during which the parabolic dunes were formed. Owing to aeolian deposits, the drainage lines were blocked, buried and dis-organized, and parabolic dunes were created by strong wind action as blowouts.

2. The blowouts become the sites of accumulation of seasonal water from adjoining dunes where clay and sandy clays were deposited. The desiccation of the pans is confirmed by the occurrence of 1-3 meters thick layers kankar pan (11-15%).

3. In conclusion, pans must be considered as virtually fossil landforms. Clearly, the origin and development of pans has taken place under climatic conditions unlike those prevailing in western Rajasthan today. The presence of a greater percentage of clay in the lower parts of the pan is clear evidence of the existence in former turns of increased rainfall and reduced evaporation. The thick layer of calcium carbonate indicates reduced rainfall and greater evaporation. Thus the origin and development of paras can be seen to reflect the influence of climatic changes in the region during the late Quaternary era and to provide further evidence of the nature of these evironmental changes.

Boulders and Gravels

During geomorphological field investigations the author had come across widespread unsorted deposits over the Mesozoic-Tertiary sequence of western Rajasthan. The occurrence of boulders and gravels is completely out of place in the present geographic setting of the Rajasthan desert. These deposits are separated from the main Arravali range by a distance of 100-150 km. As there are no integrated drainage systems at present, a number of problems arise regarding their age and origin. Such gravel and boulder spreads are heterogenous in composition and exhibit well-developed faceting, striation and grooves on the pebbles and boulders around Bap, Jaisalmer, Barmer and south-west of Bikaner. But around Jayal in Nagaur district, such deposits occuring as anomalous fossil features have puzzled many earth scientists. Mukhopadhayay and Ghose (1976), Biswas and Ghose (1980), Virendra Kumar and Laul (1981) have interpreted boulder beds around Bap, Jaisalmer and Barmer as a product of Pleistocene glaciation. On the other hand, Allchin, Goudie and Hegde (1978) considered such gravel spreads as a product of strong fluvial activity during the late Pleistocene era. Considering climatogenic factors, Agrawal et. al. (1980) a rules out the possibility of the origin of the boulder beds as a consequence of Pleicetocene glaciation.

Mode of occurence of Bap boulders and their continuity

Wide spread occurence of boulders of different sizes and shapes was noted in an triangular area of which Bap, Jaisalmer and Barmer form the apex points. On the basis of lithological similarity and on geomorphological considerations the boulders of Jaisalmer and Barmer have been correlated with those of Bap and their continuity has been traced from Deh (27°54°;72°54°30) on the north-west of Kahla in the south-west by earlier workers (Biswas and Ghose, 1978). Around Pokaran large boulders of granite lie over sandstones and shales; at places these lie embedded within recent kankar layers. From Pokaran to Hamira thick mats of pebbles and cobbles, may be found at several places e.g. near Basanpir and Bojka. The individual occurrences are spread over several kilometers. Near Hamira and Thaiyat several deposits of drumlin like heaps have been located. Bedded deposits of tillitic nature are abundant around Jaisalmer town. Angular boulders in association with cobbles, pebbles and gravels can be seen to be spread at several places, especially to the south-west of Jaisalmer town. Around Jaisurana, Bhojasar and Hamira they overlie the Jaisalmer formation. Other similar occurrences but with erratic boulders of quartzite have been recorded near Birmon-ki-Kanod and north of it where they overlie the Jaisalmer formation. However, north-east of Kanod Rann, the pebbles and cobbles along with erratics have been recorded overlying the Baisakhi formation of the upper Jurrasic age. In addition to this, such pebbles and cobbles spredds comprising mainly resistant rocks such as quartzites, vein quartz, sandstones, chert, jasper, ironstone, etc. have been found overlying the khuiala formation (lower Eocene) and the Summer formation (Pleistocene) in the Khinya-Danwar area (Virendra Kumar and Laul, 1981).

Virendra Kumar and Laul (1981) who carried out a survey in the Jaisalmer area pointed out that boulder deposits in eastern parts are heterogeneous in composition, exhibit well-developed faceting, striations and grooves on pebbles, cobbles and boulders indicating their deposition in a glacial environment. In the western parts, this spread mainly comprises quartzites, ironstones, jasper, chert, silicified wood, etc. Here the absence of erratic blocks, reduction in size of pebbles and cobbles, increase in the roundness of pebbles and cobbles, and absence of well-developed grooves and facets suggest their deposition by fluvio-glacial and fluvial agencies.

Geometry of boulders:

As has been pointed out earlier the Mesozoic-Tertiary sequences of western Rajasthan are capped at number of places by boulders and gravels comprising well-faceted and grooved pebbles varying in size from less than 2 cm to 15 cm (5-6 cm average) and comprised of quartz of various colours, quartzites, ironstones, sandstones, Dolomite, rhyolites, granites, fossil wood, basic rocks, phyllites, etc. In addition to these pebbles, a number of erratic boulders of various types of quartzites varying in size from 80 cm x 45 cm x 40 cm to 1.75 m x 0.70 m x 0.80 m have also been observed. Some of these erratics exhibit grooving in the NNE-SSW direction which is the prevailing wind direction in Rajasthan. Bioclastic limestone underlying the pebbles also show grooves trends in the same direction (Virendra Kumar and Laul, 1981).

Height correlation:

Boulders and cobbles composed of metamorphic and igneous rocks can be traced almost continuously from the south-west of Bikaner to the west of Jaisalmer, stretching over a distance of more than 250 km over hills and valleys, at various heights. However, studies show that they always occur below the 260 m contour. The highest deposits occur at a height slightly above 245 m near Hanumanji-ka-Mandir and the lowest at over 200 m near Kahla in Jaisalmer district.

Origin of boulders:

The stratigraphic position of the Bap boulder spread on the geological map of Rajasthan belongs to the upper Carboniferous period. They have been correlated with that of the Salt Range. The upper carboniferous age of Bap boulder is accepted for this type of locality. But some geologists (Biswas and Ghose, 1978; Mukhopadhyay and Ghose, 1976; Virendra Kumar and Laul, 1981) have recorded boulders and gravel in Jaisalmer and Barmer areas on the Summer formation of the early Pleistocene period and have suggested that their origin is owing due to glacial and fluvio-glacial agencies in the Pleistocene period. They have based their conclusions on the groove direction on the erratic blocks, and on the striations and faceting and computed their age on the assumption that they occur also on the early Pleistocene rocks of the Jaisalmer basin. Evidence shows that nowhere did the Pleistocene glaciation come down lower than 40°N, latitude except in mountainous regions. In India, the lowest recorded glacial deposits are near Dharmsala, about 32°N latitude at the foothills of the Himalaya. The boulders around Bap, Jaisalmer and Barmer are only about 27°N latitude at about 260-300 m altitude. Nowhere in the world the continental glaciation have been reported at this altitude and latitude during the Pleistocene. Therefore, the glacial origin of boulders in Pleistocene period is refuted.

No other explanation seems plausible except to attribute the origin of rocks and boulders to differential weathering accompanied by wind abrasion and deflation of Lathi and Jaisalmer conglomerates overlain by Jurrasic formations. Generally, wind abrasion manifests itself through (i) polishing and pitting, (ii) grooving, and (iii) shaping and faceting. It is assumed that the differential weathering has exposed conglomeratic beds on the surface for Wind erosion. The effect of wind abrasion is plainly written on the face of rocks and boulders in the form of grooving, striation and faceting. Wherever, there are strong winds, abundance of sand,and absence of vegetation, the gravels are in the form of ventifacts, which are generally confused with erratic blocks of glacial origin.

The boulders and gravels around Jaisalmer are confined to depressions, therefore, it appears that they were transported from the uplands by rainwash which is common in desert regions. The various stages of differential weathering, wind abrasion and rainwash coincide with wet and dry phases of climate. Such phases were witnesses by the Rajasthan desert during the quaternary period (Singh, 1977; Ghose, Singh and Kar, 1977).

Jayal Gravel:

On the way to the Phalodi-Bap area a ridge of gravel and boulders near Jayal had been examined. Jayal, a small town is situated 52 km east of Nagaur on Nagaur-Didwana road in flat country, formed mainly by fluvial and aeolian deposits of the Quaternary age. The most conspicuous feature of the area around Jayal is the gravel ridge, locally known as Magra. This ridge is 60-70 m high above the plain and runs north-east to south-west about 14 km in length from Katholi to Chajoli. Another part of it extends for about 10 km in the north-south directions from Katholi towards Ambali and Khatu Kalan. The width of this ridge varies from 150 m to 300 m. The summit of the ridge is rounded, with a 5° to 10° slope. The ridge has been modified into two distinct ridges separated by a valley flat. From a distance these discrete ridges still give the appearance of a continuous hill. Vegetation grown on soils and sediments is dominated by thorn and scrub.

Mode of occurrence

The gravel ridge is composed of rounded to subrounded boulders (maximum size recorded is 45 m across) and cobbles laid down in a matrix of yellowish brown silty sand and cemented by calcium carbonate. The gravel is poorly sorted, ungraded and unstratified. Carbonate coating on pebbles is common, and the carbonate concretions are soft, nodular and amorphous. Pebble to pebble contact is moderate.

Origin of Jayal gravel

In contrast to Bap boulders and other spreads, the good rounding of gravels and cobbles, presence of lenticular patches of sands and silts, and presence of calcareous cement throughout the thickness of 60-70 m suggest that the gravels are product of fluvial processes operating in a semi-arid landscape at a relief higher than that of today. The provenance of metamorphic and igneous rock material in the gravel shows that these rocks have probably been brought by streams-originating from the western slopes of the Aravallis. The evidence suggests that the gravel seem to have been deposited by braided bed load streams carrying occasional high floods during the monsoons. Such features are also very common in the Lohargarh nala in the north Aravallis. Mishra et. al. (1980), on the basis of the occurrence of lower paleolithic tools on the surface (up to 40 cm) of the gravel ridge proves that the gravels were formed in pre-lower palaeolithic times. The presence of amorphous nodular calcrete, suggest age of boulder as possibly early Pleistocene. The gravel formation, therefore, represents a palaeobraided drainage of western Rajasthan of the early Pleistocene period. Though the climate under which they were deposited was certainly semi-arid, the monsoon rainfall was definitely more than it is today, and provided high seasonal discharge in shallow streams transporting boulders and cobbles intermittently.

Laterite Deposits

The recent investigations carried out by the Geological Survey of India in the Deccan trap areas of south-eastern Rajasthan has added a new chapter to the denudation chronology of Rajasthan. The laterite deposits are found over the Deccan trap in Chittorgarh, Jhalawar and Kota districts. The laterite is grey, red, black, yellow and of variegated colours. It is highly ferruginous, vasicular, pisolitic to bouldary in form. It is a product of in situ weathering alteration of the Deccan trap and is found in the form of blanket cappings.

The existence of laterite in the area gives some clues regarding the climatic conditions under which laterization takes place. In India laterite has also been reported in Central India and in South India. The process of laterization started when the monsoonal condition in India were set after mid-Miocene period. On this basis it can be inferred that warm and moist conditions prevailed in this part of Rajasthan in the beginning of the Tertiary period. Detailed discussions of laterization with reference to climatic changes will be given in the chapter on planation surfaces.

ARCHAEOLOGICAL EVIDENCE

It is interesting to note that paleoforms such as prior channels carrying pebbles and gravels (below a depth of 3 m to 4 m in the Luni valley), fresh water clay beds below the saline lake beds at Malhar, Sambhar, Didwana and Lunkaransar, interlayered aeolian and fluvial sediments of the hillside slopes,moderate degree of calcification, and yellowish brown sand dunes, all are associated with archaeological remnants. These associations help in determining the chronological sequence of the environmental history of Rajasthan. Except for the work of Heron and La Touche, the first systematic study in south-eastern Rajasthan was undertaken by Mishra (1958). Leshnik (1966-67) made a further reconnaisance of parts of north Gujarat and Rajasthan. After 1970 a team consisting of Mishra, Rajguru and Thomas of Poona, Agrawal, Pant, Krishnamurthy of Physical Research Laboratory, Ahmedabad, and Dhir of CAZRI, Jodhpur, undertook the work of surveying the stone age sites in Jalor, Barmer, Pali, Jodhpur, Nagaur and Ajmer districts. The main aim was to study the stone age sites from the points of view of geomorphology, palaeoclimate, archaeology and ethnobotany. Rich microlith sites were discovered on the stabilized surfaces of moderately calcreted yellowish brown dunes (obstacle and barchans) at Sokarna, 8 km east of Jalor, on the western shore of Pachpadra lake in Barmer district, at Dechu in Jodhpur district, at Belaser 74 km west of Jodhpur in the same district, and in Pushkar valley in Ajmer district. According to the findings of the team the dunes do not appear to be very old, probably not earlier than the early Holocene era. Aeolian activity seems to have continued even during the early Holocene period. Middle Paleolithic tools have also been collected in a stratified contact at Jayal in Nagaur district.

The observations of archaeologists show that Rajasthan especially the south-western part has witnessed a complex palaeo-climatic history. The region does not seem to have experienced a hyper arid or humid climate during the Quaternary era. The area and has therefore, provided a congenial ecological setting for the habitation of early man since late Acheulian times.

PALAEOBOTANICAL EVIDENCES

The palynological studies in Rajasthan has been done by Agrawal (1951), Singh (1967); Singh (1971); Singh et. al. (1972); Singh (1973); Singh (1974), Vishnu-Mittre and Savithri (1973), Vishnu-Mittre (1976). Aggrawal's work formed the basis of later work by Singh and Vishnu Mittre. Agrawal constructed a generalized profile of the Sambhar lake which exemplified that shallow lacustrine deposits overlie aeolian sands which passed directly into the numerous sand dunes all round the lake. The main contribution of Singh and his co-workers has been to date the sequence of deposits and thereby to give, the probable end of major aeolian activity. A series of radio-carbon dates from charred wood fragments from deposits were obtained for dating the layers of laminated clay. The results are summarized in Table 4.4 below.

TABLE 4.4 **Radio carbon dating of Laminated Clay**

Depth	C-14 years BP based on 5730+40 years half life	TF number
1.35 - 1.50	4510 $\pm$ 110	883
1.50 - 1.60	4665 $\pm$ 100	739
1.85 - 1.95	6335 $\pm$ 315	884
2.70 - 2.85	8300 $\pm$ 135	738
3.12 - 3.20	8835 $\pm$ 140	698
3.15 - 3.25	9250 $\pm$ 130	887

Source: Singh Joshi and Singh (1972).

The salt lakes which had originated about 10,000 years ago has most probably had high salt content in their waters though perhaps as that they have today. Studies suggest that there was a recession in the level of lakes during 9500-6000 BP and a rise in lake level during 6000-5000 BP. The

TABLE 4.5. **Proposed tentative chronology of major climatic phases in late quaternary of Gujarat and Rajasthan**

Phase	Date	Evidence
Moisture phase	10,000-9,500 BP	Pollen-rich lake sedi- ments without aeolian component overlying dunes. Fresh water gastropods around lake basins. Extensive microlithic settle-ment.
Major dry phase	Pre 10,000 BP later upper palaeolithic	Dunes extend over lake basins (like Sambhar) and middle Stone Age soils. Miliolite deposited in Kutch and Kathiawar. Aggradation of rivers of the plain of Gujarat with some interruptions for fossil soil development. Little substantial evidence of human activity over wide areas.
Major wet phase	Middle Stone Age (? younger than 40,000 BP)	Major phase of weathe-ring and decalcification of dunes. Through flowing Luni river with coarse debris load. Substantial evidence of human occupation of Pushkar, Banara and elsewhere (including cur-rently hyperarid areas).
Major dry phase	Pre-middle Stone Age	Major sheets of aeolian sand with some slope-wasp material containing kankar and rolled Lower-Palaeolithical ante facts

Source: Allchin, Goudie and Hegde (1978).

progressive evolution of vegetation toward the formation of the Callgonum-Capparis series in the vicinity of Lunkaransar and of Capparis-Prosos-Acacia and Euphorbia in the vicinity of the Sambhar lake is suggestive of increasing aridity, spread of aeolian sand and extension of dune formation activity from 5000 BP onward or even earlier. The large scale desertification was caused not only by climatic factors, but also by increasing biotic factors including human influence and overgrazing (Vishnu-Mittre 1975).

On the basis of contrastingevidence, Singh, Joshi and Singh (1972) have concluded that:

In each of three salt lakes studied the lacustrine deposited in the form of laminated clay over a bed of aeolian sand which extends outwardly to meet the sand dunes surrounding the individual basin. It is clear from stratigraphic and radio carbon determinations, that the first sedimentation started around 10,000 BP after the

TABLE 4.6. **Geomorphic evidences of desertification/dedeser-tification during the late quaternary period in western Rajasthan**

Period in years B.P* (After Singh et.al. 1974)	Major geomorphic events	Geomorphic processes	Climatic phases	Desertification/Dedesertification
0-Early centuries A.D.	Increasing fluvial activities but increasing biotic stress on the land has resulted as salinity hazard along prior drainage channels and more deflation and deposition of sediments in the sandy plains and duny tracts	Biotic activities	Slight amelioration of climate	Small scale amelioration of climate but biotic deserticfication is present
Early centuries A.D. -3,800	Formation of low active dunes like branches. Shrubscoppice, etc. reactivation of the flanks and crests of the stablised dunes. Inland basins turned into saline depressions	Aeolian	Arid	Climatic desertification
3,800 - 10,000	Major dunes stablized, drainage channels revived, sandy in land basins filled with fresh water and fluvial sediments	Fluvial	Humid	Amelioration of climate leading to de-desertification
10,000-?	Formation of major dunes, disintegration of drainage systems of aeolian sand, creation of sandy inland basins	Aeolian	Arid	Climatic desertification
?	Well integrated drainage systems, vast alluvial plains	Fluvial	Humid	No evidence of desertification

Source: Ghose, et. al. (1977). *B.P means before present.

impounding of water behind closed sand basins. It is reasonable to believe that the lake basins came into being as a result of aeolian activity during an arid interval predating Holocene times. The new stablized dunes encircling the salt basin were apparently active during that period and caused the breakdown of normal drainage by the choking of river valleys with wind blown sand.

TABLE 4.7 **Broad Trends of Desertification in the LateQuaternary History of Climate in Rajasthan**

B.P.	A.D		
0	0	DESERTIFICATION II	
		Lakes start to dry and turn	SLIGHT AMELIORATION OF CLIMATE
		Salt meosphytic vegetation disappears	
2,000	B.C		
4,000	2,000	AMELIORATION OF CLIMATE	
6,000	4,000	SAND DUNES STABILIZED	
		Present Salt lake Basins filled with fresher water	
8,000	6,000		
		Vegetation comparatively more mesophytic than present	
10000	8,000		
		DESERTIFICATION I	
		SAND DUNES ACTIVE	

The tentative chronology of the climate change in Rajasthan and adjoining parts have been presented in Table 4.5, Table 4.6 and Table 4.7.

Further geomorphic evidence from Rajasthan reveals that the Rajasthan desert and adjoining areas had experienced more than one climatic phase and associated desertification and de-desertification during the late Quaternary period. Ghose Pandey and Kar (1977) are of the opinion that the signs of any recent climatic desertification are not apparent in the Rajasthan desert, although indications of more than one such desertification in the past are available. According to them, desertic conditions are being accentuated by human activity with the result that not only an ecological imbalance but possibly an ecological disaster has taken place. On the contrary, Gurdip Singh (1977), is of the opinion that desertification II, though less severe than desertification I, has continued to the present day. Since the fluctuations in the phase have been only minor, there is little reason to believe that the broad trend towards desertification in Rajasthan has been reversed.

PART TWO
MORPHOGENETIC FORM AND PROCESSES

FIVE

Aeolian Forms and Process

Earlier studies have shown that Rajasthan had experienced two major humid phases with intervening arid phases during the late Quaternary period. During the intervening arid phases, dune building processes took place in western Rajasthan and also towards the east of the Aravalli range up to Jaipur-Bharatpur. The presence of stabilized dunes east of the Aravalli range indicates that formerly desertic conditions were more widespread and the desert subsequently shrank upto the present boundary (Fig.4.1). Within the desert belt, there are also two distinct zones of aeolian and fluvial processes. The 300 mm isohyte forms the boundary between the two (Singh, 1977). West of the 300 mm isohyt, the aeolian processes are more intensive and have re-activated and degraded the stabilized dunes, thus creating new active dunes. To the east of the 300 mm isohyte, the fluvial processes are dominant and have dissected the old existing landforms with rills and gullies. The dunal forms of eastern Rajasthan have also been dissected by rills and gullies especially on the windward side of the obstacle dunes.

West of the 300 mm isohyte, there is a vast expanse of sand throughout western Rajasthan. The most dreary parts of the sandy desert are called thal and tracts relatively less sandy are known as thali. A study of landsat imageries and physical sheets of Rajasthan reveals that sand dunes and sheets are not continuous but are found in discontinuous patches. Almost everywhere sand, driven by wind, is piled up in heaps and ridges or spread out in sheets, obliterating the pre-desert fluvial topography. There are two major and two minor zones of sand dunes and sand sheets in Rajasthan.

The first major zone of sand dunes runs through Miajlar Thal, Jaisalmer, Shahgarh, Ramgarh, Tanot, Pugal and

Mahajan. The continuity of dunal fields is interrupted by stony topography around Jaisalmer, Barmer, Bap and Pokaran. The area west and south of Jaisalmer is characterized by a complex system of sand dunes. Longitudinal or siefs dunes predominate in this part on which occassionally develop transverse dunes of the barchan type. Flat floored depressions or pats occur sandwiched between two longitudinal dunes. It is these pats that provide sites for verdant oases. The length and width of the pats appear to be very much restricted in areas of rapidly moving sand. The dunes around Barmer are more or less stable sand dunes of the barchan type. Beyond Jaisalmer, both the thals of Pugal and Birsilpur contain sand dunes of the barchan type with a steep concave northern face which rises to a maximum height of 50 m.

The second major zone of medium to high and very high dunes runs through Shri Dungargarh, Sardarsahar, Nagaur and terminates around Rajgarh and Suratgarh. The old course of the Ghaggar river forms the north-western limit of this zone.

The third small dune zone runs into discontinuous patches and passes through Sanchor, Jalor, Saila, Didwana, Sikar and Jhunjhunu, joining the two major zones near Surajgarh.

The fourth patch of dunes passes through the Pushkar valley, Sambhar, and the trend continues as isolated obstacle dunes up to Bharatpur.

These four zones constitute the dunes of different types and origin and they are unevenly distributed.

In the Rajasthan desert, some investigations on the dynamics of sand dunes and temporal changes of dunes have been carried out by Pandey, Singh and Ghosh (1964), Vats et. (1976), Saxena and Singh (1976), Singh (1977), Bedi, (1982), Allchin and Goudie (1971).

The analysis of fossil dunes and their characteristics has already been discussed earlier. In the present chapter, wind regime, processes of sand movement and the accumulation and origin of different types of sand dunes are described.

WIND REGIME AND DUNE ORIENTATION

The influence of the south-western monsoonal wind is apparent on the orientation of sand dunes in the Rajasthan desert. Long narrow sand ridges of the sief type may be seen in parallel in a south-west to north-east direction, enclosing between them dune lakes known as dhands. The study of wind roses has already revealed that the dominant wind direction during summer is from the south-west to the north-east and that the wind direction in winter changes from the north-east to south-west. However, the winter winds also influence the forms. Figure 5.1 shows the extent of sand dunes in Rajasthan. It is quite clear that the maximum percentage of area covered by sand dunes is to be found in the western part of the Jaisalmer desert. This percentage decreases in the south-eastern part of the desert in the Luni basin, indicating that there exist a close relationship between wind direction and the orientation of dunes. The axial direction of the dunes of the old and new system is almost the same (Fig. 5.2), suggesting that practically no change in the wind regime has taken place in recent times.

Besides the direction of the wind, velocity also plays a dominant role in the movement of sand particles. It has been established through experiments that the fine grained sands (0.25 0.01 mm) can move with a wind velocity of 3.5-5 m/sec at the surface. The velocity required for the movement of sand particles of the different sizes as proposed by Sokolov (1884) are given in Table 5.1.

Table 5.1 shows that winds of different velocities are required for the movement of sand grains of different sizes and also to create the aeolian features of different orientation, shape and magnitude.

TABLE 5.1

Grain Size of Dunal Sand

Types of sand grains	Diameter	Wind velocity (m/sec.)
Fine grained	0.1-0.25	4.5-6.7
Medium grained	0.25-0.5	6.7-8.4
Coarse grained	0.5-1.0	9.8-11.4
Very coarse grained	1.0-2.0	11.4-13.0

(Source- Singh, 1977)

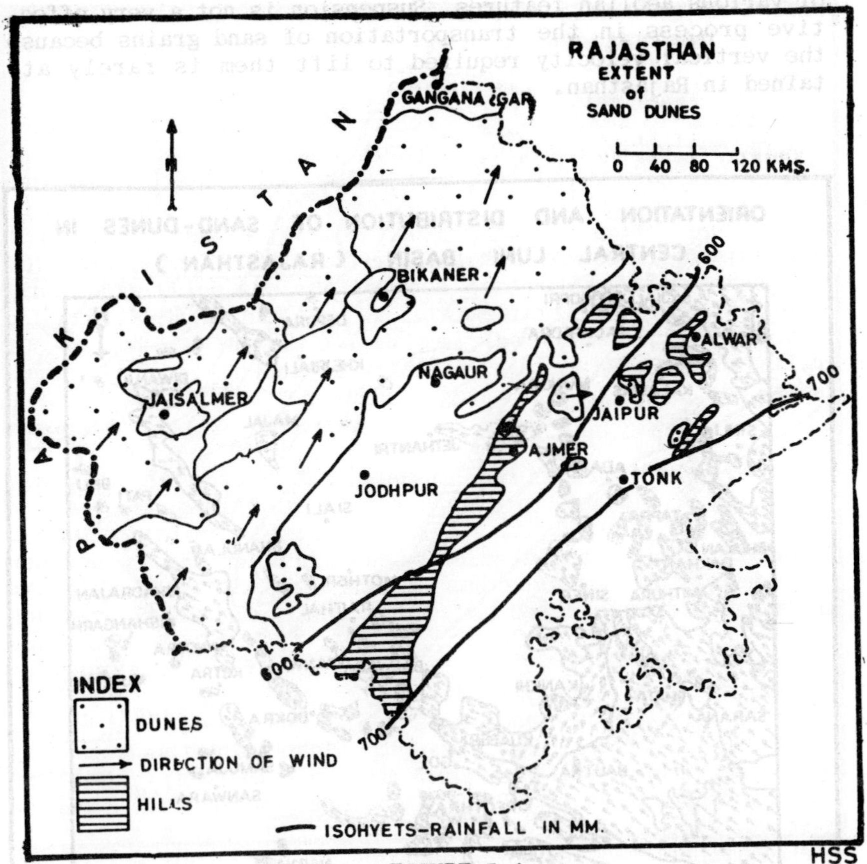

FIGURE 5.1

In the Rajasthan desert, the average size of sand grains varies from 0.12 mm to 0.14 mm. Three of classes of roundness have been identified (Singh 1977), (i) angular to subangular, (ii) subangular to subrounded, and (iii) rounded to well rounded. It has been shown that roundness in sand grains is caused by abrasion during transportation.

In the Rajasthan desert sand grains of different sizes are transported by saltation, surface-creep and suspension. Saltation is the dominant process and about 90 per cent of sand grains are transported by this method. Owing to surface creep processes, sand grains above 0.5 mm diameter are transported along the surface and are deposited in the form of various aeolian features. Suspension is not a very effective process in the transportation of sand grains because the vertical velocity required to lift them is rarely attained in Rajasthan.

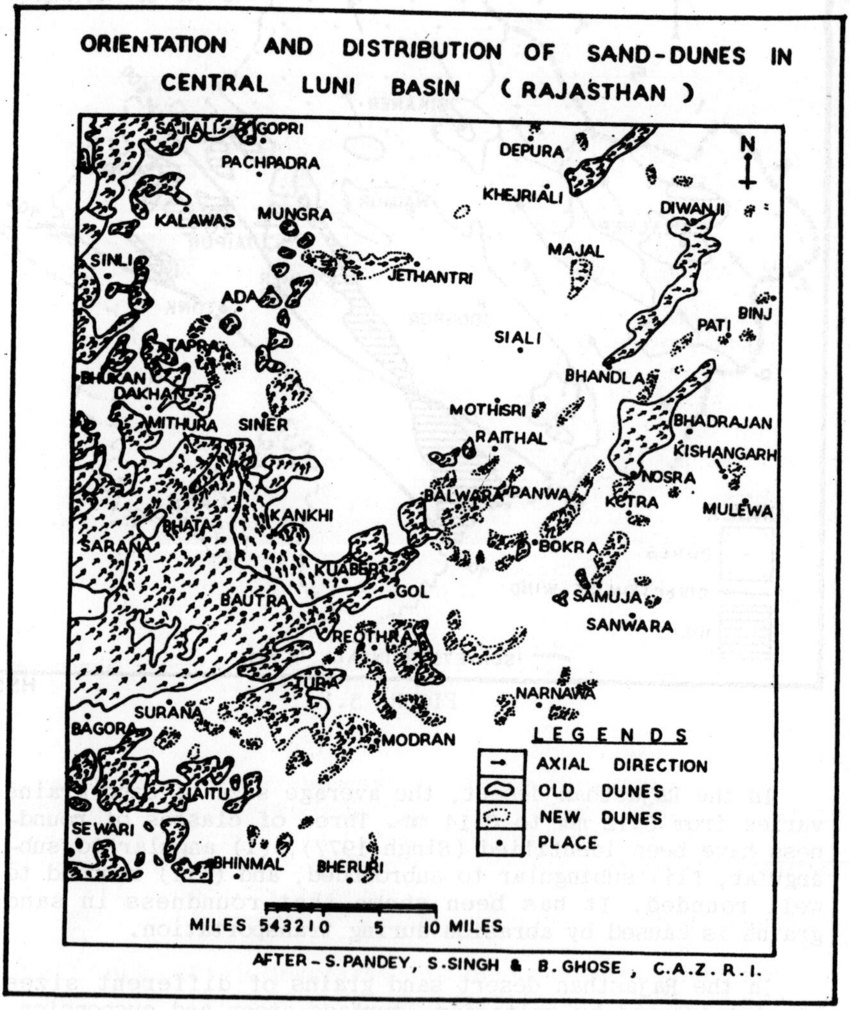

FIGURE 5.2

FORMATION AND TYPES OF SAND-DUNES

Two principles are critical to dune formation. The first principle is that a free sand patch larger than a few metres is likely to increase in size and thickness and thus becomes a mound or hill. The second principle is that a dune, like a cliff, develops of low velocity air that promotes deposition there. As soon as a mound of sand grows to moderate size, it develops a classic dune assymmetry of a gentle windward slope and a steep angle of repose face to the leeward side The sand on the windward slope is firm and compact, but it blows over the crest and cascades down the slip face; it may become loose enough to make walking difficult.

Dunes have a wide range of forms in Rajasthan, some of which were unknown until aerial photographs and landsat imageries became available for larger areas of hitherto unexplored sand regions in the extreme western and south-western part of the state around Ramgarh and Tanot.

Based on the interpretation of the landsat imageries of different years 1972, 1975 and 1979, and tophsheets of 1:50,000 scale and field studies, the following major types of dunes in Rajasthan have been identified (Fig. 5.3):

a. obstacle dunes : old dunes system;
b. parabolic sand-dunes : old dunes;
c. longitudinal dunes : old dunes;
d. transverse dunes : old dune; and
e. shrub coppice : new dune system;
f. barchan dunes : new dunes.

The sand dune characteristics of Rajasthan are shown in Fig. 5.4. The dunes are of different shapes, orientation and dimension in different parts of Rajasthan. For example, pattern of dunes is entirely different. in Jaipur district, where they trend from west to east.

Obstacle Sand Dunes

Obstacle dunes are formed on the windward and/or leeward sides of isolated and continuous chains of hills, owing to the obstruction in the path of sand laden winds. Therefore, they are known as windward and leeward obstacle dunes. The windward obstacle dunes in the shape of flat semi-cones are formed because of fixed eddy, decreased wind velocity and a calm zone on the upwind side of the hills. It was observed that the prevailing south-west winds in Rajasthan swept the sand over the hills and deposited it on the leeward side, right from the slopes of the obstruction up to a few kilometres in the direction of the down wind and thus the leeward obstacle dunes were formed.

Fig. 5.3 : Landsat MSS band 7 showing Dunal Pattern in Rajasthan Desert (1972)

It was before, pointed out that obstacle dunes were the product of an earlier arid phase in Rajasthan, probably around the preholocene period. These dunes at present are badly dissected, well-cemented and vegetated. They are separated from the hills by gullies of 25 m to 50 mm deep and 50 m to 100 m wide (Figs. 4.8 and 4.9). These dunes were stabilized in a prolonged humid phase of the early holocene period around 10,000 B.P.

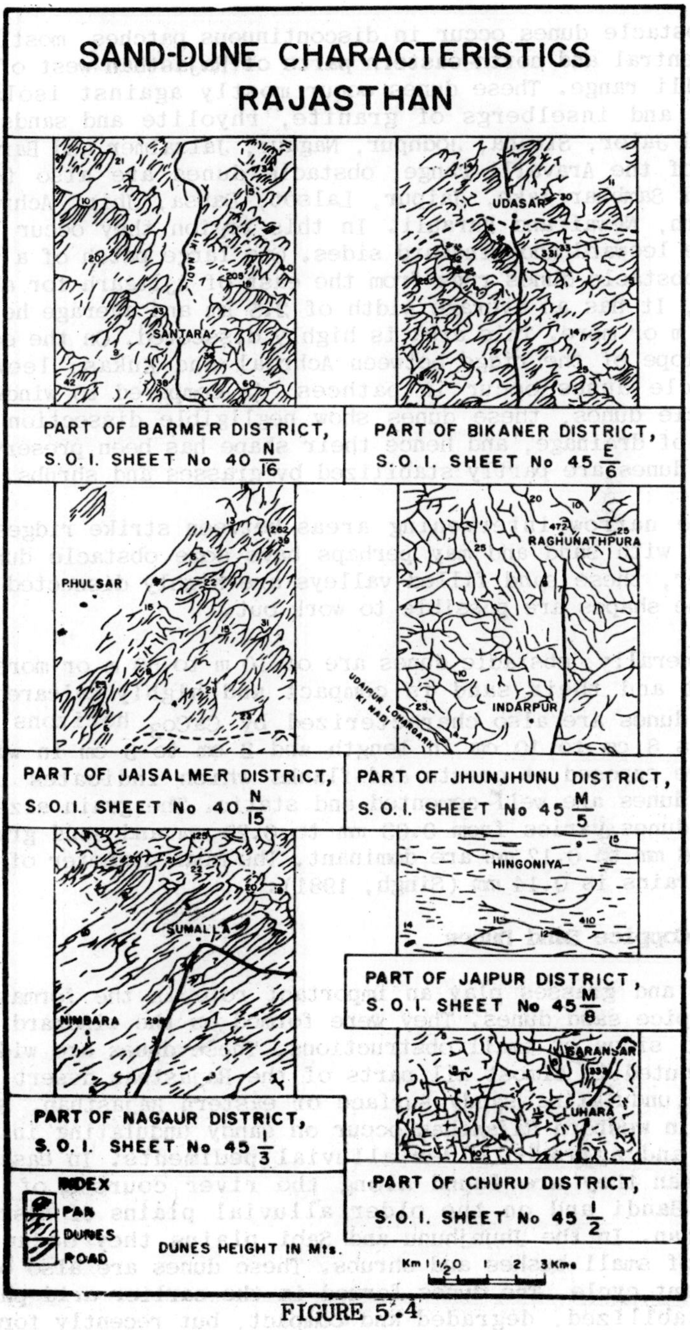
SAND-DUNE CHARACTERISTICS
RAJASTHAN
SANTARA
LIK NADI
UDASAR
PART OF BARMER DISTRICT,
S.O.I. SHEET No 40 N/16
PART OF BIKANER DISTRICT,
S.O.I. SHEET No 45 E/6
PHULSAR
RAGHUNATHPURA
UDAIPUR LOHARGARH KI NADI
INDARPUR
PART OF JAISALMER DISTRICT,
S.O.I. SHEET No 40 N/15
PART OF JHUNJHUNU DISTRICT,
S.O.I. SHEET No 45 M/5
HINGONIYA
SUMALLA
NIMBARA
PART OF JAIPUR DISTRICT,
S.O.I. SHEET No 45 M/8
BARANSAR
LUHARA
PART OF NAGAUR DISTRICT,
S.O.I. SHEET No 45 1/3
INDEX
PAN
DUNES
ROAD
DUNES HEIGHT IN Mts.
PART OF CHURU DISTRICT,
S.O.I. SHEET No 45 1/2
Km 1 1/2 0 1 2 3 Kms

FIGURE 5.4

Obstacle dunes occur in discontinuous patches, mostly in the central and north-eastern parts of Rajasthan west of the Aravalli range. These dunes occur mostly against isolated hills and inselbergs of granite, rhyolite and sandstone around Jalor, Siwana, Jodhpur, Nagaur, Jaisalmer and Barmer. East of the Aravalli range, obstacle dunes are also found around Sambhar lake, Jaipur, Lalsot, Dausa, Jhir, Achroal, Ramgarh, Newai and Karauli. In this region they occur both on the leeward and windward sides. One large patch of a wind ward obstacle dunes runs from the east of Ajitgarh for about 10 km, It has an average width of 2.5 km and average height of 50 m or more. This area is highly dissected. On the eastern slope of the ridge between Achroal and Kukas, leeward obstacle dunes occur in pathces. As compared to windward obstacle dunes, these dunes show negligible dissection because of drainage, and hence their shape has been preserved. These dunes are partly stabilized by grasses and shrubs.

The narrow intervening areas between strike ridges is filled with sand and may perhaps have some obstacle dunes. However, these sand filled valleys are highly dissected and no dune shapes are possible to work out.

Generally, obstacle dunes are of 10 m to 50 m or more in height and their sand is compact and highly calcareous. These dunes are also characterized by $CaCo_3$ horizons and nodules 8 cm to 10 cm in length and 2 cm to 5 cm in width and are exposed at crests and flanks which indicates that these dunes are well-cemented and static. The grain size of these dunes varies from 0.03 mm to 0.25 mm and sand grains of 0.06 mm to 0.12 mm are dominant. The mean diameter of the sand grains is 0.14 mm (Singh, 1981).

Shrub Coppice Sand Dunes

Shrubs and grasses play an important role in the formation of coppice sand dunes. They were formed on the windward and leeward sides of small obstructions. These dunes are widely distributed in almost all parts of the Rajasthan desert and on the undulating sandy surface of eastern Rajasthan. Such dunes in western Rajasthan occur on sandy undulating inter-dunes and aggraded older alluvial pediments. In eastern Rajasthan they are found along the river courses of the Masi, Bandi and on the older alluvial plains of eastern Rajasthan. In the Jhunjhunu and Sabi plains they occur on sides of small bushes and shrubs. These dunes are also of a different cycle. The dunes formed in the earlier arid phase are stabilized, degraded and compact, but recently formed dunes are active and their sands are loose and non-calcareous. They do not have any shape and definite orientation and are quite close to each other. These dunes do not have slip faces and the slopes on the windward side are 1° to 2°. About 49 per cent of the sand grains are of 0.06 mm to 0.007 mm diameter, this percentage is larger than that found in other dunes.

Parabolic Sand Dunes

The parabolic dunes stand out very clearly on landsat imageries particularly in band 7 (Fig. 5.2).

Parabolic sand dunes are characteristics of partially stabilized sandy surfaces that develop blowouts. The crescentric shape of parabolic dunes has a very different orientation from that of barchan dunes. A parabolic dune may become elongated by the down wind migration of a blowout until it is split into two minor longitudinal ridges paralled to the long axis of the blowout. These dunes are usually much longer and narrower than barchans and are associated with blowouts and pans (central depressions) (Fig. 4.10 and 4.11). In Rajasthan, parabolic dunes were formed by erosional and depositional processes during the prolonged arid phase. The sands were removed by the unidirectional winds through erosional processes thus creating long scoop-shaped hollows or parabolas with flank interfere with up to several kilometers towards the gentle windward slopes.

Parabolic dunes occur in more areas than other dunes and occupy almost all the districts of western Rajasthan. The height of these dunes ranges from 10 m to 25 m. They are well stabilized and vegetated. The slope of the leeward side of these dunes is steep, varying from 14° to 16° and that of the windward side is gentle ranging between 2° to 4°. The sands are calcareous and compact. The coalesced parabolic dunes are also stabilized degraded and have a good growth of vegetation. Their height ranges from 8 meters. The number of dunes in the dune chains varies from 15 to 35 and length of these chains range between 0.5 km to 10 km. the diameter of the sand grains on all the slopes varies from 0.03 mm to 0.25 mm, and 60 percent to 70 per cent of the sand grains are from 0.06 mm to 0.12 mm is size. The average length and width of these dunes are 1288 m and 226 m respectively. The coalesced parabolic dunes are covered with fresh new sand of 3 m to 4 m thickness. The existence of barchan dunes on the flanks of these dunes indicates renewed aeolian activity.

Longitudinal Sand Dunes

Longitudinal sand-dunes form huge landforms extending hundreds of kilometres in length, a kilometre or more in width and hundreds of metres in height. On landsat imageries and toposheets of 1:50,000 scale they distinctly appear in straight parallel ridges. The longitudinal dunes occur mostly in the south-western parts of Jaisalmer, Jodhpur and Barmer districts. These dunes are well developed because of constant wind direction from the south-west to north-east. The height of the dunes ranges from 10 m to 60 m. They are also well stabilized and their sand is compact and calcareious. The orientation of the dunes varies from N 40°E to N 45°E. The slope of their flanks, leeward and windward, varies from 9° to 14° upto 24° and 2° to 3°, respectively. The interduned width is from 600 m to 1200 m. The dune sand a also contains lime nodules of 2 m to 25 mm length. The sand grains are of 0.07 m to 0.15 mm diameter, the average diameter being 0.10 mm. Active longitudinal dunes are also found in areas devoid of vegetation and where sand supply is irregular. These dunes are also called sief dunes in North Africa and Saudi Arabia, and sand ridges in Australia.

There is a some controversy regarding the origin of these dunes. Glennie (1970) is of the opinion that high wind velocity favours the development of sief dunes. Bagnold (1941) suggested that the longitudinal dune is a modification f the fundamental barchan form, produced by strong cross-winds transverse to the prevailing wind direction. Singh (1981) studying the longitudinal dunes of the Rajasthan desert pointed out that these dunes had not resulted from barchan dunes but were the outcome of parabolic dunes. Early in the Holocene period when the arid phase ended and the humid phase began these dunes were stabitized. Verstappen (1968) also advocated the theory that these dunes originated from parabolic dunes. King (1960) has put forward the hypothesis that longitudinal dunes are the outcome of the separation of the trailing arms of parabolic dunes from the dune noses.

Transverse Sand Dunes

Transverse sand-dunes are associated with massive quantities of sand and relatively ineffective winds. The crests of these dunes are ridged roughly perpendicular to the effective wind. Transverse dunes occupy a relatively small area in Rajasthan. They occur in the south-western part of Bikaner district around Pugal, southern Ganganagar district around Rawatsar, and south-eastern Jodhpur district. Small pockets of transverse dunes are found in Churu and Jhunjhunu districts.

The height of these dunes ranges from 10 m to 40 m and interdunal distance from 500 m to 1000 m. The orientation of high dunes is N 50° W and that of low dunes is N 45° W. The average length and width of these dunes are 750 m and 150 m respectively. The sand grains are from 0.03 mm to 0.18 mm diameter, the average diameter being 0.11 mm. The sands are highly calcareous and compact.

Barchan Sand Dunes

Barchans or crescentric dunes are the classic aeolian landforms. Their crescentric form consists of a gently inclined windward slope and a steep lee side around which the horns of the dunes project windwards making the slip face concave to the downwind direction. They are isolated but migratery. They represent a remarkable balance between accumulation, transportation and erosion.

Barchan dunes are distributed in the central, western and south-western part of arid Rajasthan. They are well-developed in Jodhpur, Bikaner, Churu and Jhunjhunu districts and are generally found near the settlements on old dunes and in river beds. Around Churu, Bhaleri and along the Katli river course, ideal examples of barchan dune can be cited. Such dunes also occupy small areas in eastern Rajasthan where old dunes have been reactivized.

Barchan dunes are active, devoid of vegetation and of a non-calcareous character. The height of the dunes ranges from 2 m to 15 m. The orientation varies from N 38° E to N 40°E. Horizontal and vertical sand ripples, 12 cm apart, are formed on the dune crest. The grain size of the sand is from 0.06 mm to 0.12 mm, the average size being 0.12mm.

TEMPORAL CHANGES IN DUNAL FORMS

Arid areas have a very dynamic environment where changes take place quite often. Sand is one of the most mobile elements here. Its generation, mobility and accumulation is controlled by moisture and vegetation. A change in these factors brings about a corresponding change in sand deposition. Therefore, the behaviour of sand has to be studied from time to time. Such temporal changes in the dunal form can be easily monitored by studying the landsat imageries of an area for two or more different periods.

Bedi (1981) has studied the temporal changes in the aeolian landscape with the help of landsat imageries of 1975 and 1979, toposheets of 1968 and 1974 on the scale of 1:250,000 and geological maps. This example gives an idea of the temporal changes in dunal form in Rajasthan. While comparing various maps and imageries, Bedi noticed most striking changes in the sand activity between Jodhpur and Balotra. The changes that had taken place were as follows:

1. Toposheets of 1968 and 1974 showed a 3.25 km wide transverse dune complex to the south-west of Salawas.

2. The width of this dune complex had increased to 8-10 km and a fresh 4 km wide longitudinal dune complex had been superimposed on that, as seen in the landsat image of 1975. Already existing longitudinal dunes in the north-eastern part had become more prominent.

3. It was found in the landsat image of 1979 that the width of the transverse dune complex had again been reduced to 6-8 km. The activity of the longitudinal dunes had also been reduced as their zone width had come down to 1.5 km. In the northern part also the longitudinal dunes had become less conspicuous and transverse dunes had again reappeared.

On the basis of these temporal changes in dunal form and other geomorphic forms, Bedi concluded that the following environmental changes had taken place in this area.

1. Up to 1968, moderate dry aridity and wind conditions prevailed in the area. The aridity, however, started increasing from 1968 onwards as transverse dunes increased in width. By 1975 wind velocities also had increased so much that the transverse dune complex surface was stirred and a prominent longitudinal dune was superimposed. This was the deflation phase when the spread of desertic conditions took place.

2. From 1975 to 1979 the area seems to have gone through a phase. Higher precipitation during 1979 and during the preceding years brought about the stabilization of sand and the growth of vegetation. There was also a reduction in wind velocity as longitudinal dune activity considerably lessened.

SOURCE AND ORIGIN OF DUNAL SAND

Considerable controversy exists regarding the source of dunal sand. Wadia (1960) has pointed out that dune sands of Rajasthan have largely been transported by south-west winds from the Rann of Kutch and that they are of marine origin. Contrary to this, Goudie and Sperling (1977), and Agrawal and Roy (1976) consider that the sand is of aeolian origin. However, Goudie and Sperling consider that the source is from the Rann of Kutch. Agrawal and Roy carried out an SEM analysis of sand grains from Dungarpur and Junagarh and found that all the main features were indicative of aeolian deposition under a hot climate. The studies conducted by Singh (1982) on sand grains of two blocks, namely, Chotatan block which is close to the Rann of Kutch and Ahor block which is far away from the Rann of Kutch indicate that significant variation exists between the mean diameter of the grains of dune sand of these two blocks. On this basis he has concluded that the sands are of local aeolian origin.

From the study of aeolian forms and processes it has been found that in Rajasthan two systems of dunes old and new exist, which were formed under different climatic phases during the Quaternary period. The recent morphological changes in dunal form indicate that the present climate is ameliorating.

SIX

Fluvial Morphology

It had been shown earlier that the 300 mm isohyte forms the boundary between aeolian and fluvial activities. East of this isohyte, fluvial processes in the form of sheet, gully and surface wash are predominant and aeolian processes become less effective. East of the Aravalli range there is a region of predominantly fluvial activities except in the north-east where both fluvial and aeolian forms are evident.

The evolution of landforms and planation surfaces makes it evident that the existing geomorphic features in Rajasthan are the outcome of long and continuous morphogenetic processes which began after the upwarping of the Aravallis and the Central Vindhyan Plateau in the Cretaceous period. In the subsequent periods the fluvial cycles were interrupted twice owing to the intermittent upheavals of the Himalaya. After the last phase of the Himalayan orogeny in the Pleistocene period, the present cycle of erosion has been interrupted, as a result of which most of the rivers of Rajasthan mainly the Chambal and its tributaries were rejuvenated and thus are actively engaged incising their courses.

Broadly, the following drainage systems can be identified in Rajasthan (Fig.6.1):

a. the disorganized Ghaggar drainage system;
b. the Chambal-Banas drainage system;
c. the Mahi drainage system;
d. the Luni-Jawai-Sukri drainage system;
e. the Sabarmati drainage system;
f. the western Banas-Sipu drainage system;
g. the Gambhir drainage system; and
h. the inland drainage systems, which consist of:

i. the Banganga system;
ii. the Katli-Dohan-Sabi and Berah system;
iii. the Sambhar Didwana - Kuchaman drainage system; and
iv. the inland drainage of Jaisalmer-Barmer-Lik river and Gughari system.

FIGURE 6.1

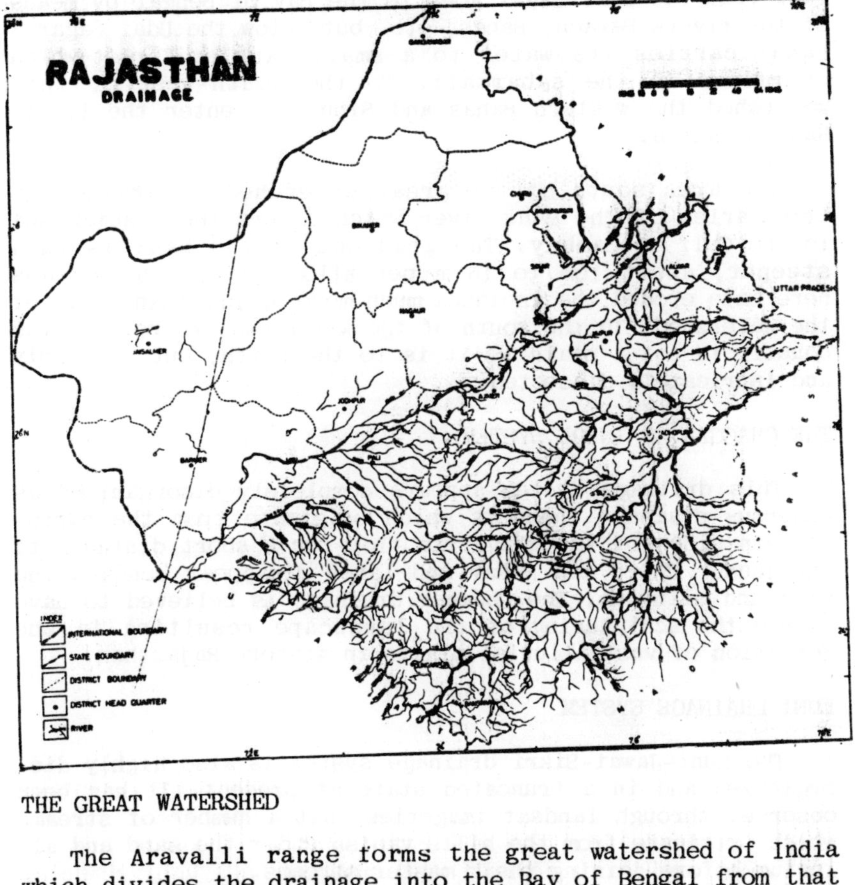

THE GREAT WATERSHED

The Aravalli range forms the great watershed of India which divides the drainage into the Bay of Bengal from that in to the Arabian Sea. It is in fact a trijunction watershed. To the north-east the flow is towards the enclosed catchment of the Sambhar lake; the Ajmer valley itself is drained westward by the Sagarmati into the Luni river; and the ridge east of Ajmer drains to the Banas and Chambal the principal rivers of Rajasthan and from them by the Yamuna and Ganges rivers to the Bay of Bengal. The watershed winds along the centre or south-east flank of the range between Ajmer and the south-west of Kumbhalgarh. At Kumbhalgarh it passes across to the north-west of the range for a few kilometers and then swings in a south-eastern direction transversely across the range, passing south of Udaipur city. To the south of this, the drainage of the range is drained by longitudinal strike valleys to the Sabarmati which enters the Gulf of Cambay. The natural overflow from the Udai Sagar goes to the Bay of Bengal by means of the rivers Berach, Banas, etc. but below the Udai Sagar a canal carries its water to a small tank at Karget which overflows to the sabarmati. To the south-west of this watershed the western Banas and Sipu also enter the little Rann of Kutch.

The area south of the great watershed is drained by tributaries of the Mahi river which enters the Arabian sea in the Gulf of Cambay. The gradient of the Mahi is much steeper, about 10m to 15 m per kilometer and the streams here have eroded the gneisses much more deeply than north of the watershed. To the south of the watershed, therefore, the country is not a plain as it is to the north, but is deeply and intricately dissected.

THE GHAGGAR DRAINAGE SYSTEM

This drainage system appears completely disorganized as numerous drainage channels which originate from the north-western sides of the Aravallis flow for a short distance to Jhunjhunu, Churu and Sikar and later disappear beneath the sand and alluvium. The Ghaggar drainage is believed to have caused the sculpturing of the landscape resulting in the formation of vast alluvial plains in western Rajasthan.

LUNI DRAINAGE SYSTEM

The Luni-Jawai-Sikri drainage system is also highly disorganized and in a truncated state at present. It has been observed through landsat imageries that a number of streams which originate from the hills vanish under the sand and alluvium before joining their master streams.

A great deal of work has been done by Ghose et. al. (1966, 1980), verma and Roy (1981) and Sood and Pathan (1982) on palaeo-drainage lines of the Luni and western Rajasthan. According to Ghose (1965):

The river Mirti for instance was one of the most important tributaries of the Luni. After rising from the Aravallis it flowed due west and after Takhatgarh it turned to the north and joined the Luni. Today it does not follow the whole length of its channel. After Takhatgarh it is completely disorganized and obliterated. There are many other tributaries like the Sukri, Bandi, the streams of Golia, Bhailan and Kundal Ramania which become disorganized since the dry climate set in.

Ghose et. al. (1980) have pointed out that the Luni system had a link with the Himalayan drainage and the older alluvial plain of the western Rajasthan was formed by the well integrated, prior drainage (Fig. 4.1).

Detailed studies have been made by Verma and Roy(1981) of the Luni basin between Dundara in the east and south of Tilwara in the west, which also suggest an integrated network of drainage joining the main channel during a more humid past. The important disorganised channels identified are the following (Fig. 6.2.) :

FIGURE 6.2

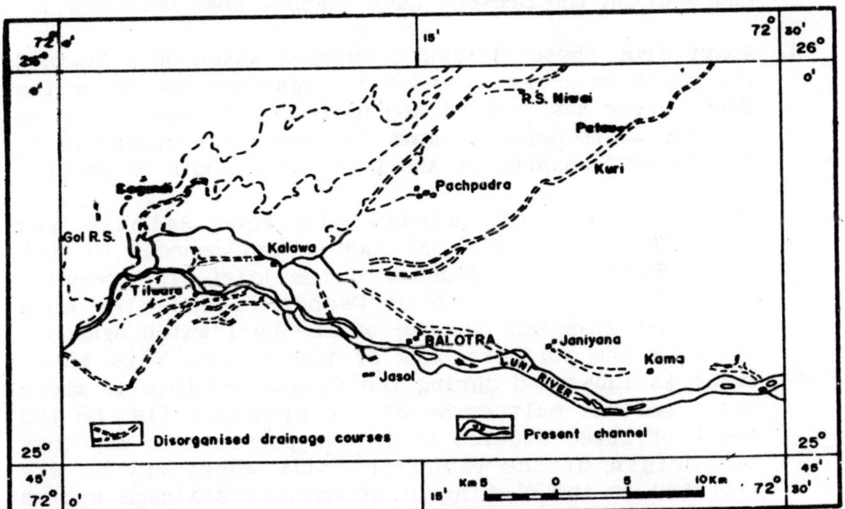

MAP SHOWING DISORGANISED DRAINAGE COURSES IN LOWER LUNI BASIN, BARMER DISTRICT, RAJASTHAN

(Arfer Verma, R.P. & Roy, P.K. 1981)

1. The abandoned course of the Jojri river which drains the highlands of Jodhpur was picked up from east of Patau to the north of Balotra. The course is almost non-existent on the ground except in some stretches in the form of isolated small bank sections.

2. Two major disorganised channels north of Balotra, one coming from Newai and other one from Bedarlai were picked up. Both these channels meet near Pachpadra town and the entire Pachpadra salt works is situated in the channel which once drained the area north of Pachpadra.

   A coursory look at the map suggests that the highlands of Shergarh were drained by this network of channels and that the rann of Thob is also an abandoned course of this network.

3. The third abandoned channel picked up with the help of aerial photographs is seen draining the area near Gol, and meeting the present Luni channel near Mekarans.

4. Apart from these abandoned courses which once drained the northern areas, a major disorganised course of the Mirti river has been picked up north of Rakhi which is now the waste Rann of Sanwarla, and which drains in to the present channel of the Luni river south of Silor.

   The zone north of Balotra (between Balotra and Pachpadra) is a zone of major confluences. In this zone meet the older courses of the Jojri, the channels draining the area north of Pachpadra and also some older disorganised courses of the Luni which branched off from the northern bank of the river. This whole tract is inundated during the floods and this is where the important salt works of the Pachpadra lie. On the basis of these studies it is now possible to say that the origin of the Pachpadra salt works may be attributed to the choking of an earlier drainage system and the concentration of salts in the stagnant drainage area.

5. A study of aerial photographs suggested that the Luni river had a series of channels taking off from its right and left banks draining into the main channel at points in the down stream after travelling for some distance. Such was the elaborate network of Luni channels seen west of Balotra, north of Jasol, around Kher

Temple, west of Bagundi, south of Tilwara and north of Kitpala and Sinli. In the last instance a vast flat carved out by the shifting course of the Luni river from east to west (broad directions), is seen north-west of Variya and north of Sinli. It is still possible to pick out one of the disorganised courses in this featureless flat.

In present times many rivulets originating from the relief around Jasol, Nakodaji, Variya die before meeting the present course of the Luni river.

Thus, if all the abandoned and disorganised courses of the Luni system picked up from aerial photographs were to be viewed as a whole then it will be amply clear that the entire highlands of Shergarh and Jodhpur were earlier drained in to the Luni river by a well-integrated drainage network, the confluence of which in present times is defined by the Pachpadra salt works and the area around it. These courses dried up during the Quaternary period owing to a major climatic shift towards aridity.

Recently, Sood and Pathan (1982) studied the palaeo-environment in Rajasthan based on landsat imageries. The analysis of multispectral and multitemporal landsat imageries in conjunction with topographical maps suggests the existence of numerous palaeo-channels west of the Aravalli range. Other notable features include rather broad floodplains (incompatible with the present day ephermeral channels), straight river courses, termination of flood plains and shifting of river courses. Sood and Pathan concluded that the disappearance of the once existing drainage system might not be a result of climatic changes as suggested by Ghose and other workers (1980) but a consequence of tectonic uplifts during the recent past.

THE CHAMBAL-BANAS DRAINAGE SYSTEM

South-eastern Rajasthan is drained by the chambal and its tributaries such as the Banas, Kali Singh and Parbati. It is a well-integrated drainage system forming dendritic and parallel to subparallel drainage patterns. The vast alluvial plains resulting from this drainage system has been highly dissected into gullies and ravines.

The Chambal River

The course of the Chambal is comprised of the following

three sections:

a. the upper valley or the course in the Vindhyan hills and the Malwa plateau;

b. the middle valley through the concentric triple scarps and the 32 km long gorge between the Gandhi sagar dam and Kota in which potholes, rapids and waterfalls are frequently found; and

c. the lower valley in the plain country before the chambal joins the Yamuna.

The upper valley:

The Chambal originates from the Janapao hill (854 m) of the Vindhyan mountain. The direction of flow is towards the north. Within 4 km of its origin the Chambal descends nearly 242 m. Thereafter the fall in height is gentle, about one metre per kilometre. The topographic features of this part seems, mature but the fluvial character of the Chambal testifies of a late youth to early mature condition. The deep-cut, incised course of the Chambal and its tributaries over the sandstone terrain indicates that vertical corrosion is a dominant process in the river.

The middle valley:

Between Rampura (24°27° : 75°31°) and Kota the chambal cuts straight through the three concentric Vindhyan scarps without any accordance to structure, though its tributaries occupy strike valleys and have exacavated in shales at the bases of sandstone scarps. Regarding the course of the Chambal in this section, it may be assumed that the Deccan Trap surface determined the direction of flow of this river. It can safely be inferred that the Chambal was developed on the Deccan Trap which overlaid the Vindhyans, and that later after cutting through the whole thickness of the Trap come to be superimposed on the buried Vindhyans. The three concentric scarps indicate three stages in which the Chambal valley was subsequently tectonically affected and which have left their imprints in the long profile of the Chambal between Rampura and Kota - "a zone of cataracts". The rapids and waterfall found in this section may be the points which starting lower down have migrated upstream to their present position (Chatterjee and Bagchi, 1957).

Potholes, terraces, waterfall, rapids and canyons are

the important erosional features between Rampura and Kota. All these features indicate the fluvial characteristics of the Chambal, hence their analysis is imperative.

Three pairs of terraces in this 'zone of cataracts' along the Chambal river have been observed during field studies. They indicate three periods of upheaval during which the area has been subjected to uplift and rejuvenation.

The most striking features indicating the youthful condition of the Chambal are the waterfall, rapids and associated potholes found along its course. The waterfall is known as Chulia and is 5 km upstream of Chambal near Bhainsrorgarh. The rapids are called Khatra and are 7 km downstream of Bhainsrorgarh (25°0° : 75°37°). The Chuli is a series of huge potholes which have been excavated by the Chambal on small faults and joints in the Kaimur sandstone, which forms the bed of the river near Rawatbhata dam site. The depth of the potholes varies between 2 m and 10 m. It was observed during the field study that in the dry season the flow of the Chambal is confined to a narrow channel, one meters to two metres wide, it swirls from pothole to pothole and finally boils through a long narrow channel cut in the quartzite.

As soon as the series of potholes ends waterfall occurs downstream. The rock formation at the side of waterfall is kaimur sandstone. The height of the fall is 15 m and its width some 10 m. It seems that this waterfall in the bed of the Chambal is not the result of any structurally weak zone, but is a knick point which starting lower down has migrated upstream to its present position.

Thirteen kilometres downstream from the Chulia waterfall are the rapids which again lend support for the theory about the youthful condition of the Chambal. The bed of the chambal at the site of the rapids is composed of the lower kaimur series. These rocks dip downstream at a low angle, in a series of great titled 'slabs, polished grooved and etched by the current of the water, resulting in a series of rapids along the course.

The erosional features make it evident that vertical corrosion is the dominant process and the deepening of the course of the Chambal is accomplished by excavation of huge potholes in the hard kaimur sandstone bed below. The lowering of the bed is so rapid that the valley has had no time to widen and, therefore, it is in the form of a deep gorge. Tributary streams are unable to keep pace with the rapid eroding course of Chambal and thus have hanging valleys.

Between Bhainsrorgarh and Kota the Chambal forms a deep gorge. The chambal gorge is about 150 m deep from the kota plateau and 32 km in length. The whole gorge section is composed of three rock formations of the Vindhyan system, namely, upper Bhander sandstone, Sirbu shale and Bhander limestone. The bed of the river through out the length is mainly formed of the Bhander sandstone except for a few patches of limestone and Sirbu shale. The general dip of rocks is extremely low and at places exposures give impression of the horizontally.

The lower valley

In the lower part of its course from Kota to the confluence with the Yamuna, the Chambal flows in a north-easterly direction parallel to the strike of the Vindhyans. The fluvio-geomorphic study of this section shows that the valley has been in existence from early times and that only the character of the bank have changed owing to the epeirogenic oscillations to which the valley has been subjected.

The incised meanders are the most significant features of the lower Chambal valley. The chambal has incised the former flood plain to a depth of nearly 40 m to 50 m. The sinousity of the Chambal seems to increase between Kota and the confluence with the Banas. The depth of upper meanders and sinousity increases beyond the confluence with the Banas further, and the same trend continues till the Chambal meets the Yamuna.

The incised course of the Chambal again has led to the conclusion that before the recent upliftment the Chambal had attained a state of maturity and was meandering freely over the wide floodplain. But the last upliftment of the region gave new power to the river to start the fluvial cycle afresh. So that now, inspite of the previous extensive valley, the Chambal has once again acquired the characteristics of a young river and as a result, meanders are now incising the region they occupied at the time of upliftment. The exposition of the older alluvium along the banks of the Chambal further supports this observation.

Another feature which indicates the youthful nature of

the lower course of the Chambal is the presence of boulders, rounded pebbles and rock islands in the river course. These features are easily discernible near the confluence of the Banas and the Chambal. Here the valley is broad and open and the amount of material brought by the rivers is so large that the channel is chocked and braided.

The longitudinal profile and erosional features of the Chambal make it evident that the river is a good example of a rejuvenated stream. Moreover, the upper Chambal may be regarded as a consequent stream on the Deccan Trap, as superimposed on the Vindhyans and strike valley in the lower course. The Chambal valley, therefore, provides an index of the morphogenetic changes the area has undergone. In the valley various phenomena associated with superimposition, rejuvenation, knick points and incision may be observed.

Valley Cross-Profiles

A study of the valley sides also throws considerable light on the evolution of the valley slopes. The shape of the valley has been determined from the data obtained during field studies. The height of the valley sides above the Chambal floor was measured at a distance of one kilometre from the centre of the valley. Such measurements were made at 20 places along the Chambal downstream from kota. The average height was then calculated. The mean height value obtained for the left bank was 66 m and for the right bank 55 m. However, the study of the obtained data reveals the asymmetrical shape of the Chambal valley.

Field studies show that the left bank of the Chambal valley is generally steeper, besides having a little higher than the right bank. The most common explanation for this asymmetry may be the formation of undercut banks and slipoff slopes attendant upon meandering and lateral erosion. These hard rocks on the left bank of the Chambal may be a possible cause of the asymmetry of the valley. At many places on the left bank of hard rocks the chambal have been exposed. Although these rocky benches are usually not continuous for any great distance (except between the Banas water gap and Dholpur), they do appear after short intervals. At the base of the Chambal scarp the upper Bhander sandstones intercalated with shales and limestones have been exposed and these may contributed have to the asymmetrical profile of the Chambal.

The asymmetry of the Chambal valley between the Banas

water gap and Dholpur may also be attributed to the inclination of the upper Bhander sandstones and Sirbu shales towards the north. Owing to the inclination of the rocks towards the north, the Chambal has a tendency to shift its course down the dip. This shifting of the river again leads to an asymmetrical cross-profile.

The valley asymmetry may also have resulted because of the unequal deposition of the alluvium on the banks of the Chambal in the previous cycle of erosion, but for want of detailed valley profiles of the old alluvium, this aspect could not be further investigated.

Another significant feature of the sides of the valley is the declivity of the channel slope and width of the Chambal course from upstream downwards. Field studies reveal that the depth of the Chambal considerably increases in the lower valley from Kota to its confluence with the Yamuna. The correlation of the channel and surface slopes reveals that the variations between them increases from Kota downstream. The profile further indicates that the difference between channel and surface slopes near Kota is 15 m while near the confluence with the Yamuna it is 45 m. This increase in difference between the channel slope and surface slope towards the lower reaches of a river is possible only when a river is in a late youth condition; such a feature indicates a reversal of the normal valley condition. In the normal cycle of erosion, the difference between the channel and surface slopes decreases, and divides widen as one proceeds downstream. But the process in the Chambal is the reverse as indicated by the profile. It seems that the Chambal has been rejuvenated in recent geological history and that the dissection of the floodplain is active owing to this phenomenon.

The Nature of Tributary Streams

The study of the tributary streams of the Chambal also sheds light on the evolution of the fluvial cycle of the region under study. After the rejuvenation of the Aravallis and the central Vindhyan plateau, the newly developing streams adjusted to the underlying geological structure but there still are some sites in the major rivers where they do not seem to be guided by the underlying structure. Such features again are important evidence of upliftment. Therefore, the tributaries of the Chambal may be divided into two major groups, (i)structurally controlled streams or subsequents, and (ii)independent to structure or superimposed streams.

The study reveals that generally all the tributaries of the major rivers such as the Kali-Sindh, Parbati and Banas are subsequent streams which are mainly controlled by the strike of the country rocks. As has been mentioned, there are two prominent strike directions in the valley, first, that of the Vindhyans and Gwalior formations running south-west to north-east along the Great Boundary Fault and second, the strike of the Mukandwara hills running north-west to south-east. The subsequent streams flow parallel to the strike of the Vindhyans and Gwaliors and change their direction according to the outcrops of hard formations. In the Bundi-Sawai-Madhopur hills and Mukandwara ranges the guidance of the country rocks is marked by evident. In the Bundi hills, owing to the long wearing away of the alternate bands of the Vindhyan sandstone and shale, the longitudinal valleys have been carved out in the softer shale and lime-stone. The Bamani, Guru, Gunjali and Ahu rivers are other examples of river flow being guided by the strike of rock formations.

The streams that flow on the Kota plateau °Dangland°, Ramgarh dome and Satur anticline are guided mainly by the dip of beds. These streams may be classed as resequent streams. On the contrary, the streams that descend from the Chambal scarp, Kota scarp and Singoli scarp are controlled by the slope and are reverse to resequent streams. These streams may be classed as obsequent streams.

Example of significant structural control of the drainage of rivers by tectonic disturbances are found at the following places in the lower Chambal valley:

1. The Banas river draining the south-eastern part of Rajasthan flows from north-west to south-east, and just before joining the Chambal it takes an abrupt turn nearly at right angles towards the south. This direction of flow is neither owing to the general slope nor to the strike of the rocks. This diversion of the Banas indicates that at one time the river might have been flowing independently, parallel to the Sawai-Madhopur-Karauli line of hills. The diversion of the river, it seems has been caused either by the tectonic movement along the boundary of the Gwaliors and the Vindhyans or by progressive stream piracy. The former appears to be the major cause of the diversion.

2. Similarly, the Mej river after draining the north-

western part of Bundi district flows from west to east, and near Ramgarh it takes an abrupt turn towards the south. In the case of the Mej it may also be inferred that once the river might have been a tributary of the Banas and was flowing parallel to the Bundi hills in the north-eastern direction, that later on its course was diverted by faulting, as happened in the case of the Banas.

From the examples given it is evident that two orientations of streams have evolved that of the tributaries of the Banas and that of the tributaries the Chambal. Structural and geomorphic evidence suggests that these two orientations were developed separately. Later during the Cretaceous period faulting (Great Boundary Fault) along the watershed (Bundi-Sawai-Madhopur hills) provided week zones and, consequently favourable sites for river capture. In this way the waters of the Banas were captured by the Chambal system.

Besides, in the course of the Chambal there are sites where the major tributary streams do not seem to be guided by outcrops of hard formations but cut across the bold ranges through narrow gorges. These sites seem to be examples of superimposition, and the course of the Kali-Sindh may be cited as one such example.

The Kali-Sindh is the second largest south-eastern tributary of the Chambal. It rises from the Vindhyans and flows northward on the Deccan Trap. After flowing for about half of its course on the trap country, the river enters the Jhalawar plateau and near Gagroan (24°37° : 76°12°) it cuts through the narrow Mukandwara hills in the dip direction forming a gorge. Here the river is definitely not guided by the underlying structure, otherwise it could have flowed eastwards parallel to the ridge and could have found a suitable outlet. It seems, here that the direction of the Kali-Sindh was established when the Deccan Trap covered the area up to the Mukandwara hills, and that the Deccan Trap surface determined the direction of the river.

Similar examples have also been observed during field studies on the Bundi-Sawai-Madhopur hills where even small streams cut across hard sandstone and quartzite.

It has already been mentioned that there is evidence of the Vindhyan plateau having been tilted north-westward. If so, the Chambal and its major south-eastern tributaries may well have developed as consequent streams on the tilted block on the youngest, that is, the Deccan Trap. Later, these streams were superimposed on the Vindhyan sandstones near the concentric triple scarps and on the Mukandwara hills. Spate (1954) is inclined to accept the view that "the Chambal, Ken, Betwa, rise within 20 miles of the Narmada, and appear as consequents on the Mesozoic surface superimposed on the scarp. The Chambal in particular cut straight across them with subsequent tributaries on the softer-shales."

Similarly, the Aravalli range was possibly arched up during the Mesozoic period along a south-west to north-east axis in the central Udaipur region. From this it may be inferred that the north-western tributaries of the Chambal river as represented by the Banas may also have been formed as consequent streams on the north-eastern flank of the Aravallis and then come to be superimposed. Varma (1961) analysing the geomorphic features of the Udaipur region has stated that:

> "the major rivers of this region still have the same course which they had at the time of rejuvenation..... the major streams did not adjust to the underlying structure and were strong enough to cut across these hard formations and their strike direction... thus these are the examples of superimposed drainage and are the evidences of rejuvenation of a peneplain."

Heron (1938) cited the example of the Banas in terms of the arching of the Aravallis. He suggested that the Banas at Rajmahal (25°52' : 75°25') wandering through the plains of Jaipur, breaches a great isolated flat-topped ridge of the Delhi quartzites to form a picturesque gorge. One can only infer that in the case of the Banas its course across the ridge was in existence before the Mesozoic peneplain was warped and cutdown vertically as the plane arched. Chatterjee and Bagchi (1957) are also inclined to accept Heron's conclusion regarding the development of the Banas and the arching of the Aravallis.

The Banas between Kathar and Dhandhi-Barwara, north of Gogunda provides an excellent example of transverse or superimposed drainage. The Banas here forms a gorge through Alwar quartzites on the south-east base of the synclinorium upstream and downstream of this gorge there are two big streams which drain water to the Banas. The course of the Banas is very sinuous, indicating its existence prior to the rejuvenation of the Aravallis. The river had developed its valley across this bold quartzitic ridge simultaneously with the arching of the Aravallis.

Besides the Banas, its tributaries such as the Berach also have sites of transverse drainage or superimposition. The Berach at its source near Udai Sagar passes through two isolated outcrops of metasediments of the Udaipur group. Downstream of Udai Sagar also, the river has cut across an elongated ridge of the Aravalli super group. These various examples of valley development in the Chambal basin indicate that the major streams are unrelated to structure and are good examples of superimposition.

On the whole the study of the tributary streams of the Chambal leads one to the conclusion that the major tributaries still have the same courses which they had at the time of the upliftment of the Aravallis and the Vindhyan plateau. With the exposure of the hard formations the weak and newly developing streams adapted their courses accordingly. On the contrary, the major rivers did not adjust to the geological structure and were strong enough to cut across these hard formations.

Migration of the Banas

The case of the migration of the Banas has been studied on landsat imageries and in the field between Rashmi in Chittorgarh district and Hamirgarh in Bhilwara district. Between these sites the Banas had migrated leaving boulders and pebbles of 20 cm to 25 cm diameter on the right bank. The existence of a gravel ridge at Uncha and of a narrow terrace between the present course of the Banas and the gravel ridge indicate that the river had rejuvenated simultaneously with its migration (Fig. 6.3 and 6.4). The terrace surface is characterized by deep black soil. From the existence of big boulders on the palaeo floodplain of the Banas it can also be inferred that fluvial conditions then were very strong as compared to the present, indicating a shift in climatic conditions from humid to less humid.

An excellent instance of the migration of the Banas and the control of lineaments can be studied in the lower reaches south of Malarana Dungar near the confluence of the Moral River. Figure 6.5 depicts palaco-channels of the Banas river which have migrated southward as inferred from the channel scar traces. The figure also shows other fluvial features such as natural old leaves, gullies, palaeo-flood plain and channel bars. A field investigations in this part of Rajasthan reveal that the area is a region of high soil moisture possibly represent old inter levee area with finer heavy sediments. It further indicating probably the greater flow regime in the Banas during recent geological period. It has been pointed out earlier by Sharma (1973 and 1976) that the Banas was formerly an independent river flowing across the Karauli plateau towards the north-east. But as a consequence of the Great Boundary Fault movement the river got diverted towards the south, taking an abrupt right-angled turn, and joined the Chambal.

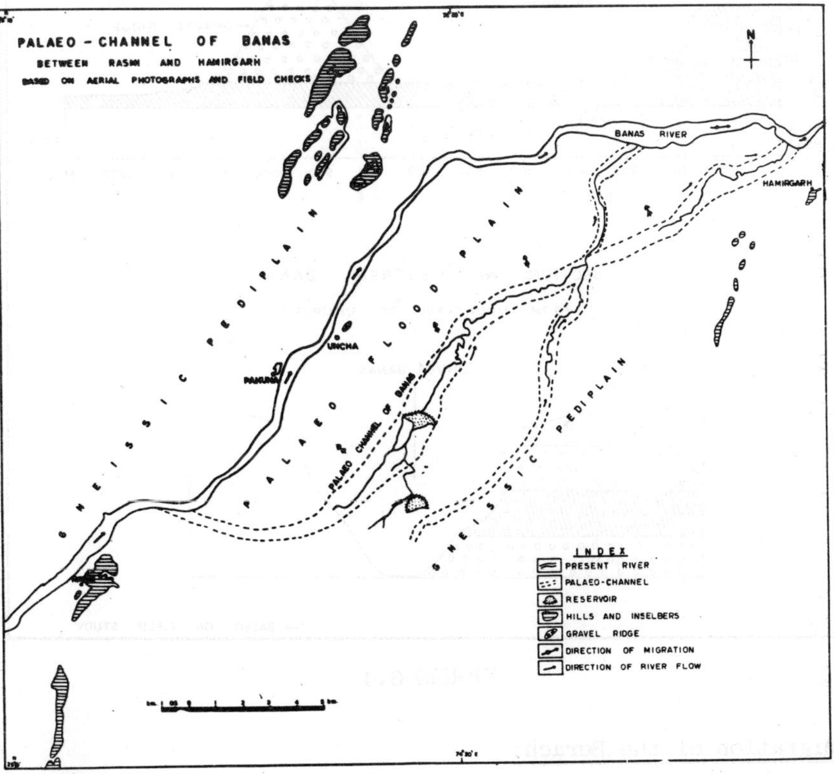

FIGURE 6.3

The above two examples of the migration of the Banas lead us to conclude that tectonic events have played a major role in the migration and that climatic changes have accentuated the effect, thus supporting the theory of Yashpal et. al. (1980).

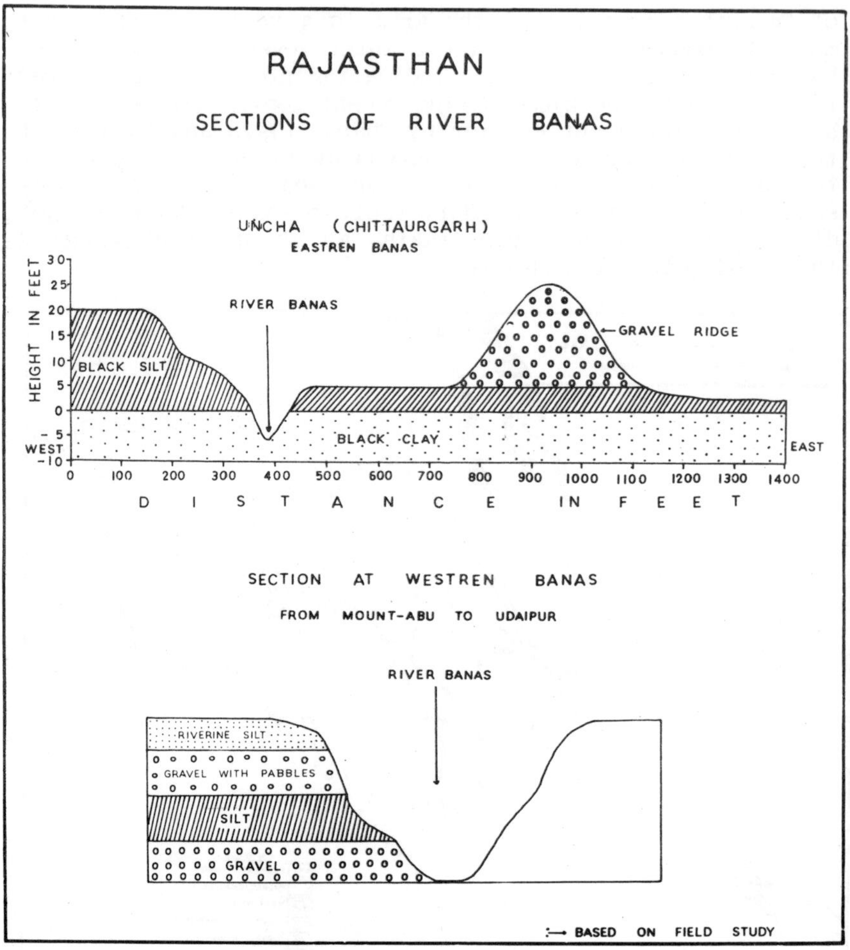

FIGURE 6.4

Migration of the Berach:

Another instance of migration is that of the Berach (tributary of the Banas) between the confluence of the Gambhir with the Berach and confluence with the Banas (Fig.6.6). Here the Berach flows parallel to the Great Boundary Fault. The river has migrated towards the north-west, leaving a palaeochannel behind. The palaeochannel was picked up on landsat imagery and subsequently verified in the field. The migration of this river can be attributed to tendency of the river to migrate towards the line of a weaker zone.

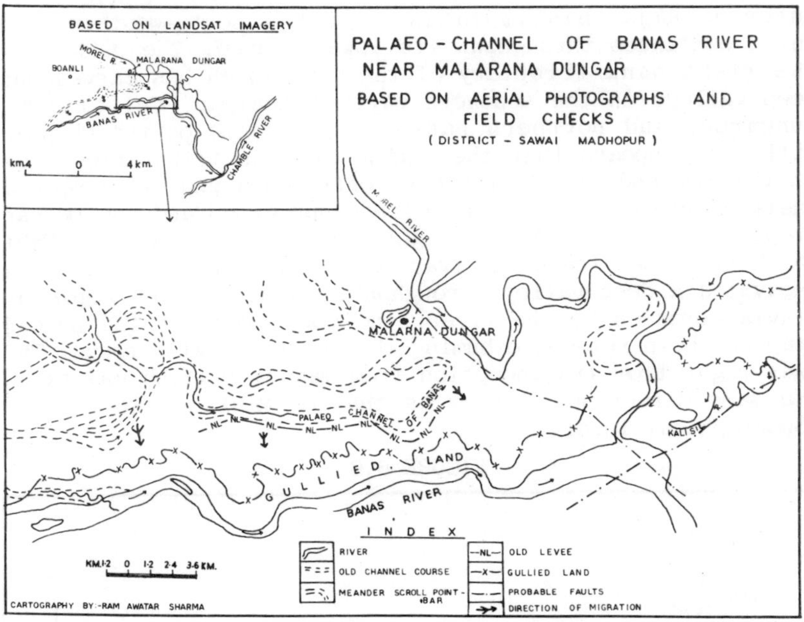

FIGURE 6.5

Waterfalls in the Chambal drainage system:

The Chambal drainage basin in Rajasthan provides a classical example of numerous waterfalls especially in the Vindhyan country from Chittorgarh to Kota. The waterfalls are generally on the margins of triple scarps. Few examples are: the water falls of the Menali river (36 m), Mahadev ka Nala, two water falls (17 m and 20 m tributary the Ruparel), Bhimlet (60 m) a tributary of Kural (Mej), Chambal (20 m), Bamani and Gunjali (make waterfalls before joining the Chambal) (figure 6.7). These examples indicate the interrupted course of these streams and upliftment of the western Vindhyan plateau. These waterfalls are knick points in the courses of the streams.

THE MAHI DRAINAGE SYSTEM

The southern part of Rajasthan is drained by the Mahi and its tributaries, namely, the Jakham, Som, Gomati, Anas and Erau. The Mahi originates in Madhya Pradesh, it enters, Rajasthan from the south in Banswara district, and then after flowing through Gujarat it joins the Gulf of Cambay. In Madhya Pradesh it flows from south to north, but after entering Rajasthan it takes an abrupt south-western turn which continues throughout its lower course. The Jakham is the right bank tributary of the Mahi which has carved out deep valleys in the southern part of Udaipur, Chittorgarh, Dungarpur and northern Banswara districts on the Chappan hills. It appears that the Jakham and its tributaries such as the Som and the Gomati have entrenched and superimposed their course over the rocks of Udaipur group and the Deccan Trap. The Erau is also a right bank tributary of the Mahi which traverses the hilly tract of Banswara district and has excavated a deep valley throughout. Owing to the vigorous fluvial erosion by the river, the top layers of the Deccan Trap has been removed exhuming the lower Aravalli rocks. The Anas right from its inception from Madhya Pradesh maintains the same direction of flow as that of Mahi but at a lower topographical level.

FIGURE 6.6

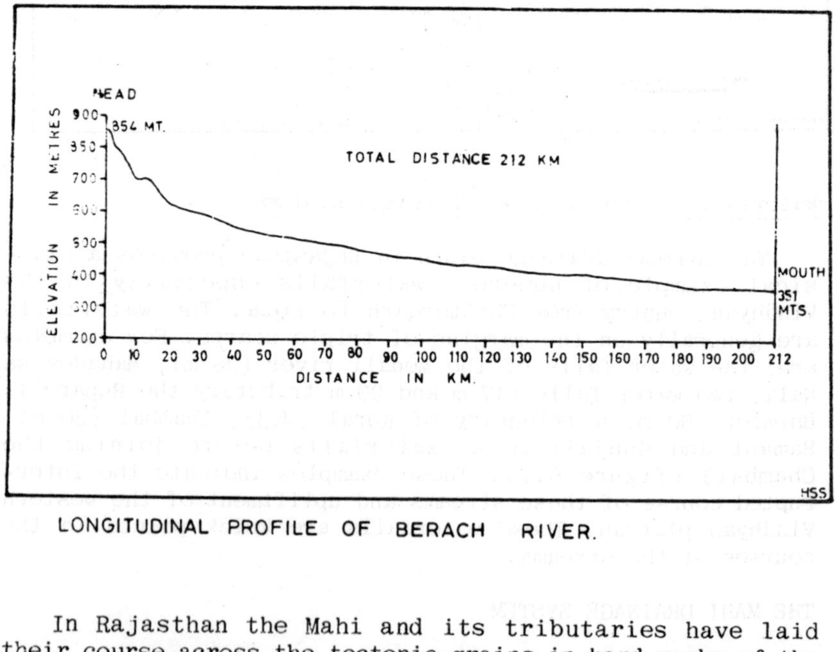

LONGITUDINAL PROFILE OF BERACH RIVER.

In Rajasthan the Mahi and its tributaries have laid their course across the tectonic grains in hard rocks of the Udaipur group and the Deccan Trap. The sharp bends of the Mahi can be attributed to the fracture pattern. From the

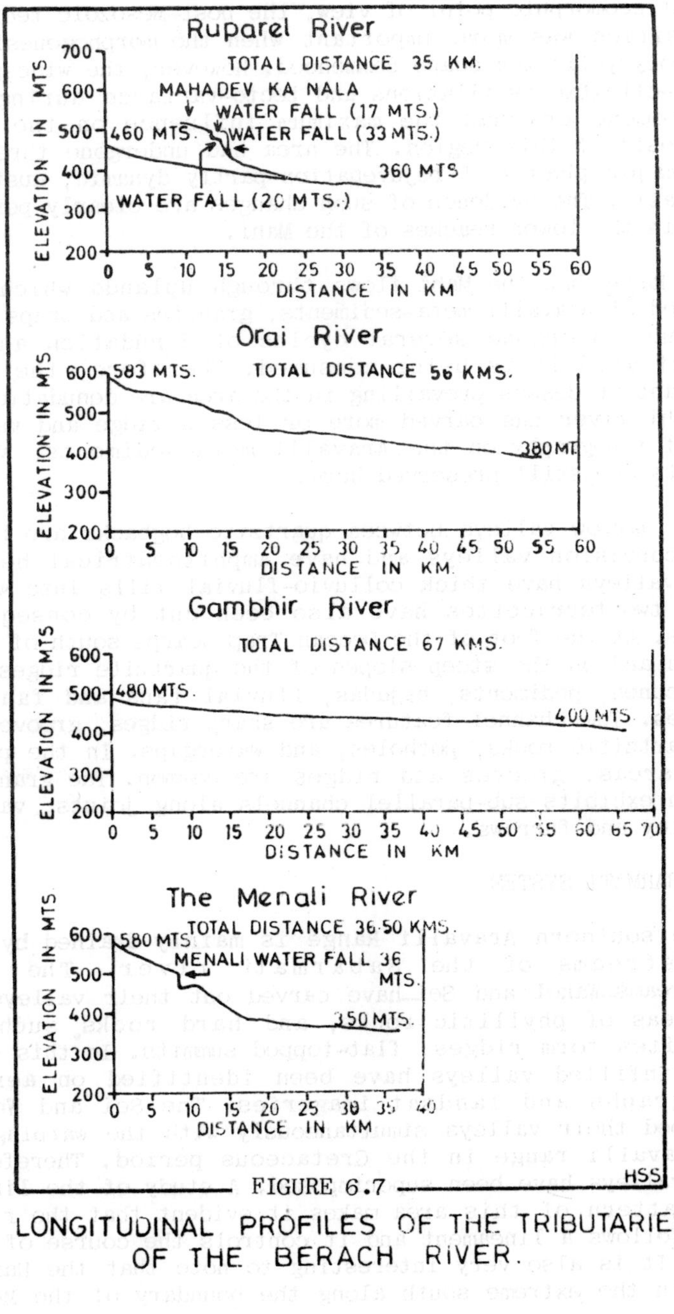

FIGURE 6.7

LONGITUDINAL PROFILES OF THE TRIBUTARIES OF THE BERACH RIVER.

fluvial geomorphic point of view, the post-Mesozoic tectonic disposition was more important when the morphogenesis of this polycyclic landscape commenced. However, the widespread palaeo-climatic oscillations and tecto-variance during the Pleistocene era had the maximum influence on the morphogenesis of this region. The area had undergone three to four major phases of rejuvenation-partly dynamic, eustatic and static. The evidence of such changes are clearly perceptible in the lower reaches of the Mahi.

In Rajasthan the Mahi flows through uplands which are composed of Aravalli meta-sediments, granites and traps. The area has undergone several cycles of denudation and at present also it is being consumed. Therefore, the most prominant processes prevailing in the area are denudational. The Mahi river has carved more or less a ridge and valley type of topography on the Aravalli meta-sediments, whose remnants are still preserved here.

The narrow valleys between quartzite hogbacks are basically corrasion valleys and have amphitheatrical heads. These valleys have thick colluvio-fluvial fills into which one or two terracettes have also been cut by consequent streams. At the foot of the Deccan Trap scarps south of Partapgarh and on the steep slopes of the quartzite ridges are talus cones, pediments, bajadas, fluvial fans and fan-cut terraces. The channel features are sharp ridges, grooves in the quartzitic rocks, potholes, and watergaps. In the phyllitic areas, grooves and ridges are common. The granitic terrain exhibits sub-parallel channels along joints, valley in valley and furrows.

THE SABARMATI SYSTEM

The southern Aravalli Range is mainly drained by the head streams of the Sabarmati river. The two headstreams_Wakel and Sei_have carved out their valleys in the areas of phyllitic rocks, and hard rocks such as quartzites form ridges, flat-topped summits. In this section, infilled valleys have been identified on aerial photographs and landsat imageries. The Sei and Wakel developed their valleys simultaneously with the warping of the Aravalli range in the Cretaceous period. Therefore, their valleys have been superimposed. A study of the lineament pattern of this area makes it evident that the river Wakel follows a lineament and it controls the course of the river. It is also very interesting to note that the Harnai river in the extreme south along the boundary of the Mewar and Idar has cut across a ridge of quartzites and formed a gorge, which indicates a case of superimposition.

THE WESTERN BANAS AND SIPU SYSTEM

The southern Aravallis and the Mount Abu mass is drained by the Banas and Sipu rivers and their tributaries. A view of both the plains can be had from the Gurushikar. The plain of these rivers is almost flat and is dotted with inselbergs. The river terraces formed of boulders, cobbles and riverine silt are discernible on the way from Abu road to Udaipur.

INLAND DRAINAGE

Rajasthan exhibits numerous examples of inland drainage which at one time were part of bigger integrated drainage systems. The whole of western Rajasthan round Jaisalmer, Pokaran, Thob and That have inland seasonal streams which flow from higher topographical levels and drain their rain-water into flat ranns. The Lik is the only big inland seasonal stream in the Barmer-Jaisalmer area.

The inland drainage of Sambhar, Didwana, Kuchaman and Nawa lakes are classic examples of enclosed basins or bolson. The Rupnagar, Khandel and Mendha are the main streams which feed waters to these lakes. The study of landsat imageries reveals that a large number of vegetation alignments are found cross cutting around the Sambhar lake and along these rivers. Therefore, the explanation of Sood and Pathan (1982) that these lineaments might be in the nature of deep-seated faults or fractures extending well into the basement appears convincing. However, the disappearance of this pre-existing drainage system as a consequence of tectonic upliftment during the recent past and not because of climatic change is doubted.

The area in the vicinity of the Sambhar and other lakes is characterized by paired terraces, pediments/bajadas, pediplains, structural hills, inselberges, sand-dunes and sand sheets.

Drainage of Ranns

In the country around Jaisalmer, Pokaran and Barmer there are several enclosed depressions which receive rain-water from the high topographic level. There are two separate basins, namely, Jaisalmer and Barmer. The Bitki,

Sukri and Ghugri rivers drain into Jaisalmer basin and the Lik drains into the Barmer basin. The pattern of surface drainage appears to be radial (around Thob) and centripetal in the ranns. There is a controversy whether the ranns around Jaisalmer, Thob and Pokaran are the results of simple disruption of drainage or not. Pandey and Chatterjee (1970) do not consider the Mitha, Kharia and Kanodwala ranns as a result of flurial action on the ground that they occupy a fault zone along the junction of Baisakhi and Jaisalmer formations. On the contrary, the author has observed fossil river deposits around Jaisalmer and Pokaran. Recently, Satyanarayana and Singh (1979) have also identified palaeochannels in this area through landsat imageries, which further support the view that Jaisalmer and Pokaran ranns are the result of disorganized drainage of the past. Allchin, Goudie and Hegde (1978) have also reported quartzitic gravel in the Jaisalmer area. However, such gravels do not found in the immediate vicinity of Jaisalmer ranns.

Inland Drainage of North-Eastern Rajasthan

The region lying between the Indus and the Ganga contains incomparable geomorphological and archaeological evidence which bears testimony to the rise and decay of many civilizations. There is considerable literature to explain the evolution of the drainage and changes in the river courses of the Ghaggar, Drishadwati and Yamuna, and to explain the climatic changes. A variety of hypotheses have been put forward by geologists, geomorphologists and archaeologists, namely Oldham (1874), Krishnan (1952), Wadia (1966), Singh (1952), Ali (1942), Indras (1967), Stein (1942), Wilhelmy (1969), Raikes (1968), Yashpal and coworkers (1980), Bakliwal and Sharma (1980), Pande (1977), Dikshit (1977) and so on. However, no work has yet been done on the evolution of the inland drainage basins of the Katli, Sabi, Dohan, Barah and Banganga. The area between the Ghaggar and the Yamuna is characterized by an inland drainage pattern. It is almost barren, desolate, highly sandy and thinly populated. It experiences an extremely arid type of climate for the rainfall is erratic.

Recently, commendable work has been done by Yeshpal et al. and Bakliwal and Sharma using remote sensing, aerial and geomorphological surveys on the lost Saraswati and the migration of the Yamuna. The hypothesis of Yashpal and others indicates that the vast width (6-8 km wide) of the Ghaggar bed can be explained by assuming that at some time in the past major tributaries flowed into it. According to these workers the Sutlej river flowed into the Saraswati. The other major river system contributing water to the Ghaggar may have been some prior channel of the Yamuna. The evidence regarding the palaeo-channels of the Sutlej-Yamuna have been identified in the following manner: The three palaeo-channels Y_1, Y_2 and Y_3 shown in Figures 4.1 and 4.2 represent courses of some sizeable ancient rivers which flowed south-east of the river Markanda and into the Ghaggar or into the Chambal. Y_1 joined the ancient bed of the Ghaggar, channel Y_2 flowed into Chautang (Drishadwati) and channel Y_3 flowed through the present course of the Yamuna down to Delhi, then connected with the small lake near Bharatpur (Ghana) and finally probably joined the Chambal. This vagrant river has been named the palaeo-yamuna. It would, mean therefore, that the palaeo-Yamuna changed its course three time before assuming the present one. In the first instance it flowed through channel Y_1 into the ancient Ghaggar. Later on it flowed through channel Y_2, which includes the present Chautang, and met the Ghaggar near Suratgarh. The third time it went southward and passed through the channel joining the Ganga through the Chambal.

The changes in the configuration of the palaeo-channels have been attributed to neo-tectonism by Yashpal and coworkers who have discarded the earlier views of Raikes attributed the vagrancy of the Yamuna for coriolis force.

The history of the migration of the Yamuna in the area lying between Delhi and Agra has been traced by Bakliwal and Sharma. The earlier course of the Yamuna has been identified as that which flowed through Hodal in the north and Bharatpur in the south. Baklival and Sharma considered accretion of sediment as the principal cause of migration.

The purpose of reviewing the history of the migration of the Yamuna and the changes in the course of the Ghaggar system is that if we accept the validity of these hypotheses mentioned, then it would not be difficult to identify the palaeo-channels of the inland streams of north-eastern Rajasthan and to link them with Ghaggar-Yamuna systems.

Inland basins and their geomorphic significance

The Katli, Sabi, Dohan, Barah and Banganga form inland drainage basin in the northern and eastern parts of Rajasthan (Fig. 6.9). These streams originate from the Aravalli hills (Torawati, Jaipur and Alwar Hills) and flow in a radial fashion. These are seasonal streams, but during the monsoon season they create floods in different parts of the region. The Katli originating from the Torawati hills loses itself in the dunes near Naurangpura and then branches off as one to three desert nallahs* up to Dhinwali and Inskhela Johra. It was observed that in the year 1960-62 the waters of this river flowed up to Nesal and Bairarsar near Rajgarh. The fluvial sand in Rajgarh and Sidhmukh indicate the possible link of this river with Drishadwati. Similarly the Dohan and Sabi becomes very notorious during monsoon and lost themself west of Nohar, Jhajjar and Najafgarh west of Delhi respectively. The Archaeological sites on the existing stream flowing from north to south through Rohtak and the Archaeological mounds west of this stream are sufficient evidence to link these streams with a larger drainage system. The Barah ends near Pahari. Its palaeo-channel may further be traced up to the south of Hodal. The Banganga, a major stream of eastern Rajasthan becomes braided near Bharatpur. The area is repleted with evidence of numerous ancient oxbow lakes and palaeo-channels suggesting its possible link either with the Chambal or the Yamuna.

If we draw a curved line from Jhajjar in the north to Bharatpur in the south, connecting Nuh and Pahari, it would coincide with the existing trend of the flow of the yamuna. Thus it may be suggested that the Dohan, Sota, Sabi, Baran and Banganga were formerly a part of Ghaggar-Yamuna system. The Katli which is an extreme westerly stream, was a tributary of the Drishadwati which was supposed to be probably the second stage of the migration of the Yamuna. If we accept the possible link of these streams with a bigger system, it may further be suggested that owing to the migration of the Yamuna as a result of neotectonic activity of the Ghagger-Yamuna Divide and subsequently owing to the imposition or onset of aridity and encroachment of sand, the courses of these streams were choked and were left as beheaded streams (Sharma, 1983).

Further field work is, however, required for the confirmation of the above hypothesis. Already some archaeological data are available on the palaeo-channels of the streams in this region, but more intensive work needs to be done to ascertain the age of these palaeo-channels and the process of beheadation.

* Torrent is called nallah in India. During the monsoon rains such nallahs become very turbulent.

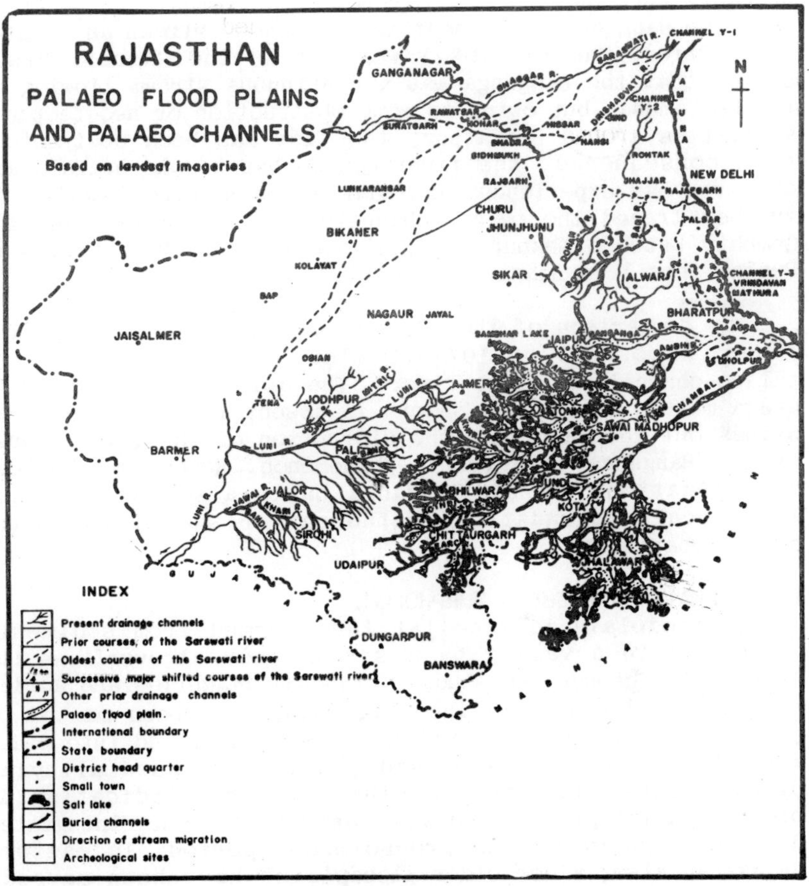

RAJASTHAN
PALAEO FLOOD PLAINS
AND PALAEO CHANNELS
Based on landsat imageries
INDEX
Present drainage channels
Prior courses, of the Sarswati river
Oldest courses of the Sarswati river
Successive major shifted courses of the Sarswati river
Other prior drainage channels
Palaeo flood plain.
International boundary
State boundary
District head quarter
Small town
Salt lake
Buried channels
Direction of stream migration
Archeological sites
GANGANAGAR
BIKANER
CHURU
JHUNJHUNU
SIKAR
ALWAR
NEW DELHI
BHARATPUR
NAGAUR
JAISALMER
JAIPUR
AJMER
JODHPUR
BARMER
PALI
JALOR
SIROHI
UDAIPUR
BHILWARA
CHITTAURGARH
TONK
SAWAI MADHOPUR
BUNDI
KOTA
DUNGARPUR
BANSWARA
JHALAWAR
SAMBHAR LAKE
N

FIGURE 6.9

The Evolution of the Gambhir-Banganga

The quaternary history of the area lying between the Gambhir and Banganga is very interesting. The study of landsat imageries with field checks brings out the fact that the Gambhir and Banganga shared the same drainage system and probably were once united to either the Yamuna or the Chambal. The Banganga at present is a beheaded stream as pointed out earlier. Sharma (1982) working in the Gambhir basin stated that the Banganga was a continuous stream flowing up to the Yamuna, but owing to some obstruction or neo-tectonic activities around Bharatpur, the flow has been obstructed. As a consequence, the river course has become braided and the area has experienced aggradation on a large scale. He further traced the palaeo-channels of the Banganga and the Gambhir between Nekpur and Kurka-Nagala Chatra villages in Rupbas tehsil of Bharatpur district.

The comparison of the 1919-20 one inch toposheet and the recent toposheet of 1971-73 also helps in tracing the palaeo-channels of these rivers. The old channel of the Banganga marked on 1919-20 sheets is absent in recent maps. It brings out the fact that the disruption of the lower course of the Banganga is a recent phenomenon. Recent geophysical investigations in the Banganga basin also indicate the large-scale aggradation in the saucer shaped depression around Bharatpur.

FLOOD PLAINS AND THEIR MORPHOLOGY

A floodplain is the flat land formed from sediments deposited in a valley by a stream when it overflows its channels*. Through the study of landsat imageries and aerial photographs the floodplains (Fig. 6.9) have been marked on the geomorphological map and have been classified into older aggraded alluvial plains and younger aggraded alluvial plains. Western Rajasthan exhibits an area indicating an older alluvial plain which was formed by the once integrated drainage system in the pre-Holocene period. In eastern Rajasthan also, wider older floodplains have been mapped. The older floodplains have been identified over the Karauli Danga land and a number of examples of floodplains resulting from the migration of various rivers have also been observed and verified in field. The floodplains which were marked on landsat imageries are incompatible with the present day ephemeral channels, which indicate that fluvial activity in the past was much more vigorous than at present.

* Encyclopedia America, Vol. II, p. 415.

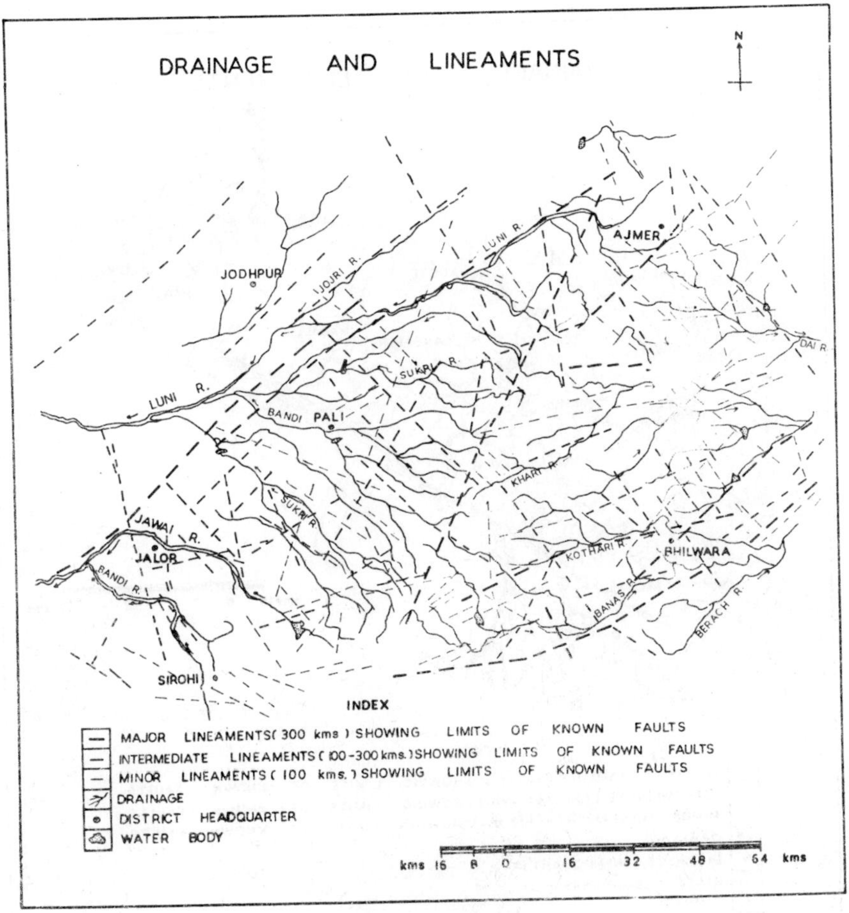
DRAINAGE AND LINEAMENTS
N
JODHPUR
AJMER
LUNI R.
JOJRI R.
DAI R.
SUKRI R.
BANDI
PALI
KHARI R.
JAWAI R.
JALOR
BANDI R.
SUKRI R.
KOTHARI R.
BHILWARA
BANAS R.
BERACH R.
SIROHI
INDEX
MAJOR LINEAMENTS(300 kms) SHOWING LIMITS OF KNOWN FAULTS
INTERMEDIATE LINEAMENTS (100-300 kms.) SHOWING LIMITS OF KNOWN FAULTS
MINOR LINEAMENTS (100 kms.) SHOWING LIMITS OF KNOWN FAULTS
DRAINAGE
DISTRICT HEADQUARTER
WATER BODY
kms 16 8 0 16 32 48 64 kms

FIGURE 6.10

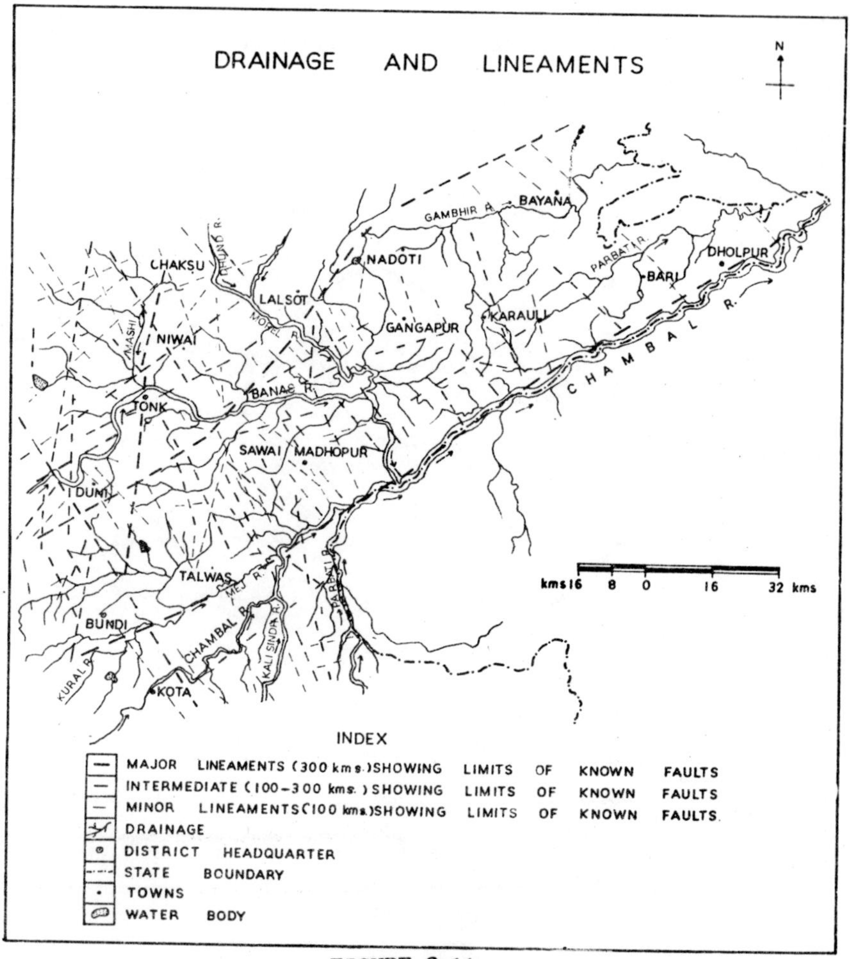
DRAINAGE AND LINEAMENTS
N
GAMBHIR R.
BAYANA
NADOTI
CHAKSU
LALSOT
MORE R.
GANGAPUR
KARAULI
PARBATI R.
BARI
DHOLPUR
CHAMBAL R.
MASHI
NIWAI
BANAS R.
TONK
SAWAI MADHOPUR
DUNI
TALWAS
MEJ R.
CHAMBAL R.
KALI SINDH R.
PARBATI R.
BUNDI
KURAL R.
KOTA
kms 16 8 0 16 32 kms
INDEX
MAJOR LINEAMENTS (300 kms.) SHOWING LIMITS OF KNOWN FAULTS
INTERMEDIATE (100-300 kms.) SHOWING LIMITS OF KNOWN FAULTS
MINOR LINEAMENTS (100 kms.) SHOWING LIMITS OF KNOWN FAULTS.
DRAINAGE
DISTRICT HEADQUARTER
STATE BOUNDARY
TOWNS
WATER BODY

FIGURE 6.11

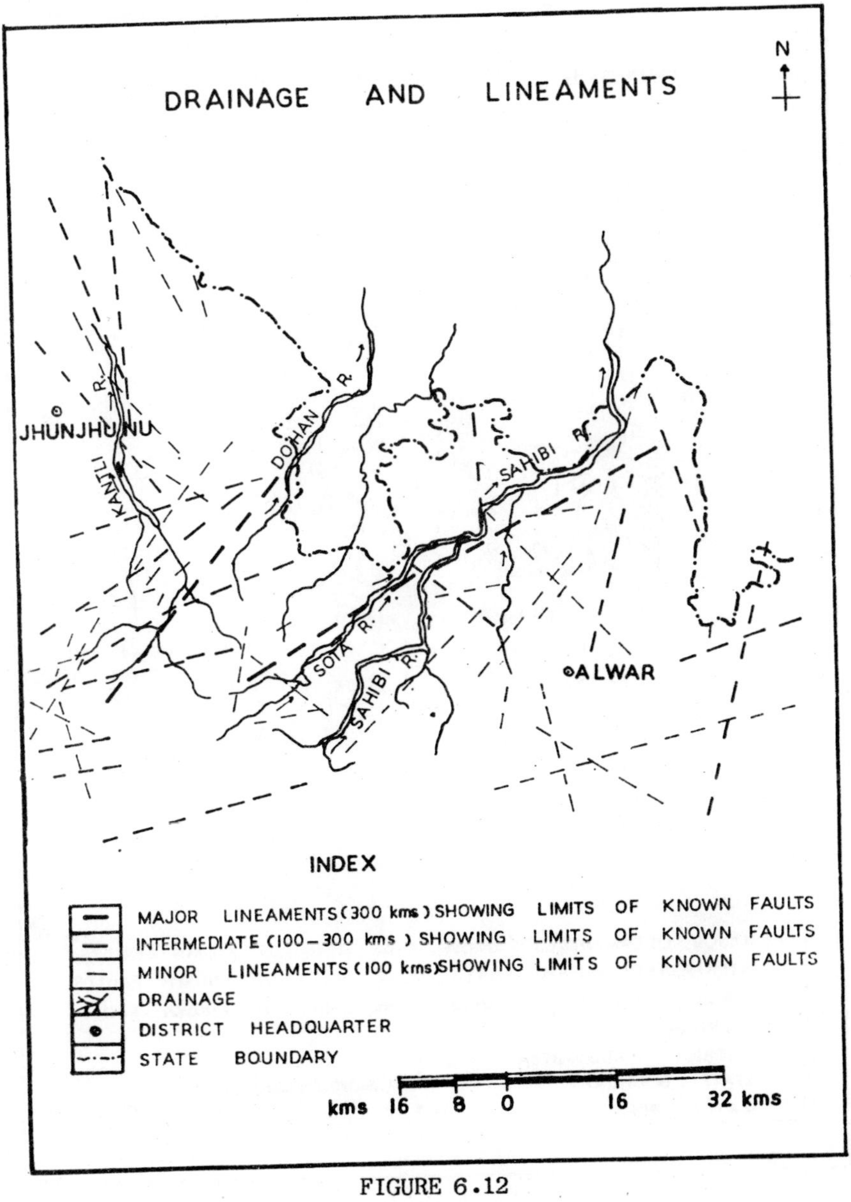
DRAINAGE AND LINEAMENTS
N
JHUNJHUNU
KANTLI R.
DOHAN R.
SAHIBI R.
SOTA R.
SAHIBI R.
ALWAR
INDEX
MAJOR LINEAMENTS (300 kms) SHOWING LIMITS OF KNOWN FAULTS
INTERMEDIATE (100–300 kms) SHOWING LIMITS OF KNOWN FAULTS
MINOR LINEAMENTS (100 kms) SHOWING LIMITS OF KNOWN FAULTS
DRAINAGE
DISTRICT HEADQUARTER
STATE BOUNDARY
kms 16 8 0 16 32 kms

FIGURE 6.12

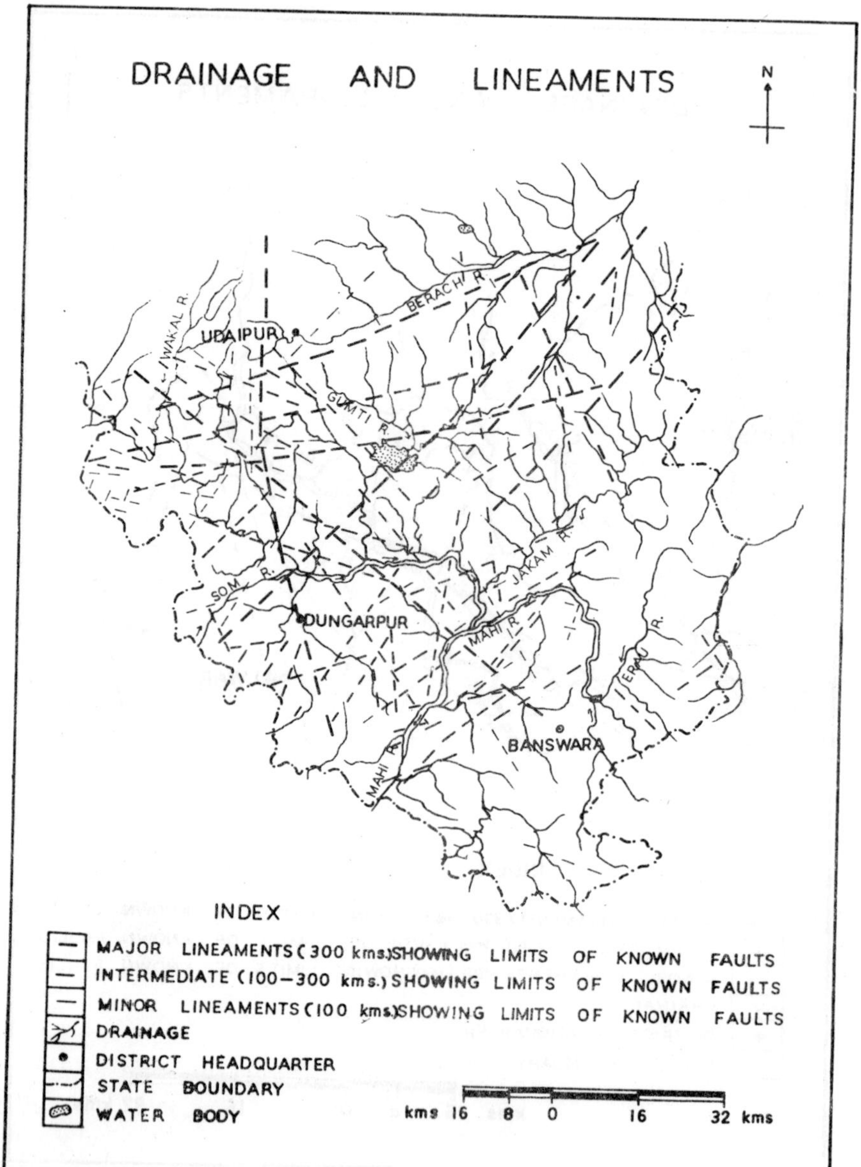
DRAINAGE AND LINEAMENTS
N
WAKAL R.
UDAIPUR
BERACH R.
GUMTI R.
SOM R.
DUNGARPUR
JAKAM R.
MAHI R.
ERAU R.
BANSWARA
MAHI R.
INDEX
MAJOR LINEAMENTS (300 kms.) SHOWING LIMITS OF KNOWN FAULTS
INTERMEDIATE (100–300 kms.) SHOWING LIMITS OF KNOWN FAULTS
MINOR LINEAMENTS (100 kms.) SHOWING LIMITS OF KNOWN FAULTS
DRAINAGE
DISTRICT HEADQUARTER
STATE BOUNDARY
WATER BODY
kms 16 8 0 16 32 kms

FIGURE 6.13

The newer floodplains are found along the Katli, Sabi, Banganga, Luni, Sukri and the Banas rivers the new floodplains along the Chambal and its tributaries, namely, the Kali Sind and the Parbati, have also been identified but the rivers have a tendency towards incision rather than towards the formation of floodplains. The breadth of the new floodplains is replenished every year during the monsoonal season through the usual flooding.

LINEAMENTS AND DRAINAGE

As has been mentioned earlier, Rajasthan exhibits a spectacular array of lineaments of varied dimensions which to a great extent control the drainage line.

Luni-Sukri and Trans-Aravalli lineaments show direct control on the drainage direction of the rivers, Luni, Sikru and Chambal, respectively (Fig. 6.10, 6.11, 6.12. 6.13). A right angle band in the Banas (western) is controlled by the fracture splay, produced at the confluence of the Banas with the Sipu near Deesa. The course of the Chambal is aligned with the trans-Aravalli lineament and the diversion of the Banas through Indargarh-Karauli Dang is also attributed to it. There are numerous streams in the Karauli Dang which are guided by lineaments. Similarly, the drainage lines of the southern Aravallis such as Wakal, Sei and other are controlled by lineaments. It is seen that the lineaments are major fractures without any sign of displacement. The Toda Bhim-Kakor lineament shows evidence of faulting and neotectonism, and controls the course of the Banas. The control of the lineaments can also be seen along the Mahi. The sharp bends of the Mahi can be attributed only to the fracture pattern. Similarly, the course of the Katli, Gambhir, Berach, Kothari, Sabi and Barah rivers and their tributaries in part or full are controlled by the lineaments, so is the migration of the river Banas near Pahuna, Banganga, Kothari, Gambhir and Luni.

SEVEN

Basin Morphometry

The validity of climatic geomorphology depends on its ability to establish a relationship between climate and process and to show that regional variations in climatic factors cause spatial differences in landforms. The significance of climate in relation to processes must be evaluated in order to make it more meaningful than the descriptive morphogenetic study. Similar to aeolian processes which are confined to the western part of Rajasthan, the fluvial systems also are controlled by climatic regimes and in particular to rainfall pattern. It is interesting to note that the fluvial processes become more effective east of 300 mm isohyte and the intensity of the processes increases with the increase in rainfall. The channel network also tends to increase in relation to increase in rainfall.

An attempt is made here to analyse the various parameters of fluvial morphometry and to show the relationship between them. In order to demonstrate the control of climatic factors on fluvial systems, morphometric parameters have been correlated with mean annual rainfall. The analysis has been made on 62 random order basins and 54 third order basins from different climatic regions of Rajasthan.

STREAM NUMBER

The quantitative method to analyse morphometric characteristics of fluvial landforms was initiated by Horton (1945). Subsequently a great deal of work has been done by several workers in India and abroad to evaluate the geomorphic characteristics of the drainage basin. The First Law of Horton is the law of stream numbers. The computed data have been presented in tabular form. It is evident from the Tables 7.1 and 7.2 that the number of streams decrease with the increasing order thereby the exponential function appears to be negative in both third order and random order basins irrespective of climatic control. But when the data is viewed from the number of stream segments of the first order it becomes evident that the basins of a humid climatic region tend to have a higher number of stream segments and that the value decreases in basins of the arid region. The number of stream segments of different orders to vary from basin to basin even in different climatic zones. That fact may be attributed to the character or underlying geological structure.

BIFURCATION RATIO

Horton suggested that the slope of fitted lines represent the mean bifurcation ratio of the series. He considered bifurcation ratio as an index of relief and dissection. A number of workers have pointed out that the bifurcation ratio tends to have values between 3 and 5 the usual value being around 4. In the area, the bifurcation ratio of the third order basin ranges from 2 to 15 with most of the values being between 3 and 6. The highest values are found in humid areas, indicating a greater degree of dissection. The mean bifurcation ratio in random order basins does not show much variations and generally falls between 3 and 4.19 in the humid region, 2.822 and 6.51 in the subhumid region, 2.8 and 4.92 in the semi-arid region, 2.85 and 3.91 in the transitional zone and 2.83 and 4.09 in the arid zone. The values of the mean bifurcation ratio of third order basins increase to as much as 9.73 in humid regions and 8.66 in subhumid regions (Table 7.1 & 7.2).[1]

STREAM LENGTH

The Second Law of Horton concerns stream length. In the present analysis mean stream length, total stream length and cumulative total stream length have been computed. Total stream length, mean stream length and cumulative stream length for all the third order basins (Table 7.1) yields different results. The highest mean stream length value (32.25) has been found in the basin of the subhumid climatic zone while the lowest value (5.0) is found in the arid region. The stream length ratios of all the basins also yield different results. The highest value is found in basins of the subhumid zone and the lowest in the arid zone.

---

1. Maps and Tables of this chapter are at the end.

It is evident from Table 7.3 and 7.4 that the values for total stream length of all the basins in different climatic regions also vary.

The value for total stream length of basin number 16 which falls in the subhumid zone is recorded as the highest 65 km and that of basin 35 which falls in the semi-arid region as 57.15 km. The total stream length values in arid and transitional zones have been found to be the least.

BASIN AREA

The Third Law of Horton is the law of basin area. It is evident from Table 7.5 that basin area values have been found to be the highest in basins of the subhumid climatic zone of third order basins. Again, the highest value of basin area (of random order basins) is to be found in subhumid and semi-arid regions. Table 7.5 and Table 7.6 also indicate that values for basin length and perimeter are also higher in subhumid and semi-arid area indicating that finger tip tributaries do have a greater chance of growth in these regions and that the basin shape is elongated.

RUGGEDNESS NUMBER

Ruggedness number is an index of dissection. It is evident from the Tables 7.5 and 7.6 that the highest ruggedness number is found in humid regions and that the value tends to decrease in arid region.

CORRELATION COEFFICIENT

The correlation between the number of streams, mean stream length, cumulative total stream length, total stream length of third order and random order basins has been determined and the regression equation, level of significance and correlation coefficient have been calculated. The values for regression equation are presented in tabular form. (Tables 7.7, 7.8, 7.9, 7.10, 7.11, 7.12, 7.13 and 7.15). The level of significance in some of the basins has been found to be highly significant.

Seventeen basin parameters have been correlated with each other and the values of correlation coefficients are presented in Table 7.15 and 7.16.

DRAINAGE DENSITY

Drainage density can be useful in the spatial prediction of drainage basin processes and in the interpretation of temporal changes in the drainage network. For these purposes it is necessary to determine the magnitude of variation in drainage density in relation to climate. In the present study the relationship between drainage density and climate has been examined at a macro climatic level.

It has been pointed out by some workers that the regression between total stream length and the basin area can be used to describe world variation in relation to the mean annual precipitation. These studies indicate that the drainage density value increases through arid to semi-arid and decreases through humid continental and subtropical areas to the values obtain for humid temperate areas (Gregory and Gardiner, 1975). The validity of the above hypothesis is tested here. Drainage density can be regarded as a response to control of climate and of catchment characteristics and yet it can also be regarded as the factor influencing runoff and sediment response of a drainage basin. The central position that drainage density occupies in fluvial systems and the sensitive function that it performs should, therefore, be the basis of determining its usefulness.

The drainage density of Rajasthan has been measured for sample drainage basins and also by grid method proposed by some workers such as Gardiner. The grid method technique is useful for spatially correlating drainage density with mean rainfall. The density calculated basinwise has been plotted on lograthim paper and the regression equation and r value calculated.

An analysis of Fig. 7.1 brings out the following facts:

1. The north-western region of Rajasthan does not have streams. It is dunal country. Therefore, it is classed as an area without drainage. Only the south-western part has drainage density mostly between 0.20 km/km^2 and 0.40 km/km^2. A higher value of density is found north-west of Jaisalmer, but this is because there are only few centripetal inland drainage lines which give rise to the higher density. Besides, the underlying sandstones also contribute towards the higher values. The rainfall west of the Aravalli range is below 300 mm.

2. The drainage density in the western Aravalli pediment zone is between 0.21 and 0.60. The rainfall in this zone is 300 mm to 400 mm.

3. The density figures increase east of the Aravallis and in the semi-arid climatic zone of Rajasthan the value tends to increase up to 0.60. The maximum grid falls between 0.41 to 0.60 while in the western Aravalli zone it is mainly between 0.21 to 0.40.

4. Higher drainage density values are found in the south-eastern part of Rajasthan. The values here range from 0.41 to 1.21 with most of the values being around 1.00. The rainfall pattern also increases in the same way.

The analysis of drainage density in relation to rainfall indicates that there is a direct relationship between drainage density and rainfall. Hence the hypothesis of earlier workers that density values increase through arid and semi-arid and decrease through humid continental and subtropical areas is not valid in the case of Rajasthan.

A study of Fig. 7.2 and 7.3 yields different results. When the drainage density of random order basins is plotted against mean annual rainfall, it shows a positive trend. The calculated value of the regression equation is as follows:

$$Y = 649.56 + 28.84x$$

The correlation coefficient value is $r = 0.6336$, which appears to be very significant. The graph also indicates that the values of drainage density of subhumid areas are scattered mainly horizontally above the plotted line. Only a few points fall below this line. The values of density of basins of transitional zones fall between the values of subhumid and arid climates. The drainage density values of humid regions always fall close to the regression line and above the values of other basins of other climatic zones.

The plotting of the density of third order basins against rainfall shows a negative trend, that is, the value of drainage density decreases with increasing rainfall. However, the pattern of scattering appears the same as that observed. In this case also for random order basins the points of arid areas are always below the line of fitness and the values of subhumid and humid basins are above, except for a few exceptions.

Therefore, the spatial analysis of drainage density and basinwise density points to the same conclusions.

In Table 7.17 drainage density values have been grouped against percentage of area of Rajasthan.

The Table 7.17 indicates that more than 33 per cent of the total area is without drainage. Besides this, 50 per cent of the area falls within the values from 0.20 to 0.60.

STREAM FREQUENCY

Stream frequency has been worked out both basinwise as well as gridwise. The drainage frequency (Fig.7.4) also shows an increase with increasing rainfall. The rainfall, as indicated earlier, progressively increases from the north-west of the Aravallis to the south-east. Minimum stream frequency between 0.02 and 0.021 is found in western Rajasthan. It tends to increase in south and south-eastern Rajasthan up to 0.15 and more. This analysis also support the earlier observations. The variation within basin or surrounding grids is because of local factors such a geological structure and vegetation cover.

The frequency of streams plotted against mean annual rainfall (Fig.7.5 and 7.6) shows the same trend as that of drainage density. The values of stream frequency of random order basins tend to increase, with a increase in rainfall and distribution of points above and below the line follow the same pattern. The points of arid region are always below the line of fitness. The regression equation and value of correlation are as follows:

$$Y = 652.62 + 16.08x$$
$$r = 06477$$

As in the case of drainage density, the graphs of stream frequency of third order basins also show negative functions. The regression equation and value of correlation are given below:

$$Y = 706.47 - 20.42x$$
$$r = 0.5316$$

The level of correlation is less significant than that of the random order basins, but pattern of scattering is the same.

The values of stream frequency and percentage of area have been given in Table 7.18:

DISSECTION INDEX

The dissection index is as index of the degree of dissection in the area. Fig. 7.7 gives the variation of dissection index from one part of Rajasthan to another. The value of dissection index is very low in the north-western part of Rajasthan. It is higher in south and south-west of the Aravallis. The dissection index does not show any particular relationship to the rainfall pattern of Rajasthan, but it is evident that dissection is maximum in the areas which either have undergone stream rejuvenation in the lower course of the Chambal or in higher rainfall zones. The maximum dissection index is found in the south Aravallis, perhaps because of the severe fluvial erosion and the higher rainfall in the Mount Abu zone. The percentage of area of dissection index is presented in Table 7.19:

Texture ratio is the combined result of drainage density and stream frequency. The length and spacing of stream collectively form a texture of drainage (Fig. 7.8 and 7.9). The texture ratio of random order and third order basins in different climatic zones are presented in Table 7.3 and 7.4

The value of texture ratio has also been plotted against mean annual rainfall (Figs. 7.8 and 7.9). The graph of random order basins (Fig.7.8) indicates that texture ratio increases with increase in rainfall. The regression equation is:

$$Y = 631.22 + 15.23x$$

and the correlation coefficient values is: $r = 0.4815$ (Tables 7.24 and 7.25).

The graph of third order basins (Fig.7.9) indicates a negative function. The points are widely scattered but generally fall above the line of fitness and in higher rainfall zones. However, the correlation coefficient appears significant, the value is: $r = 0.6050$.

The regression equation value is: $Y = 647.27 - 0.01x$.

ELONGATION RATIO

The elongation ratios of third order and random order basins in different climatic regions are also presented in Tables 7.5 and 7.6. The average elongation ratio of each climatic region has been worked out. It is evident from Table 7.5 that the ratio of third order basin increases from the humid to the arid climatic zones, while the mean elongation ratio of random order basins is highest in semi-arid regions.

The relationship of elongation ratio to mean annual rainfall has been plotted in Figs. 7.10 and 7.11. It is evident that in both the graphs the elongation ratio shows an increasing trend with increasing rainfall. The regression lines are as follows:

$$Y = 536.40 + 191.93x$$

and

$$Y = 442.00 + 270.26x$$

The correlation coefficient values are:

$$r = 0.4392$$

and

$$r = 0.4875.$$

The correlation coefficient value of third order basins appears to be more significant than of random order basins.

CONSTANT OF CHANNEL MAINTENANCE

The higher values of constant of channel maintenance are found to increase in subhumid and semi-arid areas and decrease in humid and arid zones (Tables 7.27 and 7.28). On the other hand when they are plotted against rainfall (Figs.7.12 and 7.13) they show an increasing trend with increasing rainfall. However, the correlation coefficient value ($r = 0.5143$) for random order basins appears to be significant.

CIRCULARITY RATIO

Circularity ratios also show higher values in subhumid and semi-arid areas as evident from Tables 7.5 and 7.6. Figures 7.14 and 7.15 indicate a reverse relationship between circularity and rainfall. The graph of random order basins shows a negative relationship while that of third order basins indicates a positive relationship with rainfall. The correlation coefficient value for third order basins is highly significant. Therefore, it can be concluded that the circularity ratio increases with increase in mean annual rainfall.

RELATIVE RELIEF AND RELIEF RATIO

Higher relative relief values are confined to the Aravalli mountain chain (Fig. 7.16). The maximum difference in the highest and lowest point generally occurs in this belt. Maximum relative relief (maximum 1051 m) is found in the southern Aravallis around Mount Abu and Sirohi. Relative relief is an index of dissection and vertical erosion of the topography. The area east of the Aravallis has a higher relative relief value. The relative relief of Kota plateau, Harawati plain, and desert region is generally lower because of the flat character of the topography. It has been found that relative relief does not vary with rainfall.

The relief ratio basin wise from different climatic regions has also been worked out. Table 7.5 and 7.6 indicate that the values of relief ratio are higher in humid and sub-humid areas and lower in the arid zones.

The plotting of relief ratio against rainfall (Figs. 7.17 and 7.18) indicates that in the random order basin the relief ratio increase with increasing rainfall. The regression equation is as follows:

$$Y = 625.65 + 611.52x$$

The correlation coefficient value r = 5279 is significant.

The graph of third order basins (Fig. 7.18) does not show a good relationship with rainfall. The function appears to be negative and the correlation coefficient also is not significant.

The values of percentage of area in relation to those of relative relief are presented in Table 7.20.

REGIONAL SLOPE AND BASIN SLOPE

The spatial pattern of average slope has been worked out following Wentworth's formula. The Figure 7.19 indicates that maximum slope values are found along the Aravalli range. The highest values from 6° to more than 10° are found in the southern Aravallis. Minimum slope values are found in most parts of Rajasthan. Most of areas of Rajasthan generally fall in the slope category of less than 2° to 4°. In the Alwar region of the Aravalli ranges slope values again are higher and may rise above 4° up to 6°. The slope value in the northern and central Aravallis is less than 6°:

The table 7.21 indicates that 84.78 per cent of the total area of Rajasthan fall between 0° to 2° slope.

$$\tan 0 = \frac{N}{L} \times \frac{V.1}{3361}$$

N = Number of contour crossing
L = Length of the grid
V.1 = Vertical interval
3361= Constant value sin Q.

The basin slope has also been worked out and presented in Table 7.22 and 7.23. The average slope values of random and third order basins are separately plotted against mean annual rainfall. Both the graphs (Figs. 7.20 and 7.21) indicate the same negative function. The values of the arid regions generally fall below the line of fitness. The regression equation for random order basins is:

$$Y = 731.02 - 1.51x$$

and for third order is:

$$Y = 649.38 - 0.023x$$

The correlation coefficient values are : r = 0.4939 and r = 0.5211 respectively. The correlation coefficient value for third order basins is less than 50 per cent and therefore is non-significant (Tables 7.24 and 7.25).

It can be concluded from the above analysis that the slope values decrease with increasing rainfall. However, the concentration of the points in both cases is above the line of fitness and towards higher rainfall.

RUNOFF, DISCHARGE AND SILT LOAD

There are a number of factors such as degree of slope, nature of rocks and soils, presence or absence of vegetative cover and rate of evaporation and rainfall which control surface runoff, discharge and silt load. The term runoff and discharge appear synonymous but they have different meanings. Runoff implies surface discharge of water from rainfall and snow melt down slopes, while discharge is the quantity of water passing down a stream depending on its volume and velocity expressed in cusecs or in the number of cu. ft./sec. Thus discharge is confined to stream banks. Material laid down or suspended in water with a particle diameter of 0.002 mm to 0.02mm, coarsier than that of clay and finer than sand, is termed as silt. The movement of these particles in the river is a function of the velocity of the water, which in itself is the result of the gradient of the river profile.

In Rajasthan there is no perennial source such as snow fed for any river, therefore, the life of a river surface runoff and silt load is purely a function of monsoonal rainfall. The amount of rainfall determines the volume of water in the river and other related hydrological processes.

Runoff

Runoff gives the total amount of water available in the drainage basin. Table 7.26 gives the picture of riverwise total runoff. It is evident from the table that the highest runoff is recorded in the Mahi catchment, that is, 110,780 cu.ft. The Som and Erau which are tributaries of the Mahi record 66,920 and 62433 cu.ft. runoff, respectively. The Chambal, Berach and Banas I and II record 25,930, 16,630, 18,185 and 29,190 cu. ft. respectively. The low:

runoff in the Chambal catchment is because the river has been dammed at three places to generate hydro-electricity and for canal irrigation.

The runoff data of 50 sites are represented on the map of Rajasthan by iso-runoff lines (Fig. 7.22). It is evident from the map that the whole of north-western Rajasthan is completely free from surface runoff, there is no surface drainage. The Luni is the only seasonal stream which has surface runoff and that to from 2009 cu.ft. to 3375 cu.ft. which is very low compared to that of other rivers of Rajasthan. The value of iso-runoff lines increases in the south-eastern part of Rajasthan where rainfall also is higher.

The same data has been presented in the form of a bar diagram classifying river catchment according to climatic regions. This graph (Fig. 7.23) indicates that there is a direct relationship between rainfall and runoff. In order to establish this the data or surface runoff and mean annual rainfall have been plotted on lograthim paper (Fig.7.24). It is evident from the graph that total runoff increases with increasing rainfall showing the direct positive relationship between the two factors. The regression equation is:

$$Y = 368.22 + 0.0198x$$

The correlation coefficient has also been worked out at: r = 0.5254, which appears significant.

Discharge and Silt Load

Discharge data of 8 sites are presented in Table 7.27. It is evident from the table that maximum discharge is recorded at Dhariawad in the Jakam catchment, and the minimum at Kanota in the Dund catchment. When a comparision is made between discharge and silt load, it becomes evident that the rivers which have a minimum discharge carry the maximum silt load. For example, Dund at Kanota records a high silt load (70.10).

Silt Load and Rainfall

In order to show the relationship between rainfall and silt load, five years data (1970, 1972, 1974, 1976, 1978) at two years interval, of all the eight stations have been studied (Fig. 7.25). The graph of Sodawas which is in Sabi catchment shows good correlation between rainfall and silt load. This is also in the case of Manohar Thana which is along the Parwan river. The graphs of Nagria, Shri Mahavirji, Dhariawad, Kanota, Mahuwa-Mandawar and Hingota do show some deviation from each other. It has been observed that in some years rainfall has shown a decreasing trend while the silt load has maintained an increasing trend. These facts indicate that apart from rainfall there are some other factors such as slope, nature of surface material and vegetative cover which also play an important role in the availability of silt load in the stream channel. Further, it can also be concluded that the rivers of semi-arid regions carry more silt in their flow than the rivers of humid climates. Also an analysis of the monthly data of each river is required before reaching any final conclusion.

STREAM SINUOSITY INDEX AND CLIMATE

Stream sinuosity is one of the several parameters used in morphometric studies to differentiate among various types of land units. It is commonly expressed quantitatively as a ratio between the direct length and the channel length, or as the percentage increase in channel length per unit direct length.

Mathematically, the latter is expressed as (Smithson, 1939)

$\frac{X-Y}{Y} \times 100$ When X is the average channel length in miles/kilometers

and Y is the average direct length in miles/kilometers

Leopold, Wolman and Miller (1964) have maintained that a stream showing sinuosity (defined by $\frac{X}{Y}$ over 1.5 may be considered as meandering and below 1.5 as non-meandering or sinuous. From this it may be assumed that a young river having sinuosity value over 1.5 must be attributed to rejuvenation.

In nature, all streams show significant departures from a straight line course and hence possess some degree of sinuosity. There are several factors such as hydraulic, lithologic, topographic and climatic which are responsible for the variation in the geometry of a meandering channel. However, it has been established that sinuosity index increases with the advancement of the cycle of erosion, but when direct length and channel length are plotted in different climatic regions, a distinct pattern emerges signifying the role of climate in stream sinuosity irrespective of the stage of the cycle of erosion.

In the present study an attempt has been made to correlate sinuosity index with rainfall in different climatic regions of Rajasthan. For this study, measurements were taken at 50 sites of different river basins in different climatic zones of Rajasthan. The sinuosity index has been calculated both according to Leopold's and Smithson's methods. Smithson's method has been tested by Verma and Tandon (1971) and found an effective means of appraisal of stream tortuosity. The data used in this paper were measured from 1:50,000 survey of India toposheets. The measurements have been given in the Table 7.28.

Table 7.28 indicates that sinuosity index increases from the arid to the subhumid zone and again decreases in the humid region. In terms of the critical value of Leopold all the observations pertaining to arid, semi-arid and humid areas fall below 1.5, therefore, they are non-meandering, while the values of streams of subhumid regions are above 1.5, hence they be classified as meandering streams.

When direct length is plotted against channel length in different climatic regions of Rajasthan the graph reveals that the catena/scatter of points appears to be circle-cum-loop shaped in humid zones a hyperbola in subhumid regions and like a parabola in semi-arid and arid regions (Fig.7.26). It is also evident from the graphs that the maximum scattering of points has been noticed in the humid region. In order to substantiate the variation of length in different climatic regions, the sinuosity index has been plotted against mean annual rainfall (Fig.7.27). The plotting reveals that both the variables exhibit a positive trend of increase.

The regression equation for this relationship is as follows:

$$Y = 0.98 + 0.00047X$$

It is concluded therefore that like other variables, climate also plays a significant role in determining the value of sinuosity index.

Conclusions:

The significance of climatic characteristics and particularly mean annual rainfall has already been established by a number of studies. The study of morphometric parameters of basins of Rajasthan in relation to mean annual rainfall demonstrated a significant positive relationship between them.

The study of drainage density and stream frequency which are spatially most coherently distributed not only form the most manageable basis of regionalization but also help in identifying areas of erosion hazards. Such studies can be used as indices of water yield and condition of erosion. The high drainage density and stream frequency in south eastern Rajasthan indicate high runoff. Thus any clearance of vegetation and its replacement by other forms of land use is likely to increase considerably the risk of floods. The hazards are naturally very high throughout the state. Great care is required in managing the development of the land in such areas otherwise danger of floods and erosion will be much increased with consequent effect with locally and the areas downstream.

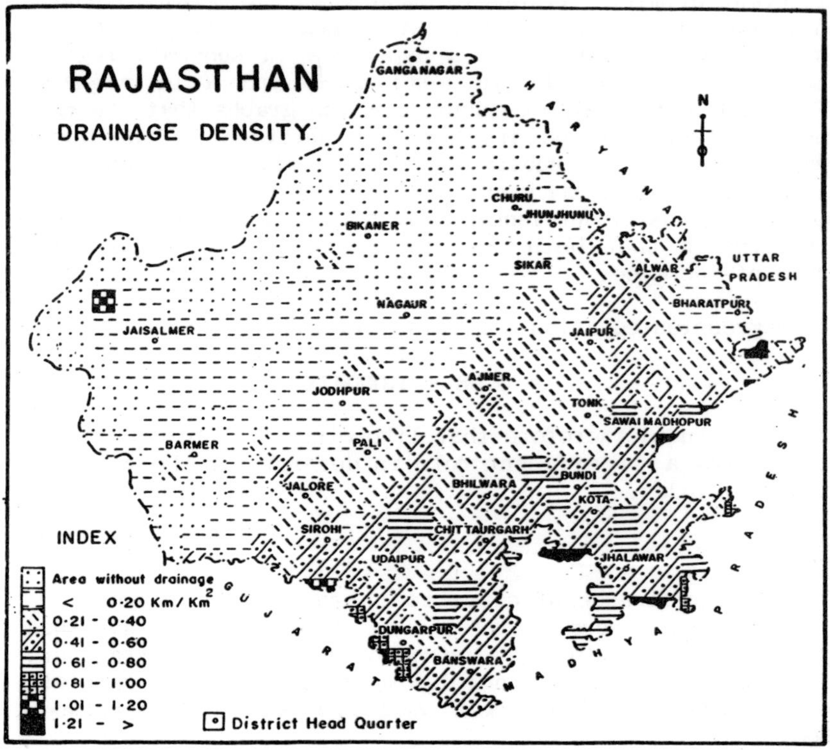
RAJASTHAN
DRAINAGE DENSITY
GANGANAGAR
BIKANER
CHURU
JHUNJHUNU
SIKAR
ALWAR
BHARATPUR
NAGAUR
JAISALMER
JAIPUR
AJMER
JODHPUR
TONK
SAWAI MADHOPUR
PALI
BARMER
BUNDI
BHILWARA
JALORE
KOTA
SIROHI
CHITTAURGARH
UDAIPUR
JHALAWAR
DUNGARPUR
BANSWARA
HARYANA
UTTAR PRADESH
MADHYA PRADESH
GUJARAT
N
INDEX
Area without drainage
< 0·20 Km/Km2
0·21 - 0·40
0·41 - 0·60
0·61 - 0·80
0·81 - 1·00
1·01 - 1·20
1·21 - >
District Head Quarter

FIGURE 7.1

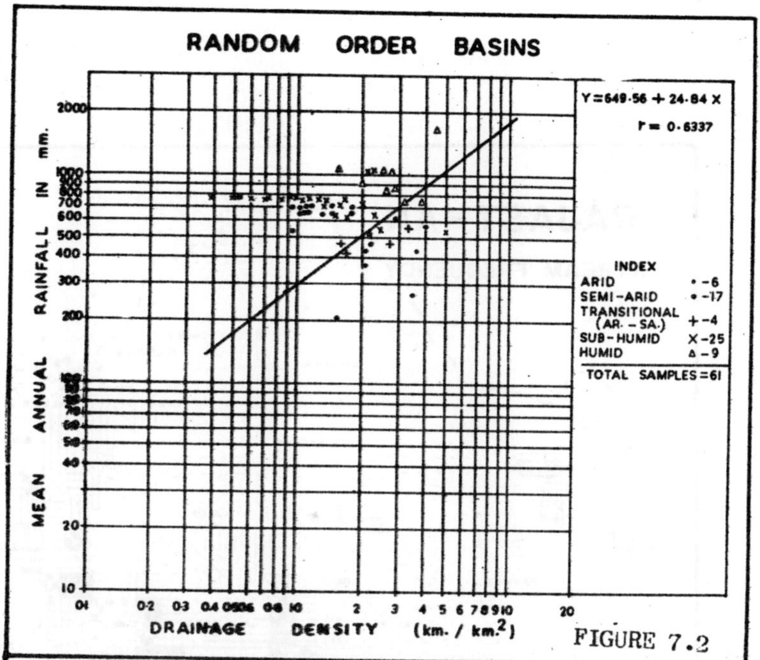

FIGURE 7.2

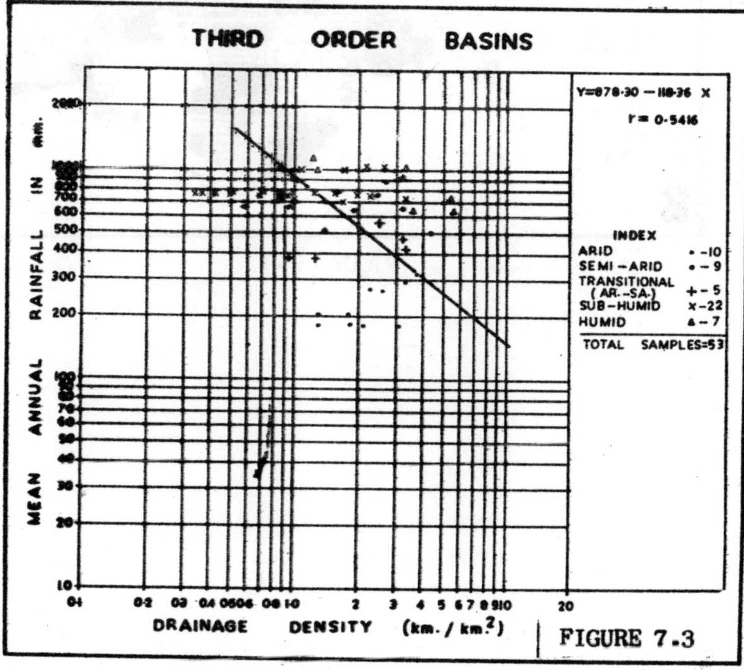

FIGURE 7.3

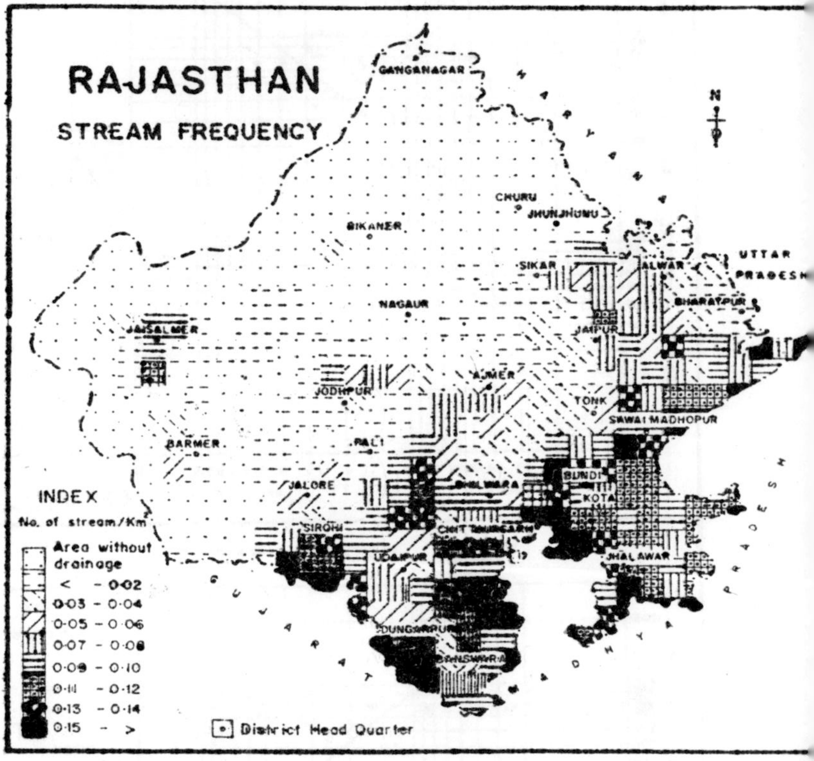
RAJASTHAN
STREAM FREQUENCY
GANGANAGAR
CHURU
JHUNJHUNU
BIKANER
SIKAR
ALWAR
UTTAR PRADESH
BHARATPUR
NAGAUR
JAISALMER
JAIPUR
AJMER
JODHPUR
TONK
SAWAI MADHOPUR
BARMER
PALI
BUNDI
KOTA
JALORE
BHILWARA
SIROHI
CHITTAURGARH
UDAIPUR
JHALAWAR
DUNGARPUR
BANSWARA
INDEX
No. of stream/Km
Area without drainage
< - 0·02
0·03 - 0·04
0·05 - 0·06
0·07 - 0·08
0·09 - 0·10
0·11 - 0·12
0·13 - 0·14
0·15 - >
District Head Quarter
N

FIGURE 7.4

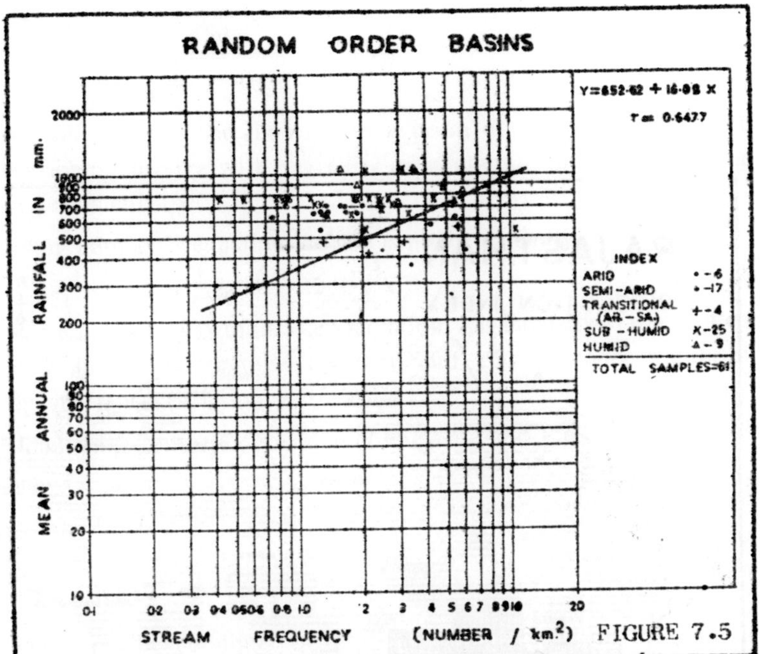

FIGURE 7.5

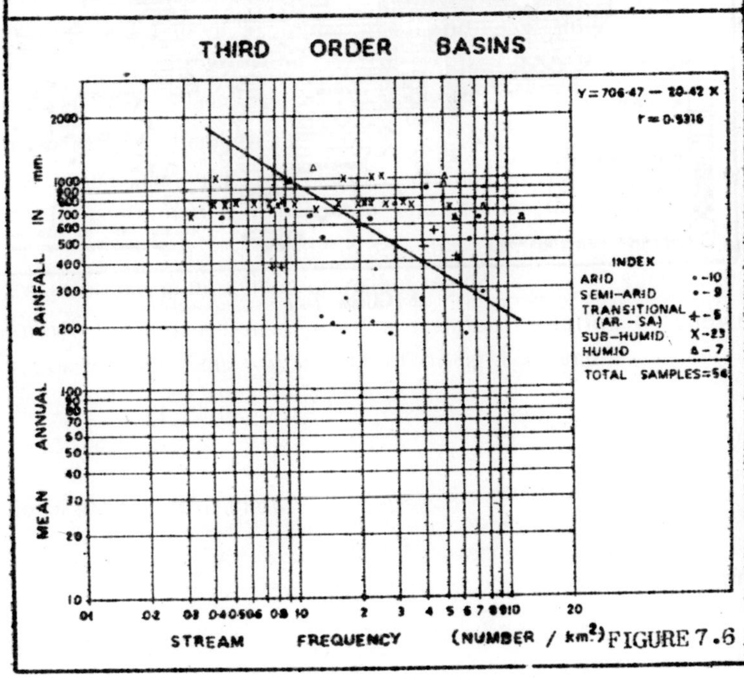

FIGURE 7.6

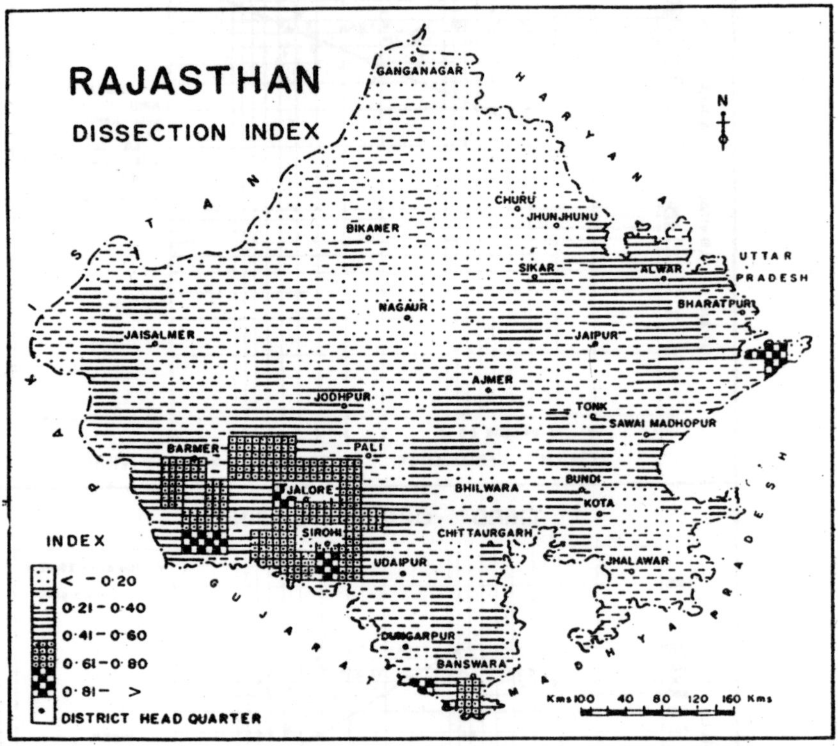
RAJASTHAN
DISSECTION INDEX
GANGANAGAR
BIKANER
CHURU
JHUNJHUNU
SIKAR
ALWAR
BHARATPUR
NAGAUR
JAISALMER
JAIPUR
AJMER
JODHPUR
TONK
SAWAI MADHOPUR
BARMER
PALI
JALORE
BHILWARA
BUNDI
KOTA
SIROHI
CHITTAURGARH
UDAIPUR
JHALAWAR
DUNGARPUR
BANSWARA
UTTAR PRADESH
N
INDEX
< -0·20
0·21-0·40
0·41-0·60
0·61-0·80
0·81- >
DISTRICT HEAD QUARTER
Kms 100 40 80 120 160 Kms

FIGURE 7.7

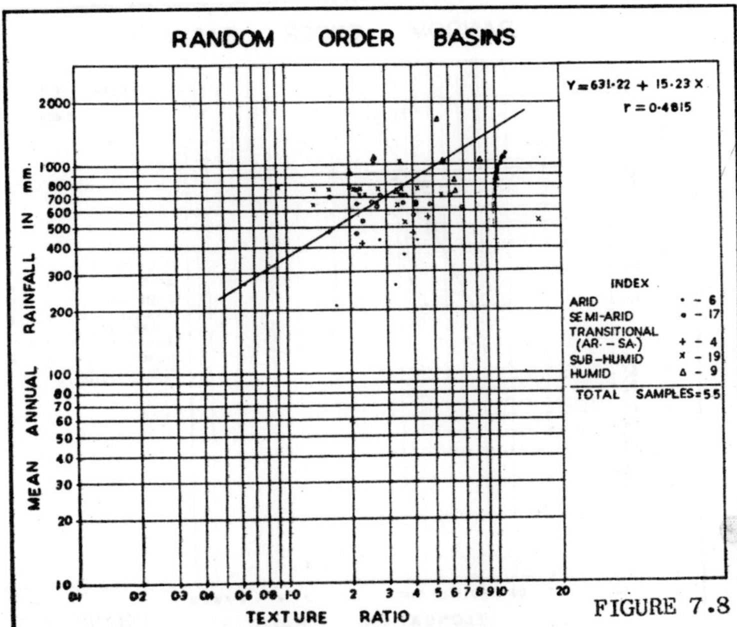

FIGURE 7.8

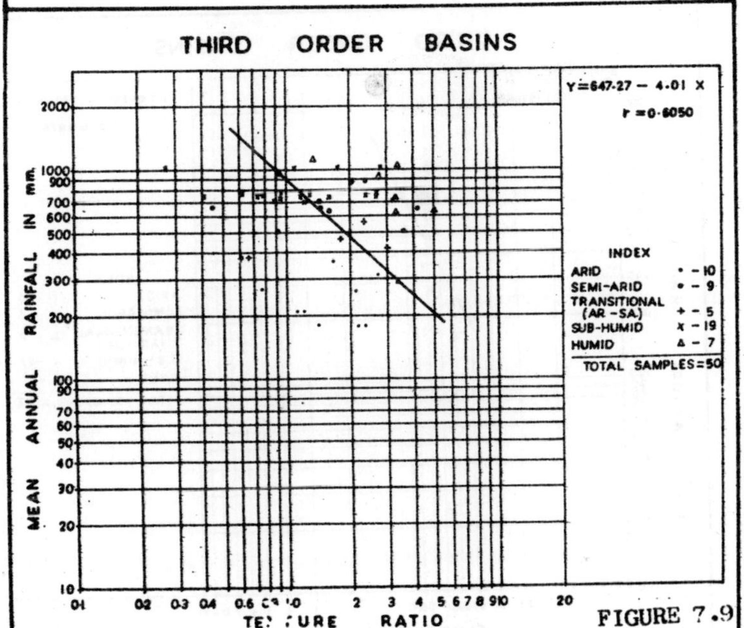

FIGURE 7.9

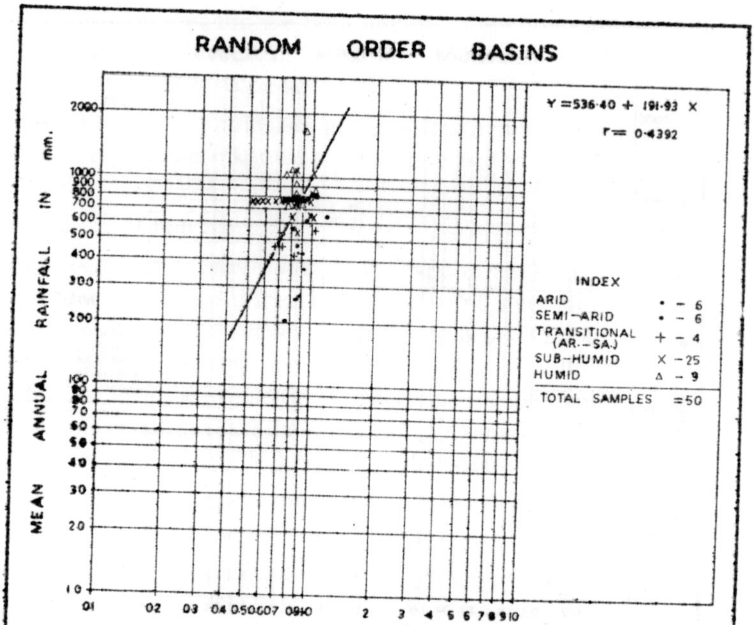

FIGURE 7.10

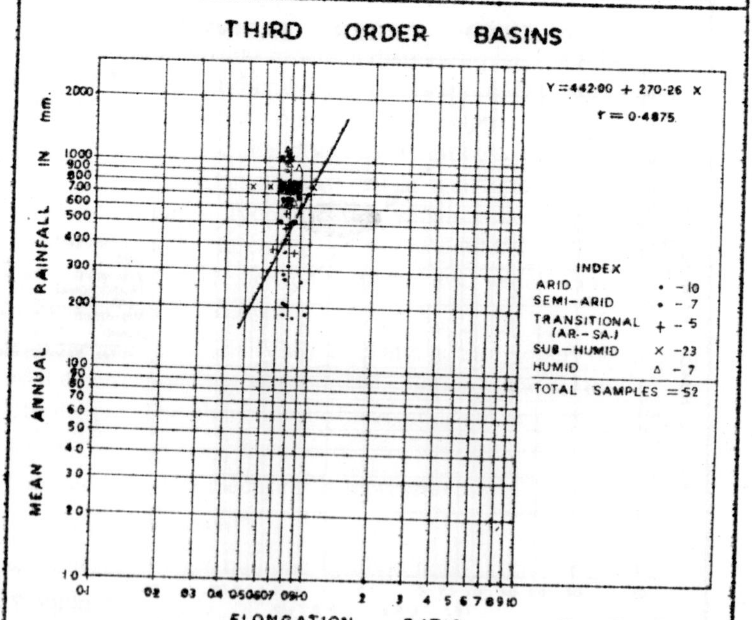

FIGURE 7.11

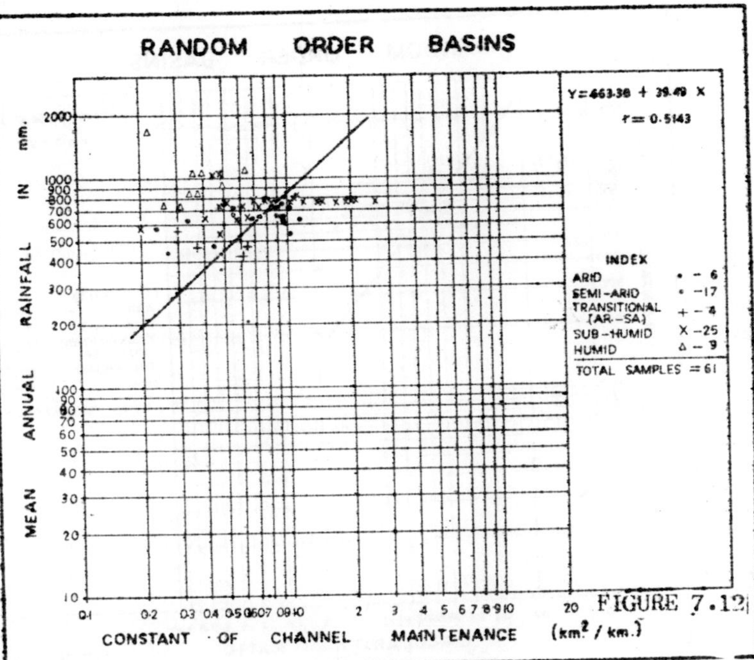

FIGURE 7.12

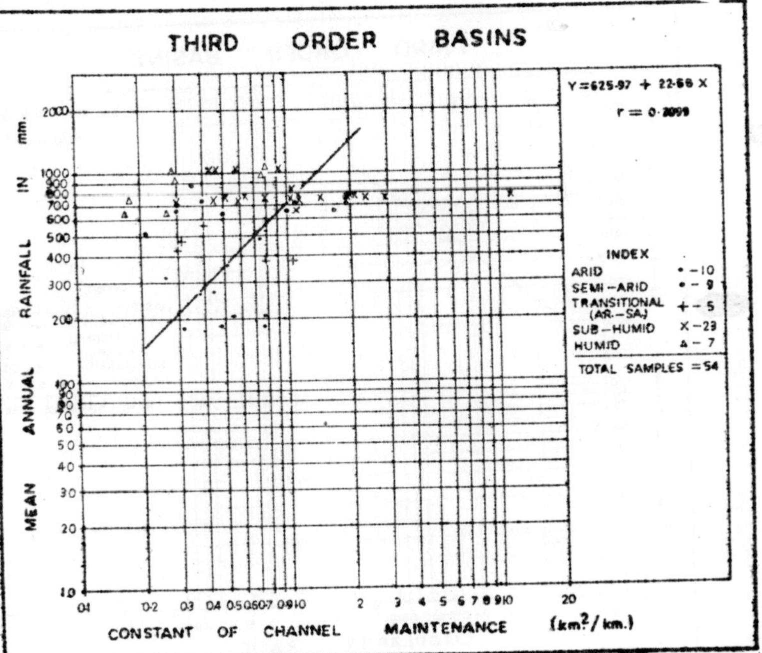

FIGURE 7.13

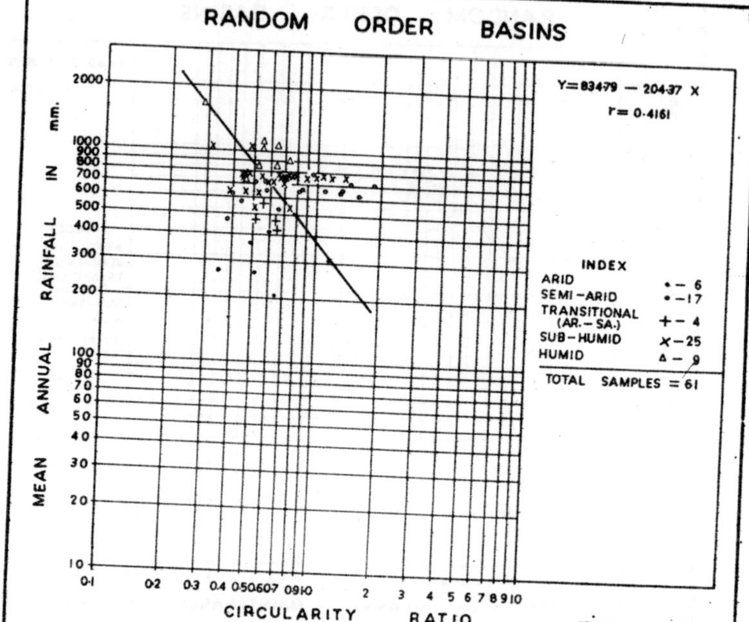

FIGURE 7.14

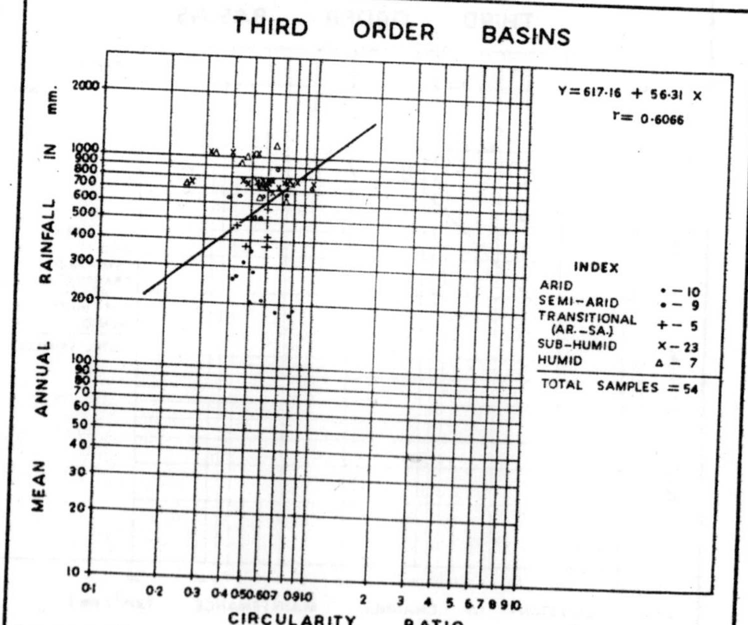

FIGURE 7.15

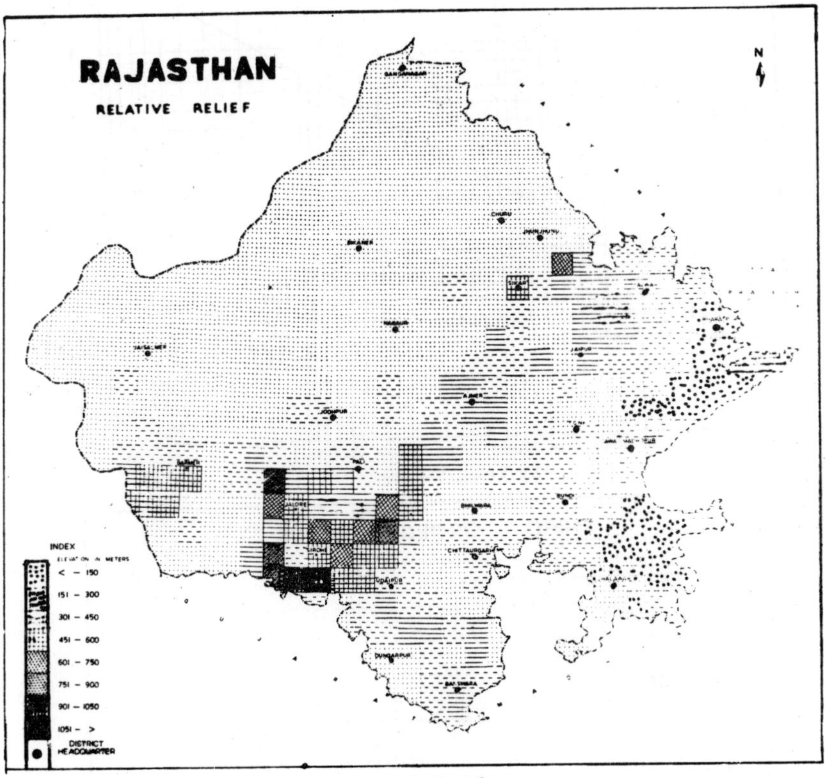
RAJASTHAN
RELATIVE RELIEF
N
INDEX
ELEVATION IN METERS
< – 150
151 – 300
301 – 450
451 – 600
601 – 750
751 – 900
901 – 1050
1051 – >
DISTRICT HEADQUARTER

FIGURE 7.16

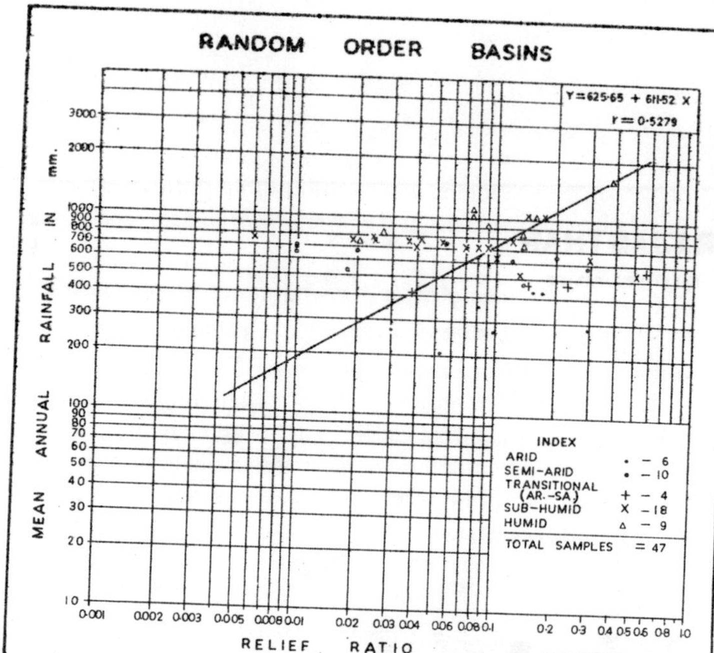

FIGURE 7.17

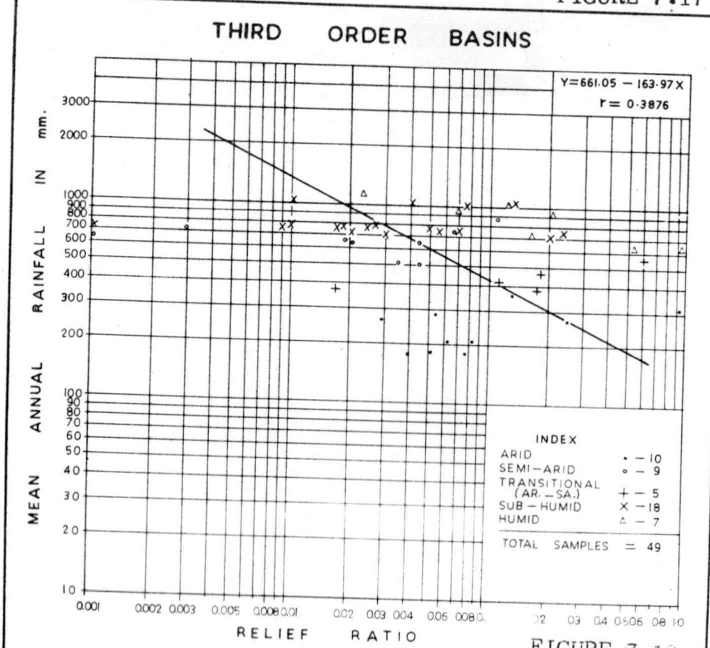

FIGURE 7.18

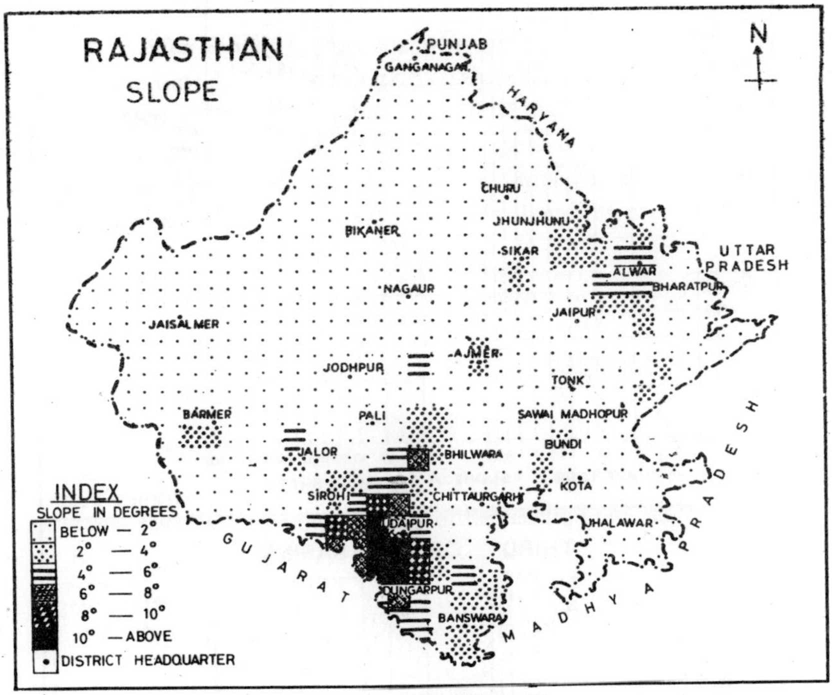
RAJASTHAN
SLOPE
N
PUNJAB
GANGANAGAR
HARYANA
CHURU
JHUNJHUNU
BIKANER
SIKAR
UTTAR PRADESH
ALWAR
BHARATPUR
NAGAUR
JAIPUR
JAISALMER
AJMER
JODHPUR
TONK
BARMER
PALI
SAWAI MADHOPUR
BUNDI
JALOR
BHILWARA
KOTA
SIROHI
CHITTAURGARH
UDAIPUR
JHALAWAR
DUNGARPUR
BANSWARA
GUJARAT
MADHYA PRADESH
INDEX
SLOPE IN DEGREES
BELOW — 2°
2° — 4°
4° — 6°
6° — 8°
8° — 10°
10° — ABOVE
DISTRICT HEADQUARTER

FIGURE 7.19

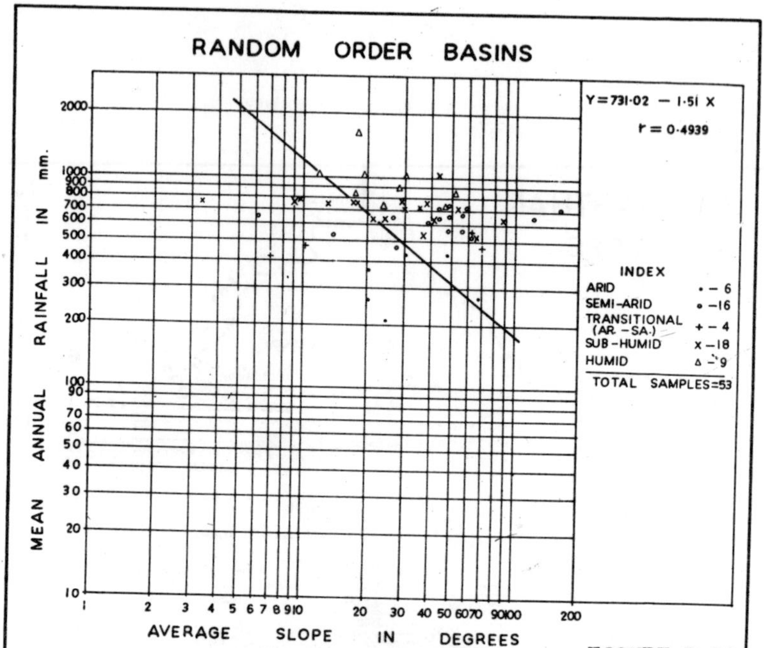

FIGURE 7.20

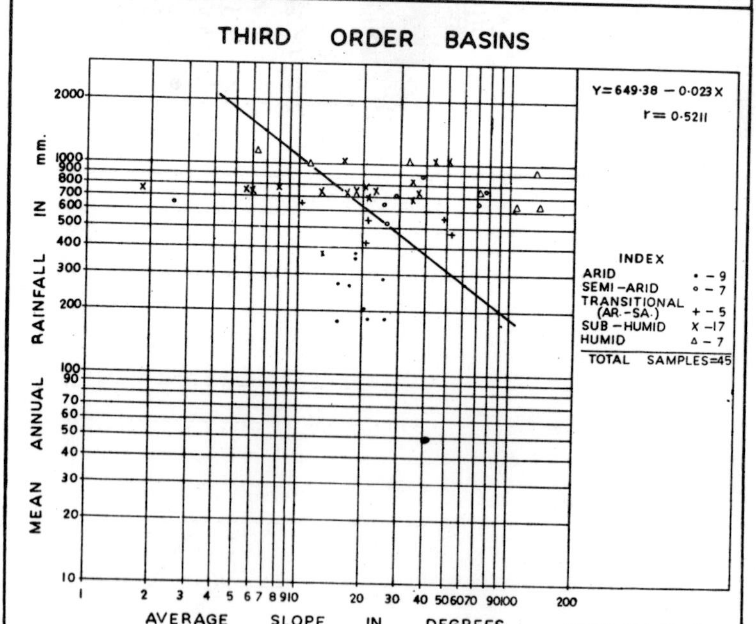

FIGURE 7.21

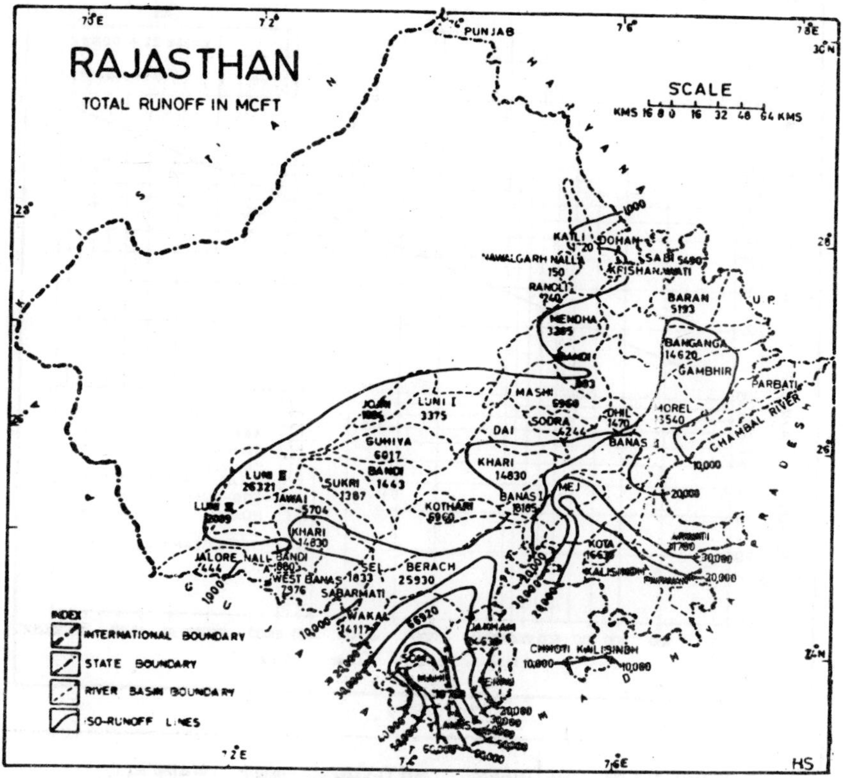

FIGURE 7.22

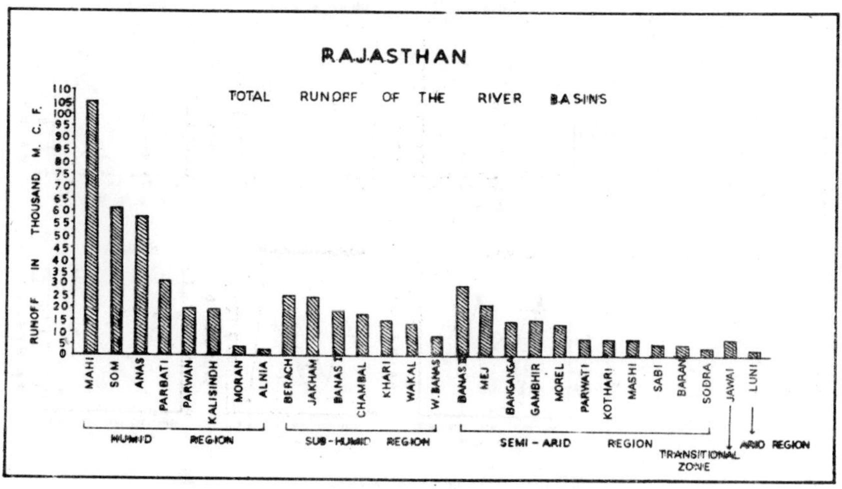

FIGURE 7.23

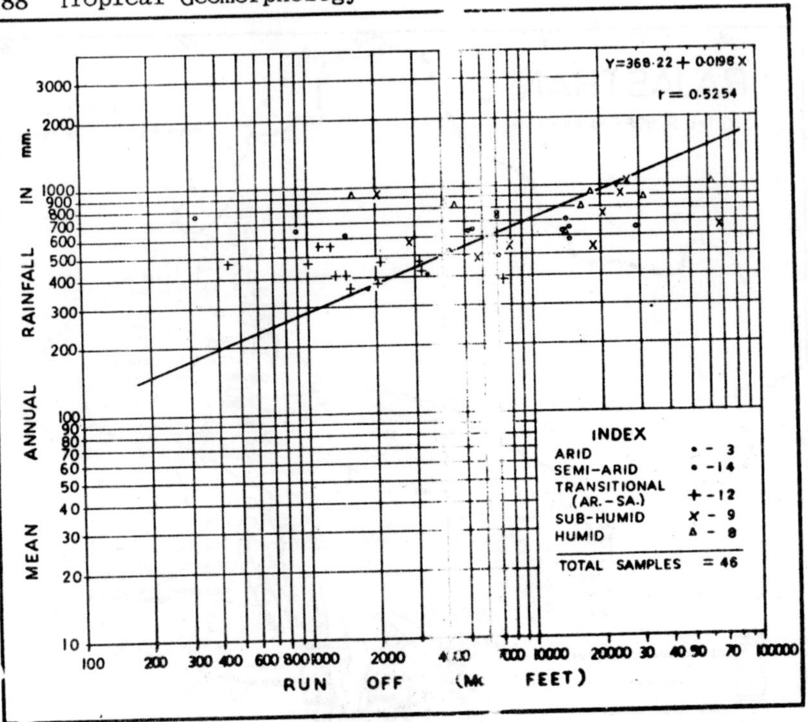

FIGURE 7.24

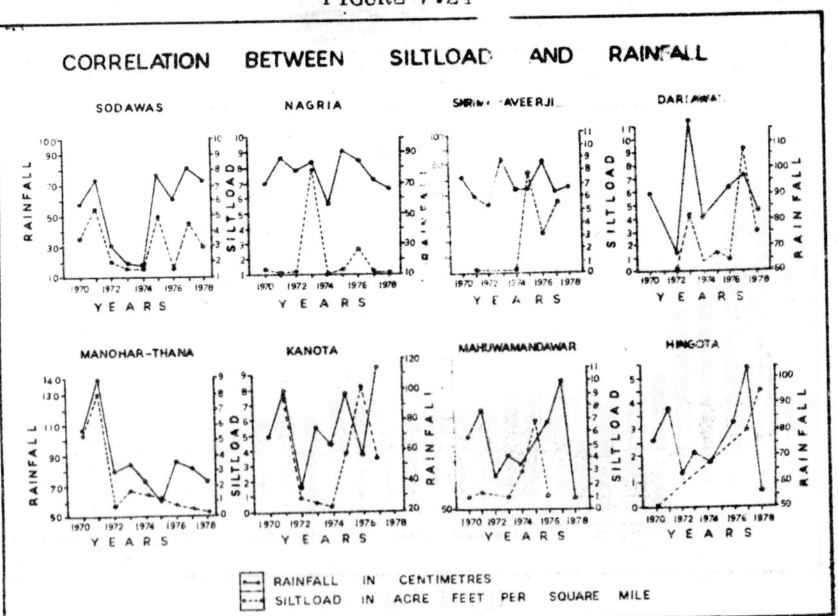

FIGURE 7.25

FIGURE 7.26

PLOTS SHOWING RELATION BETWEEN CHANNEL LENGTH AND DIRECT LENGTH IN DIFFERENT CLIMATIC ZONES OF RAJASTHAN

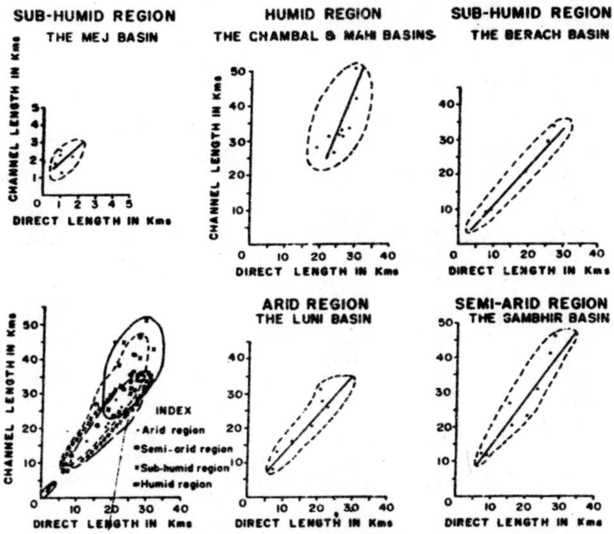

Relationship Between Tortuosity Index And Mean Annual Rainfall (Rajasthan)

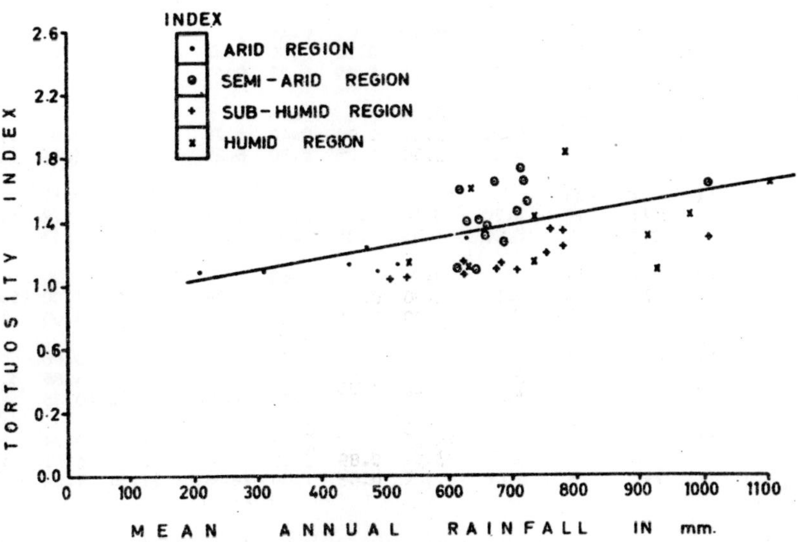

FIGURE 7.27

TABLE 7.1. **Quantitative Analysis of Fluvial Morphometry, Rajasthan (Third Order Basin)**

Climatic Region	Basin number	u	Nu	Rb	(L)u Kms	Re	$\overline{Lu}$ km	Re	CTSL km	CMSL km	Rb
1	2	3	4	5	6	7	8	9	10	11	12
Humid region	1	I	17		15.00		0.88		15.00	0.88	
		II	5	3.40	8.00	0.53	1.60	1.818	23.00	2.48	4.20
		III	1	5.00	7.50	0.93	7.50	4.687	30.50	9.98	
	2	I	20		15.25		0.76		15.25	0.76	
		II	6	3.33	4.75	0.31	0.79	1.039	20.00	1.55	4.66
		III	1	6.00	7.20	1.51	7.20	9.110	27.20	8.75	
	3	I	53		28.50		0.53		28.50	0.53	
		II	15	3.53	10.25	0.35	0.68	1.28	38.75	1.2	9.26
		III	1	15.00	9.00	0.87	9.00	13.23	47.75	10.21	
	4	I	32		3.75		0.42		13.75	0.42	
		II	6	5.33	6.10	0.44	1.01	2.40	19.85	1.43	5.66
		III	1	6.00	5.00	0.81	5.00	4.95	24.85	6.43	
	5	I	67		32		0.47		32.00	0.47	
		II	15	4.46	12.25	0.38	0.81	1.72	44.25	1.28	9.73
		III	1	15.00	11.20	0.91	11.20	14.00	55.45	12.48	
	6	I	31		13.75		0.44		13.75	0.44	
		II	5	6.20	2.00	0.14	0.40	0.90	15.75	0.84	5.60
		III	1	5.00	2.00	1.00	2.00	5.00	17.75	2.84	
	7	I	35		16.75		0.47		16.75	0.47	
		II	6	5.83	5.20	0.31	0.86	1.82	21.95	1.33	5.91
		III	1	6.00	4.75	0.91	4.75	5.52	26.70	6.08	
Subhumid region	8	I	17		10.00		0.58		10.00	0.58	
		II	4	4.25	7.20	0.72	1.80	3.10	17.20	2.38	4.12
		III	1	4.00	5.25	0.13	5.25	2.91	22.45	7.63	
	9	I	35		16.00		0.45		16.00	0.45	
		II	8	4.37	5.25	0.32	0.65	1.44	21.25	1.10	6.18
		III	1	8.00	6.25	1.19	6.25	9.61	27.50	7.35	
	10	I	25		14.00		0.51		14.00	0.51	
		II	5	5.40	2.00	0.14	0.40	0.78	16.00	0.91	5.20
		III	1	5.00	2.50	1.25	2.50	6.25	18.50	3.41	
	11	I	32		12		0.37		12	0.37	
		II	10	3.20	3.50	0.29	0.35	0.94	15.50	0.72	6.60
		III	1	10.00	4.75	1.35	4.75	13.57	20.25	5.47	
	12	I	24		24.00		1.00		24.00	1.00	
		II	4	6.00	8.00	0.33	2.00	2.00	32.00	3.00	5.00
		III	1	4.00	10.00	1.25	10.00	5.00	42.00	13.00	
	13	I	11		7.50		0.68		7.50	0.68	
		II	3	3.66	9.50	1.26	3.7	4.66	17.00	3.85	3.33
		III	1	3.00	7.00	1.25	7.00	2.20	24.00	10.85	
	14	I	19		14.00		0.74		14.00	0.74	
		II	2	9.50	12.50	0.89	6.25	8.44	26.50	6.99	5.75
		III	1	2.00	1.00	0.08	1.00	0.16	27.50	7.99	

15	I	6		6.00		1		6.00	1.00	
	II	1	6.00	12.00	2.00	12	12.00	18.00	13.00	6.00
16	I	64		41.00		0.64		41.00	0.64	
	II	12	5.33	14.00	0.34	1.17	1.82	55.00	1.81	8.66
	III	1	12.00	10.00	0.71	10.00	8.54	65.00	12.45	
17	I	17		17.00		1.00		17.00	1.00	
	II	5	3.40	7.00	0.41	1.40	1.40	24.00	2.40	4.20
	III	1	5.00	7.50	1.07	7.50	5.30	31.50	9.90	
18	I	21		15.00		0.71		15.00	0.71	
	II	6	3.50	4.00	0.26	0.66	0.92	19.00	1.37	4.75
	III	1	6.00	6.00	1.50	6.00	9.09	25.00	7.37	
19	I	15		5.50		0.37		5.50	0.37	
	II	5	3.00	4.25	0.77	0.85	2.30	9.75	1.22	4.00
	III	1	5.00	2.12	0.50	2.12	2.50	11.87	3.34	
20	I	17		4.32		0.25		4.37	0.25	
	II	4	4.25	1.12	0.26	0.28	1.09	5.50	0.53	4.12
	III	1	4.00	2.25	2.00	32.25	8.03	7.75	2.78	
21	I	15		3.75		0.25		3.75	0.25	
	II	5	3.00	1.25	0.33	2.25	1.00	5.00	0.50	4.00
	III	1	5.00	1.87	1.50	1.87	7.50	6.87	2.37	
22	I	23		6.50		0.28		6.50	0.28	
	II	4	5.75	3.00	0.46	0.75	2.67	9.50	0.93	5.85
	III	1	4.00	4.50	1.50	4.50	6.00	13.00	5.43	
23	I	26		15.12		0.58		15.12	0.58	
	II	5	5.20	6.61	0.43	1.32	2.28	21.73	1.90	5.10
	III	1	5.00	4.41	0.66	4.41	3.32	26.14	6.31	
24	I	13		13.86		1.06		13.86	1.06	
	II	5	2.60	0.94	0.06	0.19	0.18	14.80	1.25	3.80
	III	1	5.00	9.13	9.60	9.13	48.08	23.94	10.39	
25	I	23		5.10		0.23		5.10	0.23	
	II	4	5.75	3.75	0.74	0.93	4.08	8.85	1.16	5.87
	III	1	4.00	1.87	0.50	1.87	2.00	10.72	3.04	
26	I	7		4.00		0.57		4.00	0.57	
	II	2	3.50	3.62	0.90	1.81	0.27	7.62	2.38	2.75
	III	1	2.00	0.50	0.41	0.50	0.27	8.12	2.88	
27	I	20		8.82		0.44		8.82	0.44	
	II	3	6.67	2.83	0.32	0.94	0.66	11.65	1.38	4.88
	III	3.00	0.63	0.22	0.63	0.66	12.28	2.01		
28	I	7		7.56		1.08		7.56	1.08	
	II	3	2.33	2.20	0.29	0.73	0.68	9.76	1.81	2.66
	III	1	3.00	1.89	0.86	1.89	2.57	1.65	3.70	
Semiarid region 29	I	26		16.75		0.64		16.75	0.64	
	II	6	4.33	6.00	0.33	1.00	1.56	22.75	1.64	5.16
	III	1	6.00	4.20	0.70	4.20	4.20	26.95	5.84	
30	I	30		19.00		0.63		19.00	0.63	
	II	7	4.28	3.00	0.15	0.42	0.66	22.00	1.05	5.64
	III	1	7.00	3.00	1.00	3.00	7.14	25.00	4.05	

	31	I	12		9.75		0.81		9.75	0.81	
		II	2	6.00	3.00	0.30	1.50	1.85	12.75	2.31	4.00
		III	1	2.00	2.75	0.91	2.75	1.83	15.50	5.06	
	32	I	23		12.50		0.54		12.50	0.54	
		II	5	4.60	4.00	0.32	0.80	1.48	16.50	1.34	4.80
		III	1	5.00	2.75	0.68	2.75	3.43	19.25	4.09	
	33	I	44		14.75		0.33		14.75	0.33	
		II	8	5.50	7.00	0.47	0.87	2.63	21.75	1.20	6.75
		III	1	8.00	2.25	0.32	2.25	2.58	24.00	3.45	
	34	I	29		17.50		0.60		17.50	0.60	
		II	7	4.41	5.00	0.25	0.71	1.18	22.50	1.31	5.57
		III	1	7.00	9.00	1.80	9.00	12.67	31.50	10.31	
	35	I	20		24.50		1.22		24.50	1.22	
		II	3	6.66	20.25	0.82	6.75	5.53	44.75	7.97	4.83
		III	1	3.00	12.50	0.61	12.50	1.85	57.25	20.47	
	36	I	15		9.50		0.63		9.50	0.63	
		II	3	5.00	6.50	0.68	2.17	2.53	16.00	2.80	4.00
		III	1	3.00	5.50	0.84	5.50	2.53	21.50	8.30	
	37	I	24		21.00		0.87		21.00	0.87	
		II	6	4.00	5.00	0.23	0.83	0.95	26.00	1.70	5.00
		III	1	6.00	7.00	1.40	7.00	8.43	33.00	8.70	
	38	I	25		14.00		0.56		14.00	0.50	
		II	5	5.00	3.00	0.21	0.60	1.07	17.00	1.16	5.00
		III	1	5.00	8.00	2.67	8.00	13.33	25.00	9.16	
	39	I	24		18.00		0.75		18.00	0.75	
		II	4	6.00	6.50	0.36	1.62	2.16	24.50	2.37	5.00
		III	1	4.00	3.50	0.54	3.50	2.16	28.00	5.87	
Transitional zone	40	I	8		5.00		0.62		5.00	0.62	
		II	2	4.00	10.75	2.15	5.37	8.66	15.75	5.99	3.00
		III	1	2.00	0.80	0.07	0.80	0.14	16.55	6.59	
	41	I	25		11		0.44		11.00	0.44	
		II	7	3.57	4.00	0.36	0.57	1.29	15.00	1.01	5.28
		III	1	7.00	4.00	1.00	4.00	7.01	19.00	5.01	
	42	I	16		8.00		0.50		8.00	0.50	
		II	5	3.20	5.20	0.80	1.04	4.03	13.20	1.54	4.10
		III	1	5.00	4.20	0.80	4.20	4.03	17.40	5.74	
	43	I	9		4.50		0.50		4.50	0.50	
		II	2	4.50	8.50	1.88	4.25	8.50	13.00	4.75	3.75
		III	1	2.00	1.80	0.21	1.80	0.42	14.80	6.55	
	44	I	18		7.25		0.40		7.25	0.40	
		II	6	3.00	2.75	0.37	0.45	1.12	10.00	0.85	4.50
		IIII	6.00	4.00	1.45	4.00	8.88	14.00	4.85		
Arid zone	45	I	25		11.00		0.44		11.00	0.44	
		II	8	3.12	5.00	0.45	0.62	1.40	16.00	1.06	5.56
		III	1	8.00	6.50	1.30	6.50	10.48	22.50	7.56	
	46	I	6		7.50		1.25		7.50	1.25	
		II	3	2.00	2.00	0.26	0.66	0.52	9.50	1.91	2.50
		III	1	3.00	3.75	.87	3.75	5.68	13.25	5.66	

47	I	24		14.50		0.60		14.50	0.60	
	II	4	6.00	2.50	0.17	0.62	1.03	17.00	1.22	5.00
	III	1	4.00	6.80	2.72	6.80	10.96	23.80	8.02	
48	I	29		9.00		0.31		9.00	0.31	
	II	6	4.83	3.00	0.33	0.50	1.61	12.00	0.81	5.41
	III	1	6.00	3.75	1.25	3.75	7.50	5.75	4.56	
49	I	10		5.25		0.52		5.25	0.52	
	II	3	3.33	3.75	0.71	1.25	2.40	9.00	1.77	3.16
	III	1	3.00	2.00	0.53	2.00	1.60	11.00	3.77	
50	I	11		5.50		0.50		5.50	0.50	
	II	2	5.50	1.00	0.18	0.50	1.00	6.50	1.00	3.75
	III	1	2.00	0.50	0.50	0.50	1.00	7.00	1.50	
51	I	20		13.25		0.66		13.25	0.66	
	II	5	4.00	3.50	0.26	0.70	1.06	6.75	1.36	4.50
	III	1	5.00	2.50	0.71	2.50	3.57	19.25	3.86	
52	I	16		1.25		0.70		1.25	C.70	
	II	6	2.66	2.80	0.24	0.46	0.65	14.05	1.16	4.33
	III	1	6.00	4.00	1.42	4.00	6.69	18.05	5.16	
53	I	17		10.50		0.61		0.50	0.61	
	II	4	4.25	3.25	0.30	0.81	1.32	13.75	1.42	4.12
	III	1	4.00	5.50	1.69	5.50	6.79	19.25	6.92	
54	I	24		12.00		0.50		12.00	0.50	
	II	7	3.42	5.00	0.41	0.71	1.42	17.00	i.21	5.21
	III	1	7.00	4.00	0.80	4.00	5.63	21.00	5.21	

Symbols used:

	Stream order
Nu	Number of stream segments
Rb	Bifurcation ratio
(L)u	Total stream length
Re	Total stream length ratio
Lu	Mean stream length
Re	Mean stream length ratio
CTSL	Cumulative total stream length
CMSL	cumulative mean stream length.
Rb	Mean bifurcation ratio

TABLE 7.2. **Quantitative Analysis of Fluvial Morphometry Rajasthan (Random Order Basin)**

Climatic Region	Basin Number	u	Nu	Rb	(L)u	Re	$\overline{Lu}$	Re	CTSL	CMSL	Rb
1	2	3	4	5	6	7	8	9	10	11	12
Humid region	1	I	303	3.83	133.25	0.32	0.43	1.25	133.25	0.43	
		II	79	6.58	43.00	0.68	0.54	4.46	176.25	0.97	
		III	12	2.40	29.50	0.68	2.41	1.68	205.75	3.38	4.45
		IV	5	5.00	20.25	0.37	4.05	1.85	226.00	7.43	
		V	1		7.70		7.50		233.50	14.93	
	2	I	145	4.02	67.50	0.32	0.46	1.32	67.50	0.46	
		II	36	4.00	22.25	0.42	0.61	1.72	89.75	1.07	
		III	9	4.50	9.50	1.15	5.50	5.23	99.25	2.12	3.63
		IV	2	2.00	11.00	0.18	5.50	0.36	110.25	7.62	
		V	1		2.00		2.00		112.25	9.62	
	3	I	150	4.41	85.00	0.38	0.56	1.71	85.00	0.56	
		II	34	4.25	32.75	0.45	0.96	1.94	117.75	1.52	
		III	8	2.66	15.00	0.54	1.87	1.45	132.75	3.39	3.58
		IV	3	3.00	8.20	0.85	2.73	2.56	140.95	6.12	
		V	1		7.00		7.00		147.95	13.12	
	4	I	282	3.75	152.50	0.31	0.54	1.18	152.50	0.54	
		II	75	5.35	48.00	0.58	0.64	3.14	200.50	1.18	
		III	14	4.66	28.20	0.92	2.01	4.47	228.70	3.19	4.19
		IV	3	3.10	27.00	0.24	9.00	0.72	255.70	12.19	
		V	1		6.50		6.50		262.20	18.69	
	5	I	52	4.72	28.00	0.32	0.53	1.52	28.00	0.53	
		II	11	5.50	9.00	0.44	0.81	2.46	37.00	1.34	
		III	2	2.00	4.00	0.68	2.00	1.37	41.00	3.34	4.07
		IV	1		2.75		2.75		43.75	6.09	
	6	I	117	4.03	40.50	0.39	0.34	1.61	40.50	0.34	
		II	29	3.22	16.00	0.30	0.55	0.96	56.00	0.89	
		III	9	3.00	4.80	0.87	0.53	2.64	61.30	1.42	3.31
		IV	3	3.00	4.20	0.76	1.40	2.28	65.50	2.82	
		V	1		3.20		3.20		68.70	6.02	
	7	I	72	4.50	51.00	0.32	0.70	1.48	51.00	0.70	
		II	16	2.66	16.75	0.71	1.04	1.92	67.75	1.74	
		III	6	6.00	12.00	1.08	2.00	6.50	79.00	3.74	4.38
		IV	1		13.00		13.00		92.00	16.74	
	8	I	25	3.12	19.25	0.53	0.77	1.66	19.25	0.77	
		II	8	4.00	10.30	0.40	1.28	1.64	29.55	2.05	
		III	2	2.00	4.20	0.59	2.10	1.19	33.75	4.15	3.04
		IV	1		2.50		2.50		36.25	6.65	
	9	I	105	4.56	41.20	0.29	0.39	1.35	41.20	0.39	
		II	23	4.60	12.20	0.53	0.53	2.45	54.40	0.92	
		III	5	5.00	6.50	1.06	1.30	5.30	60.90	2.22	4.12
		IV	1		6.90		6.90		67.80	9.12	
Subhumid region	10	I	300	4.10	112.00	0.37	0.37	1.54	112.00	0.37	
		II	79	3.84	42.00	0.36	0.57	1.42	154.00	0.94	
		III	19	3.80	15.50	0.41	0.81	1.60	169.50	1.75	
		IV	5		6.50		1.30		176.00	3.05	3.24

	V	2	2.50	5.00	0.76	2.50	1.92	181.00	5.55				
	VI	1	2.00	3.20	0.64	3.20	1.28	184.20	8.75				
11	I	76		50.20		0.66		50.20	0.66				
	II	18	4.22	36.75	0.73	2.04	3.09	56.95	2.70				
	III	5	3.60	21.00	0.57	4.20	2.05	107.95	6.30	4.27			
	IV	1	5.00	8.75	0.41	8.75	2.08	116.70	15.65				
12	I	227		127.00		0.55		127.00	0.55				
	II	48	4.72	36.00	.28	0.75	1.36	163.00	1.30				
	III	11	4.36	22.50	0.62	2.04	2.72	185.50	3.34	4.14			
	IV	2	5.50	9.80	0.43	4.90	2.40	195.30	8.24				
	V	1	2.00	15.25	1.55	15.25	3.11	210.55	23.49				
13	I	58		38.00		0.65		38.00	0.65				
	II	15	3.86	20.00	0.52	1.33	2.04	58.00	1.98				
	III	2	7.50	19.25	0.96	9.62	7.23	77.25	11.60	4.45			
	IV	1	2.00	4.00	0.20	4.00	0.41	81.25	15.60				
14	I	92		36.00		0.39		36.00	0.39				
	II	22	4.18	14.50	0.40	0.65	1.61	50.50	1.04				
	III	5	4.40	5.25	0.36	1.05	1.61	55.75	2.09	4.52			
	IV	1	5.00	10.00	1.90	10.00	9.52	65.75	12.09				
15	I	54		29.00		0.54		29.00	0.54				
	II	14	3.85	8.00	0.28	0.57	1.06	37.00	1.11				
	III	3	4.67	7.00	0.88	2.33	4.09	44.00	3.44	3.84			
	IV	1	3.00	8.00	1.14	8.00	3.43	52.00	11.44				
16	I	94		59.5		0.63		59.50	0.63				
	II	30	3.13	21.5	0.36	0.72	1.14	81.00	1.35				
	III	7	4.28	15.0	0.70	2.14	2.97	96.00	3.49	3.[illegible]			
	IV	2	3.50	12.5	0.83	6.25	2.82	108.50	9.74				
	V	1	2.00	6.0	0.48	6.00	0.96	114.50	15.74				
17	I	170		108.00		0.63		108.00	0.63				
	II	48	3.54	62.00	0.57	1.29	2.40	170.00	1.92				
	III	12	4.00	38.00	0.61	3.16	2.40	208.00	5.08	6.[illegible]			
	IV	1	12.00	25.00	0.65	25.00	7.91	233.00	30.08				
18	I	18		10.00		0.55		10.00	0.53				
	II	4	4.45	8.00	0.80	2.00	3.63	18.00	2.55				
	III	2	2.00	2.00	0.25	1.00	0.50	20.00	3.55	2.32			
	IV	1	2.00	4.50	2.25	4.50	4.50	s24.50	8.05				
19	I	53		28.50		0.54		28.50	0.54				
	II	12	4.41	7.00	0.24	0.58	1.07	35.50	1.12				
	III	3	4.00	4.50	0.64	1.50	2.58	40.00	2.62	3.80			
	IV	1	3.00	7.00	1.55	7.00	4.66	47.00	9.62				
20	I	27		27.50		1.02		27.50	1.02				
	II	7	3.85	11.50	0.41	1.64	1.60	39.00	2.66				
	III	2	3.50	5.00	0.43	2.25	1.37	44.00	4.91	3.11			
	IV	1	2.00	8.00	1.60	8.00		52.00	12.91				
21	I	60		49.00		0.76		46.00	0.76				
	II	18	3.30	19.00	0.41	1.06	1.39	65.00	1.82				
	III	5	3.60	10.50	0.55	2.10	1.98	75.50	3.90	3.97			
	IV	1	5.00	7.50	1.40	7.50	3.57	83.00	11.42				
22	I	61		52.50		0.84		52.50	0.84				
	II	15	4.06	17.00	0.32	1.21	1.40	69.50	2.05				
	III	5	3.00	11.00	0.64	2.20	1.80	80.50	4.25	2.89			
	IV	2	2.50	16.00	1.40	8.00	3.60	96.50	12.25				

Basin	Order	No.								
	V	1	2.00	0.25	0.01	0.25	0.03	96.70	12.50	
23	I	118	4.06	54.00	0.46	0.47	1.82	54.00	0.47	
	II	29	4.83	25.00	0.60	0.86	2.86	79.00	1.33	
	III	6	3.00	15.00	0.66	2.50	2.00	94.00	3.83	3.47
	IV	2	2.00	10.00	0.70	5.00	1.40	104.00	8.83	
	V	1		7.00		7.00		111.00	15.83	
24	I	74	8.20	62.50	0.35	0.84	2.67	62.50	0.84	
	II	9	2.25	22.00	0.68	2.25	1.60	84.50	3.28	
	III	4	4.00	15.00	0.72	3.75	2.90	99.50	7.03	4.81
	IV	1		11.00		11.00		110.50	18.03	
25	I	81	4.50	73.00	0.33	0.90	1.48	73.00	0.90	
	II	18	3.60	24.00	0.50	1.33	1.80	97.00	2.23	
	III	5	5.00	12.00	1.00	2.40	5.00	107.00	4.03	4.36
	IV	1		12.00		12.00		121.00	16.63	
26	I	30	4.29	6.75	0.31	0.22	1.40	6.75	0.22	
	II	7	3.50	2.12	0.65	0.30	2.30	8.87	0.52	
	III	2	2.00	1.37	1.00	0.68	2.00	10.25	1.21	3.26
	IV	1		2.00		1.37		11.62	2.59	
27	I	247	4.20	17.82	0.22	0.29	0.91	17.82	0.2[illegible]	
	II	59	3.30	15.75	1.20	0.26	3.90	33.57	0.55	
	III	18	3.60	18.90	0.26	1.05	0.96	52.47	1.60	3.12
	IV	5	2.50	5.04	1.00	1.01	0.49	57.51	2.61	
	V	2	2.00	5.04	0.50	2.52	1.00	62.55	5.13	
	VI	1		2.25		2.52		64.80	7.65	
28	I	22	3.14	10.71	0.44	0.48	1.39	10.71	0.48	
	II	7	3.50	4.72	0.73	0.67	2.56	15.43	1.16	
	III	2	2.00	3.40	0.09	1.73	0.18	18.90	2.89	2.88
	IV	1		0.31		0.31		19.21	3.20	
29	I	25	4.17	9.75	0.30	0.39	1.26	9.75	0.39	
	II	6	3.00	3.00	0.54	0.50	1.62	12.75	0.89	
	III	2	2.00	1.62	0.68	0.81	1.38	14.37	1.60	3.06
	IV	1		1.12		1.12		15.50	2.73	
30	I	63	4.50	12.75	0.29	0.20	1.10	12.75	0.20	
	II	14	4.67	3.75	0.73	0.22	4.09	16.50	0.42	
	III	3	3.00	2.75	0.64	0.90	1.96	19.25	1.32	4.06
	IV	1		1.75		1.75		21.00	3.07	
31	I	137	2.68	24.75	0.65	0.18	1.77	24.75	0.18	
	II	51	8.50	16.25	0.30	0.32	2.60	41.00	0.50	
	III	6	3.00	5.00	0.80	0.83	2.39	46.00	1.33	
	IV	2		4.00		2.00		50.00	3.33	4.04
32	I	51	4.25	28.35	0.30	0.55	1.27	28.35	0.55	
	II	12	6.00	8.50	0.81	0.70	4.90	36.85	1.26	
	III	2	2.00	6.93	0.18	3.46	0.32	43.78	4.72	4.08
	IV	1		1.26		1.26		45.04	5.98	
33	I	18	3.60	8.82	0.36	0.49	1.28	8.82	0.49	
	II	5	2.50	3.15	1.60	0.63	4.00	11.97	1.12	
	III	2	2.00	5.04	0.21	2.52	0.42	17.01	3.64	2.70
	IV	1		1.07		1.07		18.08	4.71	
34	I	37	5.30	15.75	0.52	0.42	2.75	15.75	0.42	
	II	7	3.50	8.19	0.23	1.17	0.80	23.94	1.59	
	III	2	2.00	1.89	0.33	0.94	0.66	25.83	2.54	3.93
	IV	1		0.63		0.63		26.46	3.17	

	35	I	33	4.10	28.98	0.69	0.88	2.86	28.98	0.88	
		II	8	4.00	20.16	0.46	2.52	1.79	49.14	3.40	
		III	2	2.00	9.45	0.40	4.72	0.50	58.59	8.12	3.36
		IV	1		3.78		3.78		62.37	11.90	
Semi-arid region	36	I	65	4.06	56.25	0.41	0.86	1.69	56.25	0.96	
		II	16	4.00	23.50	0.44	1.46	1.79	79.75	2.32	
		III	4	2.00	10.50	0.66	2.62	1.33	90.25	4.94	3.01
		IV	2	2.00	7.00	0.25	3.50	0.71	97.25	8.44	
		V	1		2.50		2.50		99.75	10.94	
	37	I	42	3.81	26.00	0.69	0.61	2.67	26.00	0.61	
		II	11	3.66	18.00	0.62	1.63	2.28	44.00	2.24	
		III	3	3.00	11.20	0.35	3.73	1.07	55.20	5.97	3.49
		IV	1		4.00		4.00		59.20	9.97	
	38	I	61	4.69	30.00	0.38	0.49	1.79	30.00	0.49	
		II	13	4.33	11.50	0.78	0.88	3.40	41.50	1.37	
		III	3	3.00	9.00	0.66	3.00	2.00	50.50	4.37	4.00
		IV	1		6.00		6.00		56.50	10.37	
	39	I	63	3.93	45.50	0.46	0.72	1.81	45.50	0.72	
		II	16	5.33	21.00	0.80	1.31	4.32	66.50	2.03	
		III	3	3.00	17.00	0.35	5.66	1.06	83./50	7.69	4.08
		IV	1		6.00		6.00		89.50	13.69	
	40	I	205	5.39	74.00	0.36	0.36	2.02	74.00	0.36	
		II	38	6.33	27.75	0.55	0.73	3.42	101.75	1.09	
		III	6	3.00	15.00	0.41	2.50	1.24	116.75	3.59	4.18
		IV	2	2.00	6.25	1.00	3.12	2.00	123.00	6.71	
		V	1		6.25		6.25		129.29	12.96	
	41	I	80	3.80	55.50	0.47	0.69	1.02	55.50	0.69	
		II	21	7.00	26.50	0.35	1.26	2.50	82.50	1.55	
		III	3	3.00	9.50	0.68	3.16	2.05	91.50	5.11	4.60
		IV	1		6.50		6.50		98.00	11.61	
	42	I	151	5.03	60.00	0.33	0.39	1.72	60.00	0.39	
		II	30	7.50	20.00	0.40	0.67	2.90	80.00	1.06	
		III	4	2.00	8.00	1.00	2.00	2.00	88.00	3.06	4.13
		IV	2	2.00	8.00	0.19	4.00	0.37	96.00	7.06	
		V	1		2.00		0.19		97.50	8.56	
	43	I	68	2.72	34.00	0.51	0.50	1.40	34.00	0.50	
		II	25	1.47	17.50	1.54	0.70	2.25	51.50	1.20	
		III	17	8.50	27.00	0.09	1.58	0.79	78.50	2.78	3.67
		IV	2	2.00	2.50	2.40	1.25	4.80	81.00	4.03	
		V	1		6.00		6.00		87.00	10.03	
	44	I	37	2.46	16.00	0.78	0.43	1.93	16.00	0.43	
		II	15	3.75	12.50	1.12	0.83	4.21	28.50	1.26	
		III	4	4.00	14.00	0.14	3.50	0.57	42.50	4.76	3.40
		IV	1		2.00		2.00		44.50	6.76	
	45	I	206	4.47	105.00	0.38	0.50	1.71	105.00	0.50	
		II	46	4.60	40.00	0.30	0.86	1.39	145.00	1.36	
		III	10	5.00	12.00	0.50	1.20	2.50	157.00	2.56	4.02
		IV	2	2.00	6.00	5.00	3.00	10.00	163.00	5.56	
		V	1		30.00		30.00		193.00	35.56	
	46	I	57	4.38	26.00	0.42	0.45	1.36	26.00	0.45	
		II	13	2.60	11.00	1.05	0.84	2.73	37.00	1.29	
		III	5		11.50		2.30		48.50	3.59	3.96

		IV	1	5.00	3.50	0.30	3.50	1.52	52.00	7.09	
	47	I	79		38.00		0.48		38.00	0.48	
		II	19	4.16	10.00	0.42	0.84	1.75	54.00	1.32	
		III	5	3.80	13.50	0.84	2.70	3.21	67.50	4.02	3.97
		IV	1	5.00	3.00	0.22	3.00	1.10	70.50	7.02	
	48	I	79		45.50		0.58		45.50	0.58	
		II	20	3.90	16.00	0.35	0.80	1.37	61.50	1.38	
		III	7	2.86	10.00	0.62	1.43	1.78	71.50	2.81	2.81
		IV	2	3.50	11.00	1.10	5.50	3.84	32.50	8.31	
		V	1	2.00	4.50	0.41	4.50	0.81	87.00	12.80	
	49	I	53		47.50		0.89		47.50	0.89	
		II	17	3.11	17.00	0.36	1.00	1.11	64.50	1.89	
		III	5	3.40	13.00	0.76	2.60	2.60	77.50	4.89	3.83
		IV	1	5.00	6.00	0.46	6.00	2.30	83.50	10.49	
	50	I	92		124.00		1.35		124.00	1.35	
		II	27	3.40	20.00	0.16	0.74	0.55	144.00	2.09	
		III	8	3.37	28.00	1.15	2.89	3.87	167.00	4.96	4.92
		IV	1	8.00	8.00	0.35	3.00	2.78	175.00	12.96	
	51	I	173		91.00		0.59		91.00	0.52	
		II	46	3.76	54.00	0.59	1.17	2.25	145.00	1.69	
		III	8	5.75	24.00	0.52	3.50	2.99	173.00	5.19	3.88
		IV	2	4.00	18.50	0.66	9.25	2.64	191.00	14.44	
		V	1	2.00	5.50	0.30	5.50	0.59	197.00d	19.84	
	52	I	10		11.50		1.15		11.50	1.15	
		II	8	1.25	6.50	0.50	0.81	0.70	19.00	1.96	
		III	2	4.00	2.00	0.31	1.00	1.23	20.00	2.96	2.42
		IV	1	2.00	1.50	0.75	1.50	1.50	21.50	4.46	
Transi-tionat zone	53	I	36		19.25		0.53		19.25	0.53	
		II	9	4.00	8.50	0.44	0.94	1.77	27.75	1.47	
		III	2	4.50	4.75	0.55	2.37	2.52	32.50	3.84	3.50
		IV	1	2.00	4.50	0.94	4.50	1.89	37.00	8.34	
	54	I	786		51.50		0.67		51.50	0.67	
		II	20	3.80	22.50	0.42	1.10	1.64	73.50	1.77	
		III	4	5.00	3.50	0.15	0.87	0.79	77.00	2.64	3.20
		IV	2	2.00	2.75	0.78	1.37	1.57	79.75	4.01	
		V	1	2.00	7.50	2.72	7.50	5.47	87.25	11.51	
	55	I	33		25.50		0.77		25.50	0.77	
		II	10	3.30	9.75	0.38	0.97	1.25	35.25	1.74	
		III	4	2.50	10.50	1.07	2.62	2.70	45.75	4.36	3.26
		IV	1	4.00	10.00	0.95	10.00	3.81	55.75	14.36	
	56	I	61		30.00		0.49		30.00	0.49	
		II	17	3.58	10.20	0.34	0.60	1.22	40.20	1.09	
		III	6	2.53	3.75	0.36	0.62	1.03	43.95	1.71	2.85
		IV	2	3.00	2.25	0.60	1.12	1.80	46.20	2.83	
		V	1	2.00	3.50	1.55	3.50	3.12	49.70	6.33	
Arid region	57	I	45		22.00		0.48		22.00	0.48	
		II	12	3.75	8.50	0.38	0.70	1.45	30.50	1.18	
		III	2	6.00	2.00	0.23	1.00	1.42	32.50	2.18	3.91
		IV	1	2.00	3.00	1.50	3.00	3.00	35.50	5.18	
	58	I	68		30.50		0.44		30.50	0.44	
		II	21	3.23	11.50	0.37	0.54	1.22	42.00	0.98	
		III	6	3.50	9.00	0.78	1.50	2.77	57.00	2.48	2.93

	IV	2	3.00	3.00	0.33	1.50	1.00	54.00	3.98	
	V	1	2.00	5.00	1.66	5.00	3.33	59.00	8.98	
59	I	53	6.62	32.25	0.35	0.60	2.38	32.25	0.60	
	II	8	2.66	11.50	0.65	1.43	1.74	43.75	2.03	4.09
	III	3	3.00	7.50	0.33	2.50	1.00	51.25	4.53	
	IV	1		2.50		2.50		53.75	7.03	
60	I	15	3.80	11.50	0.55	0.80	2.12	15.25	0.80	
	II	5	2.50	5.00	1.00	1.70	2.50	23.75	2.50	2.76
	III	2	2.00	3.00	1.52	4.25	3.05	32.25	6.75	
	IV	1		16.50		13.00		45.25	19.75	
61	I	22	4.40	11.00	0.56	0.50	2.50	11.00	0.50	
	II	5	2.50	6.25	0.56	1.25	1.40	17.25	1.75	2.96
	III	2	2.00	3.50	0.64	1.75	1.28	20.75	3.50	
	IV	1		2.25		2.25		23.00	5.75	
62	I	35	3.50	16.00	0.60	0.45	2.15	16.00	0.45	
	II	10	3.33	9.75	0.35	9.97	1.19	25.75	1.42	3.27
	III	3	3.00	3.50	0.85	1.16	2.58	29.25	2.58	
	IV	1		3.00		3.00		32.25	3.58	

Note: Symbols used:

u	Stream order
Nu	Number of stream segments
Rb	Bifurcation ratio
(L)u	Total stream length
Re	Total stream length ratio
Lu	Mean stream length
Re	Mean stream length ratio
CTSL	Cumulative total stream length
CMSL	Cumulative mean stream length
Rb	Mean Bifurcation ratio

TABLE 7.3 **Quantitative Analysis of Fluvial Morphometry Rajasthan (Third Order Basins)**

Climatic Region	Basin Number	Total Number of Stream Segments	Total Stream Length	Drainage Density	Stream Frequency	Texture Ratio	Constant of Channel Maintainance
			(m)	m/m^2	$Number/m^2$	$Tu\ \frac{Nu}{Pu}$	$C=\frac{1}{Du}$
1	2	3	4	5	6	7	8
Humid region	1	23	30.50	1.28	0.96	0.92	0.78
	2	27	27.20	1.23	1.22	1.33	0.81
	3	69	47.75	5.41	7.82	3.28	0.81
	4	39	24.85	3.25	5.10	2.67	0.30
	5	83	55.95	3.37	5.01	3.35	0.29
	6	27	17.75	5.59	11.67	5.03	0.17
	7	42	26.70	3.65	5.75	3.28	0.27
Subhumid Region	8	22	22.45	1.72	1.66	1.10	0.58
	9	44	27.50	3.28	5.25	3.20	0.30
	10	33	18.50	1.95	3.49	2.38	0.51
	11	43	20.25	1.54	3.28	2.68	0.64
	12	15	42.00	1.25	0.86	0.96	0.80
	13	29	24.00	0.85	0.53	0.62	1.17
	14	19	27.50	0.99	0.79	0.88	1.01
	15	84	18.00	1.08	0.42	0.27	0.92
	16	84	65.00	2.15	2.55	2.80	0.46
	17	77	31.50	1.70	1.24	1.24	0.58
	18	69	25.00	2.58	2.31	1.75	0.43
	19	21	11.87	0.42	0.75	0.41	2.38
	20	22	7.75	0.69	1.98	2.66	1.45
	21	21	6.87	0.86	2.63	1.60	1.16
	22	28	14.00	0.50	1.01	0.73	1.97
	23	32	26.14	0.37	0.46	-	2.70
	24	19	23.94	0.52	0.41	-	1.92
	25	28	10.72	0.87	2.27	0.77	1.15
	26	10	8.12	0.51	0.64	1.30	1.96
	27	24	12.28	0.70	1.1	-	11.41
	28	11.75	11.75	0.34	0.32	-	2.94
Semiarid Region	29	33	26.95	2.43	2.98	2.6	0.41
	30	38	25.00	4.34	6.60	3.36	0.23
	31	15	15.00	1.38	1.33	0.92	0.72
	32	29	19.25	2.71	4.09	2.48	0.36
	33	53	24.00	3.24	7.16	4.18	0.30
	34	37	31.50	1.93	2.26	1.62	0.51
	35	24	57.25	0.58	0.24	0.45	1.72
	36	22	21.50	0.94	0.83	0.95	1.06
	37	72	33.00	2.24	2.10	1.16	0.44
	38	31	25.00	0.9	1.18	1.47	1.04
	39	29	28.00	0.88	0.92	1.45	1.12
Transitional zone	40	11	16.55	1.26	0.84	0.66	0.79
	41	33	19.00	3.28	5.69	3.00	0.30
	42	22	17.40	3.18	4.02	1.76	0.31
	43	12	14.80	0.95	0.77	0.60	1.05
	44	25	14.00	2.50	4.47	2.32	0.40

Ari reg	45	34	22.50	2.58	3.89	2.12	.38
	46	10	13.25	2.25	1.70	0.76	.44
	47	29	23.80	1.94	2.37	1.65	.51
	48	36	15.75	3.28	7.50	3.39	.30
	49	14	10.95	1.80	2.30	1.13	.55
	50	14	7.00	3.09	6.19	2.33	.32
	51	26	19.25	2.08	2.81	2.16	.48
	52	23	18.05	1.30	1.66	1.42	.76
	53	22	19.25	1.29	1.47	1.22	.77
	54	32	21.00	3.81	5.81	2.66	.26

TABLE 7.4 **Quantitative Analysis of Fluvial Morphometry – Rajasthan (Random Order Basins)**

Climatic Region	Basin Number	Total Number of Stream Segments	Total Stream Length (m)	Drainage Density m/m^2	Stream Frequency Number/m^2	Texture Ratio Tu $\frac{Nu}{Pu}$	Constant of Channel Maintainance C= $\frac{1}{Du}$
1	2	3	4	5	6	7	8
Humid region	1	400	234.00	2.83	4.84	9.85	0.35
	2	193	212.25	3.14	5.40	6.20	0.31
	3	296	147.95	2.72	3.60	5.60	0.36
	4	375	262.20	2.50	3.58	8.28	0.40
	5	66	84.75	3.82	2.98	3.34	0.26
	6	159	68.70	2.58	5.97	6.29	0.38
	7	95	92.75	1.56	1.59	2.62	0.64
	8	35	36.25	1.98	1.91	1.97	0.50
	9	134	66.80	4.43	8.90	5.25	0.22
Subhumid region	10	400	184.20	4.93	10.70	16.00	0.20
	11	100	116.70	0.92	0.79	2.04	1.08
	12	289	210.55	2.05	2.81	5.47	0.48
	13	76	81.25	1.24	1.16	2.08	0.80
	14	120	65.75	1.37	2.51	4.21	0.72
	15	25	52.00	2.41	3.34	2.72	0.41
	16	231	114.50	2.43	2.08	3.67	0.48
	17	156	233.00	1.52	1.40	3.42	0.66
	18	31	24.50	1.72	1.75	1.35	0.58
	19	37	47.00	3.18	2.16	2.70	0.46
	20	7	52.00	0.91	0.65	0.92	1.08
	21	23	83.00	2.06	2.08	2.58	0.48
	22	88	96.75	1.35	1.17	2.33	0.74
	23	134	111.00	2.21	3.11	3.46	0.45
	24	105	110.50	1.61	1.28	2.24	0.62
	25	28	121.00	1.64	1.89	2.72	0.52
	26	40	11.62	0.69	2.39	3.50	1.45
	27	332	119.07	0.97	2.70	–	1.03
	28	32	19.21	0.53	0.86	–	1.89
	29	34	15.50	0.84	1.85	1.60	1.19
	30	81	21.00	1.13	4.37	1.37	0.84
	31	197	53.00	0.59	2.18	2.25	1.70
	32	66	45.04	0.71	1.04		1.40
	33	26	18.08	0.39	0.56	–	2.59
	34	47	26.46	0.50	0.90	–	1.97
	35	44	62.37	0.49	0.43	–	2.04
Semi-arid region	36	88	99.75	1.04	0.92	2.12	0.96
	37	57	59.20	2.18	2.10	2.15	0.45
	38	78	56.50	0.94	1.30	2.31	1.06
	40	252	128.75	2.88	5.63	6.76	0.34
	41	105	98.00	4.02	4.31	4.00	0.24
	42	188	97.50	1.00	1.93	4.82	0.99
	43	113	87.00	1.05	1.36	4.10	0.95
	44	57	44.50	1.10	1.41	2.53	0.90
	45	265	193.00	1.00	1.38	3.48	0.99
	46	76	52.00	1.09	1.60	3.70	0.90
	47	104	70.50	1.13	1.66	2.81	0.88

	48	108	87.00	1.44	1.79	3.60	0.69
	49	76	52.00	1.09	1.60	3.70	0.90
	50	128	175.00	1.76	1.29	4.12	0.56
	51	230	197.00	1.78	2.08	5.97	0.56
	52	1	21.50	0.94	0.92	1.68	1.08
Transi-tional zone	53	48	37.00	1.66	2.16	2.31	0.60
	54	103	87.25	2.66	3.14	4.03	0.37
	55	48	55.75	1.56	1.34	1.62	0.64
	56	87	49.70	3.30	5.78	4.72	0.30
Arid zone	57	60	35.50	3.61	6.10	4.21	0.27
	58	98	59.00	2.07	3.44	3.65	0.48
	59	65	53.75	2.08	2.52	2.81	0.48
	60	27	45.25	0.88	0.52	0.63	1.13
	61	30	23.00	1.54	2.01	1.73	0.64
	62	49	32.25	3.50	5.32	3.26	0.28

TABLE 7.5 **Quantitative Analysis of Fluvial Morphometry Rajasthan (Third Order Basins)**

Climatic Region	Basin Number	A	Lb	Z	Pu or	z	Rc $\frac{Au4}{L^2}$	H=2-z	Re=H/LB	Rh=LB	Rn=$\frac{D-H}{5280}$
1	2	3	4	5	6	7	8	9	10	11	12
Humid region	1	23.74	9.90	390	25.00	320	0.47	70	0.80	0.070	0.0169
	2	22.09	8.25	289	20.25	270	0.67	19	0.78	0.023	0.0044
	3	08.82	8.65	382	21.00	240	0.25	142	0.77	0.164	0.1454
	4	7.64	5.25	250	14.60	140	0.45	110	0.88	0.209	0.0677
	5	16.56	10.00	349	24.75	220	0.33	129	0.78	0.129	0.0823
	6	3.17	3.15	700	7.35	400	0.73	300	0.74	0.952	0.3176
	7	7.30	4.85	600	12.80	334	0.55	266	0.84	0.548	0.1838
Subhumid region	8	20.36	8.25	478	20.00	416	0.40	62	0.77	0.075	0.0201
	9	8.37	5.96	540	13.75	420	0.55	120	0.73	0.201	0.0745
	10	9.45	5.80	251	13.85	240	0.61	11	0.76	0.01	0.0040
	11	13.10	6.75	238	16.00	220	0.64	18	0.75	0.0267	0.0052
	12	33.51	11.40	455	30.00	434	0.46	21	0.83	0.01	0.004
	13	27.93	8.60	403	24.00	360	0.60	43	0.88	0.05	0.006
	14	27.69	10.70	477	25.00	440	0.55	37	0.74	0.03	0.006
	15	16.61	11.00	447	25.50	400	0.32	47	0.73	0.04	0.009
	16	30.10	11.25	590	27.50	440	0.51	150	0.76	0.13	0.061
	17	18.51	7.50	460	18.50	440	0.67	20	0.78	0.02	0.006
	18	10.82	7.05	554	16.00	480	0.53	74	0.72	0.10	0.029
	19	27.82	7.10	-	-	-	0.54	-	0.84	0.009	-
	20	11.08	4.85	-	-	-	0.56	-	0.76	0.001	-
	21	7.96	3.85	-	-	-	0.48	-	0.82	0.017	-
	22	27.60	9.15	-	-	-	0.79	-	0.64	0.001	-
	23	68.92	15.56	-	-	-	0.95	-	0.53	-	-
	24	45.84	12.29	-	-	-	0.73	-	0.53	-	-
	25	12.31	5.50	-	-	-	0.60	-	0.72	0.031	-
	26	15.69	5.45	-	-	-	0.71	-	0.82	0.024	-
	27	15.85	5.10	-	-	-	0.56	-	0.79	-	-
	28	33.89	7.94	-	-	-	0.79	-	0.83	-	-
Semiarid region	29	11.05	6.50	564	16.00	520	0.54	44	0.78	0.067	0.0202
	30	5.75	4.15	299	11.30	280	0.56	19	0.86	0.045	0.0156
	31	11.22	7.0	305	16.25	280	0.3	25	0.73	0.035	0.0065
	32	7.08	5.00	295	11.65	240	0.65	55	0.74	0.110	0.0282
	33	7.40	4.90	220	12.65	198	0.58	22	0.82	0.044	0.0135
	34	16.32	9.0	474	22.70	286	0.39	188	0.80	0.208	0.0687
	35	98.00	22.25	280	52.50	236	0.44	44	0.75	0.019	0.0048
	36	22.68	8.40	478	20.00	416	0.71	62	0.75	0.07	0.011
	37	14.73	7.40	533	26.50	360	0.26	173	1.14	0.23	0.070
	38	26.13	8.00	-	21.00	-	0.74	-	-	0.001	0.001
	39	31.52	5.11	-	20.00	-	0.99	-	-	0.003	0.003
Transitional zone	40	13.04	6.25	529	16.50	420	0.60	109	0.84	0.174	0.0260
	41	5.79	4.50	331	11.00	280	0.60	51	0.77	0.113	0.0316
	42	5.47	5.30	358	12.50	260	0.93	98	0.75	0.184	0.0590
	43	15.56	9.40	298	20.00	282	0.48	16	0.67	0.017	0.0028
	44	5.59	4.45	643	10.75	375	0.60	268	0.76	0.602	0.1268
Arid zone	45	8.72	5.50	326	16.00	310	0.42	16	0.92	0.029	0.0078
	46	5.88	5.50	283	13.00	145	0.43	138	0.75	0.250	0.0588
	47	12.23	7.30	357	17.50	260	0.50	97	0.76	0.132	0.0365
	48	4.80	4.25	263	10.60	240	0.52	23	0.79	0.054	0.0142
	49	6.07	5.00	309	12.30	268	0.50	41	0.78	0.082	0.0139
	50	2.26	2.25	309	6.00	300	0.78	9	0.84	0.040	0.0052
	51	9.25	3.90	260	12.00	230	0.80	30	0.97	0.076	0.0118
	52	13.80	6.80	273	16.10	238	0.66	35	0.75	0.051	0.0086
	53	14.89	7.70	228	17.90	180	0.58	48	0.74	0.062	0.0117
	54	5.50	4.75	657	12.00	220	0.47	437	0.80	0.092	0.3153

TABLE 7.6 **Quantitative Analysis of Fluvial Morphometry (Random Order Basins)**

Climatic Region	Basin Number	A	Lb	Z	Pu or	z	$Rc=\frac{AUX4}{L^2}$	H=2-z	$Re=\frac{}{Lb}$	$Rh=\frac{H}{Lb}$	$Rn=\frac{D-H}{5280}$
1	2	3	4	5	6	7	8	9	10	11	12
Humid region	1	82.57	1200	305	40.60	140	0.64	162	1.07	0.137	0.0884
	2	35.74	1250	393	31.10	220	0.46	172	0.79	0.137	0.1022
	3	54.31	1450	495	35.00	260	0.55	235	0.76	0.162	0.1210
	4	104.65	1740	460	45.25	326	0.64	134	0.82	0.077	0.0634
	5	22.13	700	234	19.75	220	0.71	14	0.89	0.020	0.0101
	6	26.60	950	226	25.25	200	0.52	26	0.84	0.027	0.0127
	7	59.43	1425	471	36.25	360	0.56	111	0.81	0.077	0.0327
	8	18.25	675	382	17.75	320	0.72	62	0.83	0.091	0.0232
	9	15.05	865	689	25.50	360	0.56	329	0.93	0.380	0.2760
Subhumid region	10	37.35	925	1293	25.00	820	0.74	473	0.86	0.511	0.4416
	11	125.98	1910	467	49.00	420	0.65	47	0.81	0.024	0.0081
	12	102.67	2025	563	52.75	420	0.56	143	0.82	0.070	0.0555
	13	65.28	1475	316	36.40	255	0.61	61	0.78	0.041	0.0143
	14	47.66	1125	242	28.50	220	0.73	22	0.80	0.019	0.0057
	15	21.53	800	932	26.50	700	0.38	232	1.05	0.29	0.105
	16	54.93	1350	689	36.50	509	0.51	180	0.86	0.13	0.070
	17	164.54	2080	755	67.50	500	0.45	225	1.03	0.10	0.060
	18	14.24	730	580	18.50	507	0.52	73	0.80	0.10	0.020
	19	21.69	925	516	25.50	462	0.41	54	0.87	0.05	0.022
	20	56.67	1390	559	40.00	385	0.44	174	0.91	0.12	0.029
	21	40.25	1260	621	32.50	406	0.47	215	0.82	0.17	0.083
	22	71.64	1235	555	36.00	440	0.69	115	0.92	0.09	0.029
	23	50.13	1400	572	45.00	370	0.31	202	1.02	0.14	0.080
	24	68.58	1350	500	37.50	440	0.61	60	0.88	0.04	0.018
	25	63.76	2325	603	38.50	400	0.54	203	0.52	0.08	0.072
	26	16.69	61	-	-	-	0.65	-	0.88	0.006	-
	27	121.90	1405	-	-	-	1.35	-	0.78	-	-
	28	36.16	838	-	-	-	0.47	-	0.72	-	-
	29	18.34	550	-	-	-	0.77	-	0.87	0.054	-
	30	18.53	630	-	-	-	0.65	-	0.76	0.0036	-
	31	90.01	1840	-	-	-	1.04	-	0.58	0.024	-
	32	63.03	794	-	-	-	1.15	-	0.99	-	-
	33	45.84	1084	-	-	-	0.87	-	0.62	-	-
	34	52.09	851	-	-	-	0.99	-	0.85	-	-
	35	127.51	1745	-	-	-	0.46	-	0.65	-	-
Semi-arid region	36	95.61	1500	545	41.50	460	0.69	85	0.88	0.056	0.0167
	37	36.69	975	576	26.50	440	0.48	136	0.86	0.139	0.0561
	38	59.98	1440	307	33.75	280	0.66	27	0.74	0.018	0.0048
	39	68.91	1015	511	38.50	300	0.58	211	1.20	0.207	0.0515
	40	44.70	1225	350	37.35	200	0.40	150	0.96	0.122	0.0818
	41	33.76	1000	589	26.25	300	0.44	289	0.83	0.289	0.2200
	42	97.04	1200	-	39.00	-	0.801	-	-	0.013	0.030
	43	82.64	920	-	27.50		1.273	-	-	0.018	0.033
	44	40.18	710	-	22.50		1.092	-	-	0.020	0.030
	45	191.96	2885	-	76.00		0.417	-	-	0.004	0.027
	46	47.28	750	-	20.50		1.413	-	-	0.011	0.015
	47	62.28	125	-	37.00		0.571	-	-	0.006	0.016
	48	60.13	900	-	30.00		0.839	-	-	0.016	0.040
	49	98.65	1035	-	28.00		1.580	-	-	0.009	0.015
	50	98.86	1000	-	31.00		1.292	-	-	0.008	0.026
	52	22.78	45	-	12.50		1.832	-	-	0.004	0.003

Transi-tional zone	53	2.17	810	343	20.75	312	0.64	31	0.81	0.038	0.0097
	54	46.90	1000	452	25.50	220	0.63	232	0.81	0.232	0.1168
	55	35.59	1300	481	29.50	297	0.51	184	0.72	0.141	0.0543
	56	15.05	560	663	18.40	340	0.55	323	1.04	0.576	0.2018
Arid region	57	9.82	525	330	14.25	240	0.60	90	0.86	0.171	0.0615
	58	28.47	910	356	26.80	281	0.49	75	0.93	0.082	0.0294
	59	25.79	800	402	23.10	278	0.60	124	0.91	0.155	0.0488
	60	51.15	1510	146	42.50	100	0.35	46	0.89	0.030	0.0076
	61	14.90	710	318	17.00	280	0.64	38	0.76	0.053	0.0110
	62	9.21	550	354	15.00	300	0.51	54	0.86	0.098	0.0357

TABLE 7.7 **Number of streams V/s Stream Order (Third Order Basins)**

Climatic Region	Basin Number	Regression Equation	Level of significance (%)	Correlation Coefficient
Humid region	1	Y = 16.02 - 4.98X	20	-1
	2	Y = 28.00 - 9.50X	20	-1
	3	Y = 75.00 -26.00X	5	-1
	4	Y = 44.00 -15.50X	10	-1
	5	Y = 93.62 -32.58X	5	-1
	6	Y = 42.31 -14.99X	10	-1
	7	Y = 48.00 -17.00X	10	-1
Subhumid region	8	Y = 23.31 - 7.99X	10	-1
	9	Y = 48.62 -16.98X	10	-1
	10	Y = 37.00 -13.00X	10	-1
	11	Y = 45.31 -15.49X	10	-x
	12	Y = 32.62 -11.48X		
	13	Y = 15.00 - 5.00X		
	14	Y = 25.31 - 8.99X		
	15	Y = 5.99 - 1.66X		
	16	Y = 88.62 -31.48X		
	17	Y = 23.62 - 7.98X		
	18	Y = 29.31 - 7.99X		
	19	Y = 21.00 - 7.00X		
	20	Y = 23.31 - 7.99X		
	21	Y = 21.00 - 7.00X		
	22	Y = 31.31 -10.99X		
	23	Y = 35.62 -12.48X		
	24	Y = 18.31 - 5.99X		
	25	Y = 31.31 -10.99X		
	26	Y = 9.31 - 2.99X		
	27	Y = 17.00 - 9.50X		
	28	Y = 9.62 - 2.98X		
Semiarid region	29	Y = 36.00 -12.50x	10	-1
	30	Y = 41.62 -14.48x	10	-1
	31	Y = 16.00 - 5.50x	10	-1
	32	Y = 31.62 -10.98x	10	-1
	33	Y = 60.62 -21.48x	10	-1
	34	Y = 40.31 -13.99x	10	-1
	35	Y = 27.00 - 9.50x	10	-1
	36	Y = 20.31 - 6.99x		
	37	Y = 33.31 -11.49x		
	38	Y = 34.31 -11.99		
	39	Y = 32.62 -11.48		
Transitional zone	40	Y = 10.62 - 3.48x	5	-1
	41	Y = 35.00 -12.00x	10	-1
	42	Y = 22.31 - 7.49x	5	-1
	43	Y = 12.00 - 4.00x	10	-1
	44	Y = 25.31 - 8.49x	5	-1
Arid region	45	Y = 35.31 -11.99x	5	-1
	46	Y = 8.31 - 2.49x	2	-1
	47	Y = 32.62 -11.48x	10	-1
	48	Y = 40.00 -14.00x	10	-1
	49	Y = 13.62 - 4.48x	5	-1
	50	Y = 14.62 - 4.98x	10	-1
	51	Y = 27.62 - 9.98x	10	-1
	52	Y = 22.62 - 7.48x	5	-1
	53	Y = 23.31 - 7.99x	10	-1
	54	Y = 33.62 -11.48x	10	-1

TABLE 7.8 **Number of streams V/s Stream Order (Random Order Basins)**

Climatic Region	Basin Number	Regression Equation	Level of significance (%)	Correlation Coefficient
1	2	3	4	5
Humid	1	Y = 383.40 - 67.80x	5	-1
	2	Y = 130.60 - 25.00x	5	-1
	3	Y = 138.00 - 32.90x	2	-1
	4	Y = 265.20 - 63.40x	1	-1
	5	Y = 57.00 - 16.20x	1	-1
	6	Y = 109.20 - 25.80x	1	-1
	7	Y = 79.50 - 22.30x	1	-1
	8	Y = 21.00 - 5.30x	1	-1
	9	Y = 116.00 - 33.00x	10	-1
Subhumid region	10	Y = 238.83 - 49.19x	1	-1
	11	Y = 84.50 - 23.80x	1	-1
	12	Y = 207.20 - 49.80x	1	-1
	13	Y = 65.00 - 18.40x	1	-1
	14	Y = 102.50 - 29.00x	1	-1
	15	Y = 60.50 - 17.00x		
	16	Y = 67.50 - 21.60x		
	17	Y = 193.50 - 54.30x		
	18	Y = 19.50 - 5.30x		
	19	Y = 58.50 - 16.50x		
	20	Y = 30.00 - 8.30x		
	21	Y = 69.75 - 19.00x		
	22	Y = 56.70 - 13.30x		
	23	Y = 109.50 - 26.10x		
	24	Y = 78.00 - 22.40x		
	25	Y = 89.50 - 25.30x		
	26	Y = 26.00 - 6.40x		
	27	Y = 192.66 - 39.57x		
	28	Y = 14.00 - 2.40x		
	29	Y = 27.50 - 7.60x		
	30	Y = 69.50 - 19.70x		
	31	Y = 135.70 - 32.10x		
	32	Y = 218.00 - 80.00x		
	33	Y = 20.00 - 5.40x		
	34	Y = 40.00 - 11.30x		
	35	Y = 36.50 - 10.20x		
Semi-arid region	36	Y = 60.20 - 14.20x	1	-1
	37	Y = 47.00 - 13.10x	1	-1
	38	Y = 67.00 - 19.00x	1	-1
	39	Y = 74.50 - 21.00x	1	-1
	40	Y = 183.60 - 44.40x	1	-1
	41	Y = 90.00 - 25.50x	1	-1
	42	Y = 136.00 - 32.80x		
	43	Y = 69.70 - 15.70x		
	44	Y = 44.00 - 11.90x		
	45	Y = 189.20 - 45.40x		
	46	Y = 63.00 - 17.60x		
	47	Y = 93.00 - 26.80x		
	48	Y = 74.00 - 17.40x		
	49	Y = 61.00 - 16.80x		
	50	Y = 105.00 - 29.20x		
	51	Y = 162.40 - 38.80x		
	52	Y = 13.25 - 3.20x		

Transitional zone	53	Y = 40.00 - 11.20x	1	-1
	54	Y = 71.00 - 16.80x	2	-1
	55	Y = 37.50 - 10.20x	1	-1
	56	Y = 57.90 - 13.50x	1	-1
Arid region	57	Y = 50.50 - 14.20x	1	-1
	58	Y = 65.50 - 15.30x	1	-1
	59	Y = 56.50 - 16.10x	2	-1
	60	Y = 21.00 - 5.70x	5	-1
	61	Y = 24.00 - 6.60x	2	-1
	62	Y = 39.49 - 10.90x	1	-1

TABLE 7.9 **Main stream Length V/s Stream Order (Third Order Basins)**

Climatic Region	Basin Number	Regression Equation	Level of significance (%)	Correlation Coefficient
1	2	3	4	5
Humid region	1	Y = - 3.34 + 3.33X	5	+1
	2	Y = - 0.19 + 3.22X	20	+1
	3	Y = - 5.09 + 4.24X	20	+1
	4	Y = - 2.49 + 2.30X	10	+1
	5	Y = - 6.59 + 5.36X	10	+1
	6	Y = - 0.66 + 0.80X	10	+1
	7	Y = - 2.30 + 2.16X	10	+1
Subhumid region	8	Y = - 1.64 + 2.09X	10	+1
	9	Y = - 3.35 + 2.90X	20	+1
	10	Y = - 0.90 + 1.01X	10	+0.5
	11	Y = - 2.55 + 2.20X	20	+0.5
	12	Y = - 4.69 + 4.51X		
	13	Y = - 2.77 + 3.18X		
	14	Y = + 2.38 + 0.14X		
	15	Y = - 1.75 + 11.0X		
	16	Y = - 5.47 + 4.70X		
	17	Y = - 3.25 + 3.25X		
	18	Y = - 2.88 + 2.66X		
	19	Y = - 0.66 + 0.885X		
	20	Y = - 21.12 + 16.02X		
	21	Y = - 0.21 + 0.83X		
	22	Y = - 2.40 + 2.12X		
	23	Y = - 1.75 + 1.92x		
	24	Y = - 5.99 + 4.55X		
	25	Y = - 0.63 + 0.82x		
	26	Y = 0.89 - 0.035X		
	27	Y = 0.500 + 0.095X		
	28	Y = 0.40 + 0.41X		
Semi-arid region	29	Y = - 1.66 + 1.80x	5	
	30	Y = - 1.01 + 1.18X	5	
	31	Y = - 0.30 + 0.99X	2	
	32	Y = - 0.86 + 1.11X	5	
	33	Y = - 0.77 + 0.96x	5	
	34	Y = - 5.01 + 4.22X	20	
	35	Y = - 4.48 + 5.65X	1	
	36			
	37	Y = - 2.15 + 2.45X		
	38	Y = - 3.23 + 3.06X		
	39	Y = - 4.41 + 3.73X		
Transitional zone	40	Y = + 2.06 + 0.10X	Non significant	
	41	Y = - 1.89 + 1.78x	10	
	42	Y = - 1.81 + 1.86x	5	
	43	Y = + 0.86 + 0.66X	Non significant	
	44	Y = - 2.03 + 1.82X	10	
Arid region	45	Y = - 3.54 + 3.03X	20	+1
	46	Y = - 0.66 + 1.27X	5	+0.5
	47	Y = - 3.55 + 3.11X	5	+1
	48	Y = - 1.92 + 1.72X	10	+1
	49	Y = - 0.27 + 0.76X	1	+1
	50	Calculation is not poss.	-	-0.25
	51	Y = - 0.60 + 0.94X	10	+1
	52	Y = - 1.58 + 1.66X	10	+0.5
	53	Y = - 2.63 + 2.465X	10	+1
	54	Y = - 1.81 + 1.77X	10	+1

	50	Y = 0.22 + 0.70X		
	51	Y = - 1.36 + 1.79X		
	52	Y = 0.80 + 0..2X		
Transitional zone	53	Y = - 1.28 + 1.344	Highly significant	+1
	54	Y = - 1.702 + 1.334X	10	+0.9
	55	Y = - 0.077 + 1.467X	20	+1
	56	Y = - 0.729 + 0.663	1	+1
Arid region	57	Y = - 0.70 + 0.796X	2	+1
	58	Y = - 1.261 + 1.017	1	+0.85
	59	Y = 0.02 + 0.692X	Highly Significant	+0.7
	60	Y = - 4.895 + 3.93X	2	+1
	61	Y = - 0.045 + 0.59X	Highly significant	+1
	62	Y = - 0.595 + 0.794X	2	+1

TABLE 7.10 **Main Stream Length V/s Stream Order (Random Order Basins)**

Climatic Region	Basin Number	Regression Equation	Level of significance (%)	Correlation Coefficient
1	2	3	4	5
Humid region	1	Y = - 2.318 + 1.766X	1	+1
	2	Y = - 0.489 + 0.803X	10	+0.9
	3	Y = - 1.793 + 1.471X	1	+1
	4	Y = - 2.282 + 2.004X	1	+0.9
	5	Y = - 0.455 + 0.790X	Highly Significant	+1
	6	Y = - 0.789 + 0.663X	1	+0.9
	7	Y = - 5.32 + 3.796X	1	+1
	8	Y = 0.15 + 0.60X	Highly Significant	+1
	9	Y = - 2.795 + 2.030X	2	+1
subhumid region	10	Y = - 0.625 + 0.593	Highly significant	+1
	11	Y = 49.79 - 14.852X	20	+1
	12	Y = - 5.4111 + 3.367X	1	+1
	13	Y = - 0.685 + 1.834X	10	+0.8
	14	Y = - 4.30 + 2.298X	5	+1
	15	Y = - 4.85 + 2.07X		
	16	Y = - 1.75 + 1.63X		
	17	Y = - 11.22 + 7.49X		
	18	Y = - 0.71 + 1.09X		
	19	Y = - 3.97 + 2.55X		
	20	Y = - 2.60 + 2.33X		
	21	Y = - 2.49 + 2.13X		
	22	Y = - 0.81 + 0.56X		
	23	Y = - 2.24 + 0.64X		
	24	Y = - 5.76 + 4.09X		
	25	Y = - 4.48 + 3.45X		
	26	Y = - 0.33 + 0.38X		
	27	Y = - 0.51 + 0.51X		
	28	Y = + 0.67 + 0.07X		
Subhumid region	29	Y = 0.05 + 0.26X		
	30	Y = - 0.95 + 0.68X		
	31	Y = - 0.96 + 0.74X		
	32	Y = 0.26 + 0.49X		
	33	Y = 0.22 + 0.37X		
	34	Y = 0.69 + 0.004X		
	35	Y = 0.20 + 1.10X		
Semi-arid region	36	Y = 0.473 + 0.569X	1	+0.85
	37	Y = - 0.59 + 1.232X	1	+1
	38	Y = - 2.085 + 1.870X	1	+1
	39	Y = - 1.64 + 2.02X	1	+1
	40	Y = - 1.67 + 1.42X	Highly Significant	+1 +1
	41	Y = - 1.945 + 1.938X	Highly significant	+1
	42	Y = - 15.03 + 5.58X		
	43	Y = - 1.48 + 1.16X		
	44	Y = - 1.00 + 1.07X		
	45	Y = - 1.02 + 2.71X		
	46	Y = - 0.89 + 1.06X		
	47	Y = - 0.63 + 0.95X		
	48	Y = - 1.20 + 1.25X		
	49	Y = - 1.61 + 1.69X		

TABLE 7.11 **Cumulative total stream length v/s stream order (Third order Basins)**

Climatic Region	Basin Number	Regression Equation	Level of significance (%)	Correlation Coefficient
1	2	3	4	5
Humid region	1	Y = 7.31 + 7.76X	1	+1
	2	Y = 8.82 + 5.99X	2	+
	3	Y = 19.06 + 9.63X	1	+1
	4	Y = 8.36 + 5.56X	1	+1
	5	Y = 20.45 + 11.72X	1	+1
	6	Y = 11.75 + 2.00X	-	+1
	7	Y = 11.85 + 4.97X	2	+1
sub-humid region	8	Y = 4.10 + 6.22X	2	+1
	9	Y = 10.06 + 5.76X	1	+1
	10	Y = 11.62 + 2.27X	1	+1
	11	Y = 7.62 + 4.14X	1	+1
	12	Y = 14.62 + 9.02X		
	13	Y = 0.38 + 8.27X		
	14	Y = 9.12 + 6.77X		
	15	Y = 6.00 + 12.00X		
	16	Y = 29.62 + 12.02X		
	17	Y = 9.62 + 7.27X		
	18	Y = 9.62 + 5.02X		
	19	Y = 2.67 + 3.18X		
	20	Y = 2.47 + 1.70X		
	21	Y = 2.04 + 1.58X		
	22	Y = 3.12 + 3.27X		
	23	Y = 9.93 + 5.53X		
	24	Y = 7.43 + 5.05X		
	25	Y = 2.58 + 2.82X		
	26	Y = 2.46 + 2.06X		
	27	Y = 3.41 + 3.75X		
	28	Y = 5.52 + 2.06X		
Semi arid region	29	Y = 11.95 + 5.10X	2	+1
	30	Y = 16.00 + 3.00X	-	+1
	31	Y = 6.87 + 2.89X	1	+1
	32	Y = 9.31 + 3.38X	2	+1
	33	Y = 10.87 + 4.64X	5	+1
	34	Y = 9.81 + 7.01x	2	+1
	35	Y = 9.37 + 16.39X	2	+1
	36	Y = 3.62 + 6.02X		
	37	Y = 14.62 + 6.02X		
	38	Y = 7.62 + 5.52X		
	39	Y = 13.50 + 5.00X		
Transitional zone	40	Y = 0.86 + 5.78X	10	+1
	41	Y = 7.00 + 4.00X	-	+1
	42	Y = 3.42 + 4.72X	1	+1
	43	Y = 0.42 + 5.17X	5	+1
	44	Y = 3.62 + 3.39X	2	+1
Arid region	45	Y = 5.00 + 5.75X	1	+1
	46	Y = 4.31 + 2.88X	2	+1
	47	Y = 9.11 + 4.66X	5	+1
	48	Y = 5.50 + 3.37X	1	+1
	49	Y = 2.62 + 2.89X	5	+1
	50	Y = 4.81 + 0.76X	5	+1
	51	Y = 10.37 + 3.02X	2	+1
	52	Y = 7.65 + 3.40X	2	+1
	53	Y = 5.75 + 4.37X	2	+1
	54	Y = 7.62 + 4.52X	1	+1

TABLE 7.12 **Cumulative Total Stream Length V/sStream Order (Random Order Basin)**

Climatic Region	Basin Number	Regression Equation	Level of significance (%)	Correlation Coefficient
1	2	3	4	5
Humid	1	Y = 131.89 + 21.02X	1	+1
	2	Y = 62.80 + 11.00X	Highly Significant	+1
	3	Y = 80.15 + 14.91X	Highly Significant	+1
	4	Y = 136.64 + 27.76X	Highly Significant	+1
	5	Y = 24.58 + 5.14X	1	+1
	6	Y = 18.86 + 13.18X	1	+1
	7	Y = 38.83 + 13.44X	Highly Significant	+1
	8	Y = 15.90 + 5.52X	1	+1
	9	Y = 34.47 + 8.64X	1	+1
subhumid region	10	Y = 115.20 + 13.59X	Highly Significant	+1
	11	Y = 35.32 + 22.05X	Highly Significant	+1
	12	Y = 116.45 + 19.94X	Highly Significant	+1
	13	Y = 26.34 + 14.91X	1	+1
	14	Y = 28.38 + 9.45X	1	+1
	15	Y = 21.5 + 7.6X		
	16	Y = 40.33 + 137.5X		
	17	Y = 76.5 + 41.3X		
	18	Y = 6.75 + 4.55X		
	19	Y = 22.75 + 6.00X		
	20	Y = 493.52 - 12.15X		
	21	Y = 97.75 + 12.15X		
	22	Y = 3690.51 + 12.15X		
	23	Y = 127.1 + 12.9X		
	24	Y = 54.5 + 13.9X		
	25	Y = -266 + 146.4X		
	26	Y = 5.37 + 1.60X		
	27	Y = 15.46 + 9.33X		
	28	Y = 8.82 + 2.89X		
	29	Y = 8.37 + 1.88X		
subhumid region	30	Y = 8.37 + 1.88X		
	31	Y = 23.3 + 6.55X		
	32	Y = -57.76 + 38.51X		
	33	Y = 5.76 + 3.28X		
	34	Y = 14.49 + 3.40X		
	35	Y = 22.36 + 10.91X		
Semi-arid region	36	Y = 53.30 + 10.45X	Highly Significant	+1
	37	Y = 18.40 + 11.08X	Highly Significant	+1
	38	Y = 322.47 + 8.86X	Highly Significant	+1
	39	Y = 54.00 + 6.90X	10	+1
	40	Y = 69.31 + 13.19X	Highly Significant	+1
	41	Y = - 158.53 + 82.36X	5	+1
	42	Y = - 53 + 39.1X		

	43	Y = -937.7 + 334.7X		
	44	Y = 13.0 + 7.95X		
	45	Y = 138.04 + 6.51X		
	46	Y = 96.00 d+ 54.75X		
	47	Y = 29.75 + 11.1X		
	48	Y = 10.24 + 23.28X		
	49	Y = 83.0 + 60.5X		
	50	Y = 158.5 - 2.4X		
	51	Y = 614.6 + 258.0X		
	52	Y = 10.25 + 3.10X		
Transitional zone	53	Y = 14.85 - 5.71X	Non signi- ficant	+1
	54	Y = 50.48 + 7.77X	1	+1
	55	Y = 15.24 + 10.13X	Highly Si- gnificant	+1
	56	Y = 28.39 + 4.54X	Highly Si- gnificant	+1
Arid region	57	Y = 19.47 + 4.26X	1	+1
	58	Y = 26.60 + 6.90X	Highly Si- gnificant	+1
	59	Y = 27.00 + 7.20X	Highly Si- gnificant	+1
	60	Y = - 54.03 + 29.36X	2	+1
	61	Y = 8.13 + 3.95X	Highly Si- gnificant	+1
	62	dY = 12.74 + 5.23X	1	+1

TABLE 7.13 **Total stream length V/s stream order (Third order basins)**

Climatic Region	Basin Number	Regression Equation	Level of significance %	Correlation Coefficient
1	2	3	4	5
Humid region	1	Y = 17.62 - 3.73X	10	-1
	2	Y = 17.07 - 4.00X	20	-0.5
	3	Y = 35.37 - 9.73X	10	-1
	4	Y = 17.01 - 4.36X	10	-1
	5	Y = 39.36 - 10.39X	10	-1
	6	Y = 17.62 - 5.85X	10	-1
	7	Y = 20.90 - 6.00X	10	-1
subhumid region	8	Y = 12.21 - 2.36X	1	-1
	9	Y = 18.87 - 4.85X	10	-0.5
	10	Y = 17.62 - 5.73X	10	-0.5
	11	Y = 14.00 - 3.62X	10	-0.5
	12	Y = 28.00 - 7.00X		
	13	Y = 8.50 - 0.25X		
	14	Y = 22.12 - 6.48X		
	15	Y = 23.00 - 7.00X		
	16	Y = 52.62 - 15.48X		
	17	Y = 20.00 - 4.75X		
	18	Y = 17.31 - 4.49x		
	19	Y = 6.79 - 1.42X		
	20	Y = 4.70 - 1.06X		
	21	Y = 4.17 - 0.94X		
	22	Y = 6.62 - 0.98X		
	23	Y = 19.10 - 5.19X		
	24	Y = 12.66 - 2.34X		
	25	Y = 6.78 - 1.60X		
	26	Y = 6.16 - 1.73X		
	27	Y = 12.26 - 4.08X		
	28	Y = 9.53- 2.82X		
semi-arid region	29	Y = 26.20 - 8.27X	10	-1
	30	Y = 24.31 - 7.99X	10	-1
	31	Y = 12.12 - 3.48X	10	-1
	32	Y = 16.12 - 4.85X	10	-1
	33	Y = 20.50 - 6.25X	1	-1
	34	Y = 19.00 - 4.25X	20	-0.5
	35	Y = 31.06 - 5.99X	5	-1
	36	Y = 11.12 - 1.98X		
	37	Y = 25.00 - 7.00X		
	38	Y = 14.31 - 2.99X		
Transitional zone	39	Y = 23.81 - 7.24X		
	40	Y = 9.67 - 2.08X	Non significant	-0.5
	41	Y = 13.31 - 3.49X	10	-1
	42	Y = 9.60 - 1.90X	5	-1
	43	Y = 7.61 - 1.34X	Non significant	-0.5
	44	Y = 7.87 - 1.60X	20	-0.5
Arid region	45	Y = 12.00 - 2.25X	20	-0.5
	46	Y = 8.12 - 1.85X	20	-0.5
	47	Y = 15.61 - 3.84X	20	-0.5
	48	Y = 10.25 - 2.62X	5	-0.5
	49	Y = 6.87 - 1.60X	1	-1
	50	Y = 7.31 - 2.49X	10	-1
	51	Y = 17.12 - 5.35X	10	-1
	52	Y = 13.22 - 3.60X	10	-.05
	53	Y = 11.37 - 2.48X	20	-0.5
	54	Y = 15.00 -4.00 X	10	-1

TABLE 7.14 **Total Stream Length V/s Stream Order (Random Order Basins)**

Climatic Region	Basin Number	Regression Equation	Level of significance (%)	Correlation Coefficient
1	2	3	4	5
Humid region	1	Y = 128.78 - 27.42X	1	-1
	2	Y = 175.12 - 44.22X	5	-0.9
	3	Y = 83.75 - 18.05X	1	-1
	4	Y = 146.34 - 31.30X	1	-1
	5	Y = 31.08 - 8.06X	1	-1
	6	Y = 38.56 - 8.34X	1	-1
	7		Y = 52.83 -	
	1	-0.8		
	8	Y = 23.12 - 5.63X	Highly Significant	-1
	9	Y = 43.82 - 10.85X	1	-0.8
Subhumid region	10	Y = 97.09 - 18.97X	1	-1
	11	Y = 64.17 - 14.00X	Highly significant	-1
	12	Y = 117.02 - 24.97X	2	+0.9
	13	Y = 45.28 - 10.27X	1	-1
	14	Y = 38.70 - 8.91X	Highly significant	-0.8
	15	Y = 29.0 - 6.4X	-	-
	16	Y = 57.7 - 11.0X	-	-
	17	Y = 145.5 - 34.9X	-	-
	18	Y = 11.87 - 2.3X		
	19	Y = 28.5 - 6.7X		
	20	Y = 29.25 - 6.5X		
	21	Y = 54.75 - 13.3X		
	22	Y = 51.00 - 10.55X		
	23	Y = 56.55 - 11.35x		
	24	Y = 67.97 - 16.14X		
	25	Y = 79.00 - 19.50X		
	26	Y = 7.11 - 1.68X		
	27	Y = 23.15 - 3.53X		
	28	Y = 72.87 - 2.23X		
	29	Y = 10.67 - 2.72X		
	30	Y = 13.7 - 3.4X		
	31	Y = 158.6 - 49.75X		
	32	Y = 31.91 - 8.28X		
Subhumid region	33	Y = 9.84 -2.13X		
	34	Y = 19.51 - 5.16X		
	35	Y = 37.4 - 8.62X		
Semi-arid region	36	Y = 57.15 - 12.40X	1	-1
	37	Y = 33.00 - 7.28X	Highly significant	-1
	38	Y = 32.72 - 7.44X	2	-1
	39	Y = 52.97 - 12.24X	1	-1
	40	Y = 72.95 - 15.70X	1	-1
	41	Y = 65.50 - 16.40X	Highly Significant	-1
	42	Y = 58.2 - 12.9X	-	-
	43	Y = 38.7 - 7.1X	-	-
	44	Y = 22.00 - 4.3X		
	45	Y = 93.8 - 14.4X		
	46	Y = 12.5 - 5.7X		
	47	Y = 41.5 - 10.15X		

	48	Y = 43.5 - 8.7X		
	49	Y = 53.00 - 12.85X		
	50	Y = 130.00 - 34.00X		
	51	Y = 100.55 - 20.65X		
	52	Y = 14.00 - 3.45X		
Transitional zone	53	Y = 21.25 - 4.80X	1	-1
	54	Y = 49.62 - 10.72X	1	-0.7
	55	Y= 25.33 - 4.56X	5	-0.4
	56	Y = 28.22 - 6.09X	1	-0.9
Arid region	57	Y = 24.72 - 6.34X	1	-0.8
	58	Y = 29.65 - 5.95X	1	-0.9
	59	Y = 36.70 - 9.31X	2	-1
	60	Y = 12.98 - 0.67X	5	+0.2
	61	dY = 12.50 - 2.70X	20	-1
	62	Y = 19.36 - 4.52X	1	-1

TABLE 7.15 **Values of Correlation Coefficients (Third Order Basins)**

Climatic Region	1	2	3	4	5	6	7	8	9
Humid region	0.904	0.927	0.574	0.961	0.927	0.647	0.894	0.993	0.905
Subhumid region	0.990	0.541	0.545	0.822	0.882	0.669	0.997	0.996	0.799
Semi-arid region	0.993	0.630	0.659	0.940	1.126	0.615	0.995	0.998	0.901
Transitional zone	0.904	0.813	0.603	0.790	0.850	0.470	0.973	0.966	0.715
Arid region	0.740	0.601	0.853	0.895	0.756	0.533	0.889	0.796	0.715

Climatic Region	10	11	12	13	14	15	16	17
Humid region	0.833	0.530	0.982	0.882	0.955	0.942	0.613	0.697
Subhumid region	0.004	0.850	0.965	1.056	0.660	0.710	0.555	0.904
Semi-arid region	0.871	0.640	1.040	0.514	0.624	0.578	0.565	0.896
Transitional zone	0.974	0.360	0.998	0.896	0.844	0.758	0.621	0.928
Arid region	0.519	0.565	0.558	0.648	0.558	0.598	0.556	0.855

Note:
Abbreviations used:

1. Basin area v/s basin length
2. Relief ratio v/s texture ratio
3. Circularity ratio v/s elongation ratio
4. Drainage density v/s stream frequency
5. Texture ratio v/s constant of channel maintenance
6. Elongation ratio v/s relief ratio
7. Basin area v/s basin parameter
8. Basin length v/s basin-parameter
9. Drainage density v/s texture ratio
10. Relief ratio v/s circularity ratio
11. Elongation ratio v/s constant of channel maintenance
12. Ruggedness number v/s relief ratio
13. Ruggedness number v/s constant of channel maintenance
14. Ruggedness number v/s texture ratio
15. Ruggeness number v/s stream frequency
16. Relief ratio v/s bifurcation ratio
17. Texture ratio v/s bifurcation ratio

TABLE 7.16 **Values of Correlation Coefficients (Random Order Basins)**

Climatic Region	1	2	3	4	5	6	7	8	9
Humid region	0.880	0.362	0.564	0.955	0.635	0.444	0.945	0.930	0.584
Subhumid region	0.942	0.976	0.647	0.992	0.192	0.909	0.943	0.996	0.993
Semi-Arid region	0.882	0.489	0.577	0.667	0.763	0.713	0.956	0.857	0.663
Transitional zone	0.732	0.855	0.691	0.952	0.944	0.889	0.599	0.987	0.982
Arid region	0.994	0.858	0.709	0.990	0.922	0.623	0.995	0.993	0.911

Climatic Region	10	11	12	13	14	15	16	17
Humid region	0.863	0.477	0.984	0.732	0.403	0.950	0.537	0.886
Subhumid region	0.700	0.736	1.000	0.881	0.982	0.986	0.975	0.956
Semi-arid region	0.792	0.660	0.951	0.856	0.429	0.664	0.759	0.628
Transitional zone	0.692	0.889	0.992	0.916	0.891	0.575	0.994	0.777
Arid region	0.912	0.362	1.034	0.823	0.590	0.946	0.954	0.781

Note :
Abbreviation used :

1. Basin area v/s basin length
2. Relief ratio v/s texture ratio
3. Circularity ratio v/s elongation ratio
4. Drainage density v/s stream frequency
5. Texture ratio v/s constant of channel maintenance
6. Elongation ratio v/s relief ratio
7. Basin area v/s basin perimeter
8. Basin length v/s basin perimeter
9. Drainage density v/s texture ratio
10. Relief ratio v/s circularity ratio
11. Elongation ratio v/s constant of channel maintenance
12. Ruggedess number v/s relief ratio
13. Ruggedness number v/s constant of channel maintenance
14. Ruggedness number v/s texture ratio
15. Ruggedness number v/s stream frequency
16. Relief ratio v/s bifurcation ratio
17. Texture ratio v/s bifurcation ratio

TABLE 7.17 **Drainage density and Area.**

Drainage Density in km/km^2	Area as a Percentage of Total Area of Rajasthan
Area without drainage	33.71
less than 0.20	34.83
0.21-0.40	27.22
0.41-0.60	3.08
0.61-0.80	0.40
0.81-1.00	0.40
1.01-1.20	0.34
1.21- and above	0.32
Total	100.00

TABLE 7.18 **Stream Frequency and Percentage of Area**

Stream Frequency In km/km^2	Area as Percentage of Total Area of Rajasthan
Area without drainage	33.71
less than 0.02	20.91
0.03 - 0.04	11.67
0.05 - 0.06	5.37
0.07 - 0.08	6.67
0.09 - 0.10	8.50
0.11 - 0.12	6.51
0.13 - 0.14	3.96
9.15 and above	3.10
Total	100.00

Table 7.19 **Dissection Index and Area**

Dissection Index	Area as a Percentage out of Total area of Rajasthan
Less than 0.20	28.26
0.21 - 0.40	39.66
0.41 - 0.60	24.55
0.61 - 0.80	6.38
0.81 and more	1.15

TEXTURE RATIO

TABLE 7.20 **Relative Relief and area**

Relative Relief (m)	Area as a Percentage of Total Area of Rajasthan
Less than 150	64.65
151 - 300	20.38
301 - 450	8.81
401 - 600	3.75
601 - 750	1.24
751 - 900	0.77
901 - 1050	0.20
1051 and more	0.20
Total	100.00

TABLE 7.21 **Slope and Area**

Slope in Degree	Area as a Percentage of Total Area of Rajasthan
Less than 2°	84.78
2° - 4°	8.82
4° - 6°	3.31
6° - 8°	1.04
8° -10°	1.08
10° and more	0.97
Total	100.00

Table 7.22 **Quantitative Analysis of Fluvial Morphometry average Slope in Degrees (Third Order Basins)**

Climatic Region	Basin Number	Slope in Degrees
Humid region	1	11.16
	2	6.45
	3	72.07
	4	-
	5	32.98
	6	-
	7	-
Subhumid region	8	15.93
	9	33.62
	10	7.97
	11	20.39
	12	1.81
	13	7.76
	14	20.83
	15	-
	16	52.52
	17	-
	18	49.22
	19	16.95
	20	12.85
	21	6.05
	22	8.20
	23	-
	24	-
	25	5.75
	26	23.35
	27	-
	28	-
Semi-arid region	29	76.15
	30	-
	31	25.76
	32	38.67
	33	69.55
	34	24.78
	35	10.12
	36	12.75
	37	36.20
	38	2.59
	39	29.15
Transitional zone	40	12.96
	41	21.52
	42	52.54
	43	-
	44	48.47
Arid region	45	17.45
	46	15.24
	47	18.60
	48	24.99
	49	20.30
	50	15.37
	51	25.27
	52	21.21
	53	-
	54	-

TABLE 7.23 **Quantitative Analysis of Fluvial Morphometry Average Slope in Degrees (Random Order Basins)**

Climatic Region	Basin Number	Slope in Degrees
Humid region	1	51.34
	2	23.63
	3	29.40
	4	19.10
	5	45.90
	6	17.68
	7	11.85
	8	28.05
	9	19.89
Subhumid region	10	72.71
	11	9.68
	12	30.08
	13	9.12
	14	3.25
	15	88.31
	16	36.11
	17	23.37
	18	20.48
	19	16.77
	20	28.96
	21	-
	22	53.05
	23	43.43
	24	35.00
	25	38.05
	26	12.93
	27	-
	28	-
	29	9.13
	30	17.00
	31	17.5
	32	-
	33	-
	34	-
	35	-
Semi-arid region	36	48.85
	37	27.44
	38	13.84
	39	26.76
	40	38.80
	41	56.23
	42	-
	43	6.08
	44	49.02
	45	35.44
	46	43.99
	47	59.40
	48	43.73
	49	40.06
	50	46.49
	51	63.50
	52	-
Transitional zone	53	6.96
	54	65.68
	55	10.14
	56	62.94
Arid region	57	47.09
	58	20.20
	59	30.65
	60	60.72
	61	24.76
	62	20.42

TABLE 7.24 **Relation between Climatic Variable and Morphometric Parameters (Third Order Basins)**

S.No.	Parameters	Regression Equation	Correlation Coefficient
1	Mean annual rainfall v/s Stream frequency	Y=706.47-20.42x	0.5316
2	Mean annual rainfall v/s Circularity ratio	Y=617.16+56.31x	0.6066
3	Mean annual rainfall v/s Constant of channel maintenance	Y=625.97+22.68x	0.3099
4.	Mean annual rainfall v/s Drainage density	Y=878.30-118.36x	0.5416
5	Mean annual rainfall v/s relief ratio	Y=661.05-163.97x	0.3876
6	Mean annual rainfall v/s Texture ratio	Y=647.27-4.01x	0.6050
7	Mean annual rainfall v/s Elongation ratio	Y=442.00+270.26x	0.4875
8	Mean annual rainfall	Y=649.38-0.023	0.5211

TABLE 7.25 **Relation between Climatic Variables and Morphometric Parameters (Random Order Basins)**

S.No.	Parameters	Regression Equation	Correlation Coefficient
1	Mean annual rainfall v/s Stream frequency	Y=652.62+16.08x	0.6477
2	Mean annual rainfall v/s Circularity ratio	Y=834.79-204.37x	0.4161
3	Mean annual rainfall v/s Constant of channel maintenance	Y=663.38+39.49x	0.5143
4	Mean annual rainfall v/s Drainage density	Y=649.56+24.84x	0.6337
5	Mean annual rainfall v/s Relief ratio	Y=625.65+611.52x	0.5279
6	Mean annual rainfall v/s Texture ratio	Y=631.22+15.23x	0.4815
7	Mean annual rainfall v/s Elongation ratio	Y=536.40+191.93x	0.4392
8	Mean annual rainfall v/s Average sloope	Y=731.02-1.51x	0.4939
9	Mean annual rainfall v/s Total runoff	Y=368.22+0.0198	0.5254

TABLE 7.26 **Total Runoff from different river catchments in Rajasthan**

S. No.	Name of the Basin	Total Runoff from catch-ment in	S. No.	Name of the Basin	Total Runoff from catch-ment in
1	Kantli	1720	31	Kali Sindh	18432
2	Dohan	1115	32	Chhotti Kali Sindh	1616
3	Nawalgarh Nalla	150	33	Jakham	24610
4	Krishnawati	340	34	Son	66920
5	Ranali	240	35	Moran	3561
6	Sabi	5490	36	Ahas	62433
7	Baran	5193	37	Mahi	110780
8	Banganga	14620	38	Bhadan Velirak	4845
9	Gambhir	14010	39	Wakal	14117
10	Parbati	4000	40	Sei	1833
11	Mendha	3205	41	Sabarmati	1833
12	Mashi	6960	42	Jogri	1804
13	Morel	13540	43	Luni-I	3375
14	Dhil	1470	44	Luni-II	2637
15	Kali Sindh	1952	45	Luni-III	2009
16	Sodra	4244	46	Guhiya	3017
17	Dai		47	Sukri Hemawas	1387
18	Khari	14830	48	Bandi Hemawas	1443
19	Kothari	6980	49	Sukli (Bankli to Luni	1063
20	Banas River-I	18185	50	Sukri (upto Bañkli)	1461
21	Banas-II	29190	51	Mithri	1236
22	Berach	25930	52	Jawai	5704
23	Chambal	16630	53	Khari upto Luni	2821
24	Chakhan	1640	54	Sukri Upto Luni	1535
25	Mej	20550	55	Bandi	883
26	Eru	2010	56	Sabi	316
27	Alma	3230	57	Western Banas	7974
28	Parbati	31780	58	Sukli	2047
29	Kunu	4014	59	Other Nallas of Jalore	444
30	Parwan	14637		Total	648237

TABLE 7.27 **Hydrological data of Different Sites in Rajasthan**

Site Basin	Discharge No.	(cu. metre/ sec in cumecs)	Runoff (M. C. metre)	Total Silt Load (hectare /metres)
Berach of	8	1424.07	395.07	35.544
Chittaurgarh	11	-	-	-
Dhund at	13	-	-	-
Kanota	17	227.00	31.41	71.10
Banganga at	16	-	-	-
Hingota	19	1464.30	57.76	21.66
Sabi at	17	610.32	114.17	36.50
Sodawas	20	-	-	-
Jakham at	9	-	-	-
Dhariawad	12	2551.43	516.03	19.54
Ghambir at shree	16	-	-	-
Mahavirji	19	735.60	812.00	233.27
Parwan at Manohar	1	1528.72	242.60	102.31
Thana	4	-	-	-
Banas at	14	-	-	-
Negaria	18	3032.15	1481.16	127.61

TABLE 7.28 **Sinuosity Index and Climate**

Stream Segment Measurement	Channel Length in km"X"	Direct length in km "Y"	Tortuosity Index* $\frac{X}{Y}$ (Leopold 1964)	Stream Tortuosity $\frac{X-Y}{Y} \times 100$ (Smithson's, 1939)
ARID REGION (THE LUNI BASIN)				
I	31.0	21.0	1.47	47.61
II	0.25	7.5	1.10	10.0
III	10.00	8.0	1.25	25.00
IV	27.75	21.5	1.29	29.06
V	20.5	18.5	1.10	10.81
VI	8.0	7.25	1.10	10.34
VII	13.0	8.5	1.52	52.94
VIII	16.0	13.0	1.23	23.07
IX	26.0	23.0	1.13	13.04
X	33.25	29.0	1.14	14.65
Total	193.75	157.25	12.33	236.52
Mean	19.37	15.72	1.23	23.65
SEMI-ARID REGION (THE GAMBHIR BASIN)				
I	27.00	15.50	1.74	74.19
II	46.50	28.25	1.64	64.60
III	41.25	27.00	1.52	52.77
IV	12.50	7.50	1.66	66.66
V	23.25	20.50	1.13	13.49
VI	12.00	8.75	1.37	37.14
VII	36.00	26.00	1.38	38.46
VIII	35.50	25.50	1.39	39.21
IX	20.25	16.00	1.26	26.56
X	31.00	23.50	1.31	31.91
XI	9.50	6.50	1.46	46.15
Total	294.75	205.00	15.86	491.06
Mean	26.79	18.63	1.44	44.64
SUBHUMID REGION (THE BERACH BASIN)				
I	25.00	19.00	1.31	31.57
II	34.00	27.50	1.23	23.63
III	21.00	19.00	1.10	10.52
IV	9.50	9.00	1.05	5.55
V	29.50	25.50	1.15	15.68
VI	32.00	29.50	10.08	8.47
VII	10.00	9.50	1.05	5.26
VIII	33.50	30.50	1.09	9.83
IX	4.00	3.50	1.14	14.28
Total	198.5	173.00	10.20	124.79
Mean	22.05	19.22	1.13	13.86
SUBHUMID REGION (THE MEJ BASIN)				
I	1.65	2.25	1.36	36.36
II	1.10	1.30	1.19	18.18
III	2.25	3.00	1.33	33.34
IV	0.75	1.85	2.45	146.67
V	0.60	1.50	2.50	150.00

VI	1.00	2.35	2.35	135.00
VII	1.00	2.60	2.60	160.00
VIII	1.19	2.12	1.95	78.15
Total	9.54	16.97	15.73	757.70
Mean	1.19	2.12	1.96	94.71
HUMID REGION (THE CHAMBAL AND MAHI BASINS)				
I	44.50	24.00	1.85	85.41
II	28.00	19.25	1.45	45.45
III	33.25	25.75	1.29	29.12
IV	40.25	28.25	1.42	42.47
V	31.00	26.75	1.15	15.88
VI	26.50	24.25	1.09	9.27
VII	24.50	22.00	1.11	11.36
VIII	33.50	28.50	1.17	17.54
IX	38.00	22.50	1.68	68.88
X	42.00	32.50	1.29	29.23
XI	31.50	25.75	1.22	22.33
XII	50.75	30.509	1.66	66.39
Total	423.75	310.00	16.38	443.33
Mean	35.31	25.83	1.36	36.94

EIGHT

Morphology and Origin of Ravines

The problem of ravine erosion has received much attention in tropical countries. The loss of productive croplands and valuable grazing areas, high rates of sedimentation in areas of down stream from ravines, the creation of new channels that were difficult to dam for irrigation and declining water tables resulting from channel entrenchment provided economic incentives for over two decades of research aimed at the problem of ravine origin in India.

The problem of origin of ravines has generated more interest for geomorphologists and soil and water conservationists than any other research problem. In the last two decades or so significant progress in the development of cumulative generalization has been made that may led to general theory formulation and the time has now come to review the recent trends of research in ravine geomorphology mainly in tropical regions.

Ravines, are in a sense, geomorphological curiosity. Their geomorphological evolution is by no means so simple as has sometimes been assumed. Observers view ravines as anomalous features of the landscape and their existence indicate instability of landsurface and landslope.

Research into the origin and dynamics of ravines has been significant in the science of geomorphology because the subject of ravine embodies many of the important consideration of the wider field. The geomorphic response to general environmental changes, the nature of connections between geomorphic and hydroclimatic systems, the distribution of the processes of erosion, transportation and deposition and the chronology of adjustment are foci of investigative interest shared by the specific research into ravine and the general endeavours of geomorphological work (Graf, 1983).

RAVINES

The word "ravine" denotes gullied lands containing systems of gullies running more or less parallel to each other. In the Amercian south west, the term arroyo is generally used, which is synonimous to Ravine. An arroyo is defined as a "trench with a roughly rectangular cross section excavated in valley bottom alluvium with a major stream channel on the floor of the trench" (Graft, 1983). "A gully is a recently extended drainage channel that transmit ephemeral flow, has steep sides and head scarp, a width greater than 1 ft. depth greater than 2 ft." (Brice, 1966).

A ravine is a channel of ephemeral flow denuded and guided essentially by the process of rejuvenated streams, has steep side and head scraps, width and depth always greater than a gully (Sharma, 1968).

Normally the gully used to be a current word for ravine, but its modern usage related to the fact that ravines are larger than gullies and the characterization that gully initiates along the animal trails, roads, hill slopes and path on the agricultural uplands. On the contrary, a ravine forms along the river side and encroaches upon the catchment area by scarp recession undergoing various phases of evolution like swallow holing, piping, collapsing and recessing. Thus, ravine formation is a process of major river erosion and gully formation is a function of catchment area.

DISTRIBUTION OF RAVINES IN RAJASTHAN

The National Remote Sensing Agency, Hyderabad on the basis of landsat data (1985) has estimated nearly 4.3 million hectares of land under ravines in India. Out of this .4 million hectares are found in Rajasthan. The ravine lands of Rajasthan are mainly concentrated in the eastern Rajasthan mainly along the Chambal, Kali Sindh, Parvati, Parwan, Alnia, Talera, Ghora Pachar, Mangali, Mej, Morak, Banas, Banganga, Sabi, Mahi, Gambhir, Sota, Parwati (Bharatpur) and other inland streams in the piedmont areas of the Aravallies (Fig. 8.1). In the north eastern Rajasthan around Jaipur, Thana Gazi, Achroal, Benada, Newai and Lalsot ravines are found developing on the stablised obstacle dunes.

In order to have the better appraisal of the problem of ravine erosion a detailed description of the physiographic features of the Chambal region is given in the following pages.

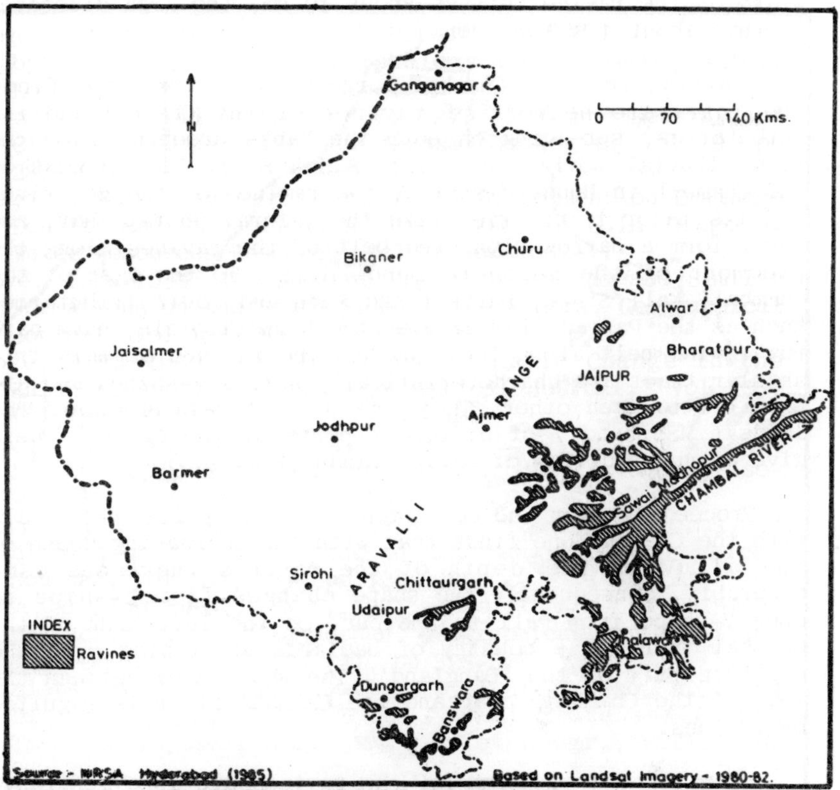

RAVINES LANDS OF RAJASTHAN

FIGURE 8.1

Ravines of the Chambal and Its Tributaries

The Chambal is one of the major tributaries of the Yamuna. The belt of ravine along the Chambal stretches from Kota city to the confluence of the Chambal with the Yamuna over a distant of about 482 km. The ravines extend along the banks and are broadly separated by the watershed of the rivers on both sides. From Kota to Dholpur, the Bundi-Ranthambhor line of hills and beyond Dholpur, the ravines of the Choti Parbati form the northern boundary of the Chambal ravines. The southern boundary fluctuates with the tributaries like the Kali-Sindh and Parbati. Further on, the south eastern boundary of the Chambal ravines is well marked by the ravines of the Kunwari river. Taking the average width of the ravine belt as about 10 km, the area dissected is thus about 4,820 sq. km.

Starting from the Chambal gorge downstreams, one finds the intricate network of ravines of the Chambal and its tributaries, spreading on both the banks creeping into the rich alluvial soil (Figs. 8.2, 8.3 & 8.4). To the west of the Chambal in Bundi district, the ravines of the Mej river and its major tributaries like the Talera. Ghorapachar, and Kural form a narrow elongated belt of the ravines along the piedmont of the northern Bundi hills. To the east of the Chambal, Kali Singh, Parbati and Kunu and their tributaries such as the Parwan flowing over the Harwati plain, have narrow ravine belt along their banks. The ravines of this triangular tract are characteristically narrow V-shaped and are parallel to each other. They are 3 to 4 metres deep. The landsat MSS band 7 of different parts of the chambal Basin give a synoptic view of ravine lands (Fig. 8.5).

Proceeding from the confluence of the Banas near Pali with the Chambal one finds that with the increasing depth of the alluvium the depth of the ravines increases considerably. Consequently the shape changes from V-shape to open V-shape from Pali to Dholpur. On the left bank of the Chambal there is a country of badlands stretching over the southern part of the 'Dangland'. The whole area between the foot of the Chambal Scrap and the Chambal river is engulfed by ravines.

From Dholpur downwards, the ravines are more destitute and are bare of vegetation. They are characteristically V-shaped with vertical walls and flat bottoms. The depth varies between 30 to 60 metres. This dissected tract is a long narrow strip between the Chambal and the Yamuna, and about half the area is occupied by the deep and far extending ravines of the Yamuna. They eat their way from the banks into the adjacent areas along the Chambal, dissected into numerous platforms, knolls and flood plain bluffs. Towards west the belt of ravines is nearly 3 km on either sides of the Chambal but on the east they almost unit with the ravines of the Yamuna and make a bewildering of ravines and gullies.

The whole country between Bhind and Etawah is mostly dissected, irregular and is made up of steep rides, low sloping hills, deep trenches and broad cut incised meanders. During the rainy season these streams become flooded and are a formidable barrier to crossing. The area in the neighbourhood of the junction of the Yamuna and the Chambal represents perhaps as wild and picturesque view as is no where to be found in the plains. As far as eye can reach one can see here dissected ravines and green valley of accasia trees. Besides, the ravines in this zone comprise so large an area of the total land that little area fit for agriculture has been left out.

Fig. 8.2 : Landsat MSS band 7 showing, alwar hills in the north western parts, meandering yamuna in the north east and gullies of the gambher, Banganga and other inland streams in Alwar hills.

Almost the entire zone of Chambal ravine is underlain by upper vindhyan (lower Bhander) sirbu shales. These shales are generally soft, thinly laminated and fissile. There are several silty bands. The most important intercalation is of limestones which are stromalotitic at several places. The shales itself is calcareous forming partings of various horizons giving rise to Kankar Pan. The nature and composition of soil is dependent on underlying lithology from which it is derived. It is known that calcareous shales will have low permeability and high porosity but presence of Kankar will increase the permeability. Silty bands will have higher porosity and help in greater runoff. Thus, the consolidation due to calcareous cementing in this semi arid to sub humid climate causes vertical joint formation and block slippage during rains and floods in the Yamuna-Chambal ravine zones.

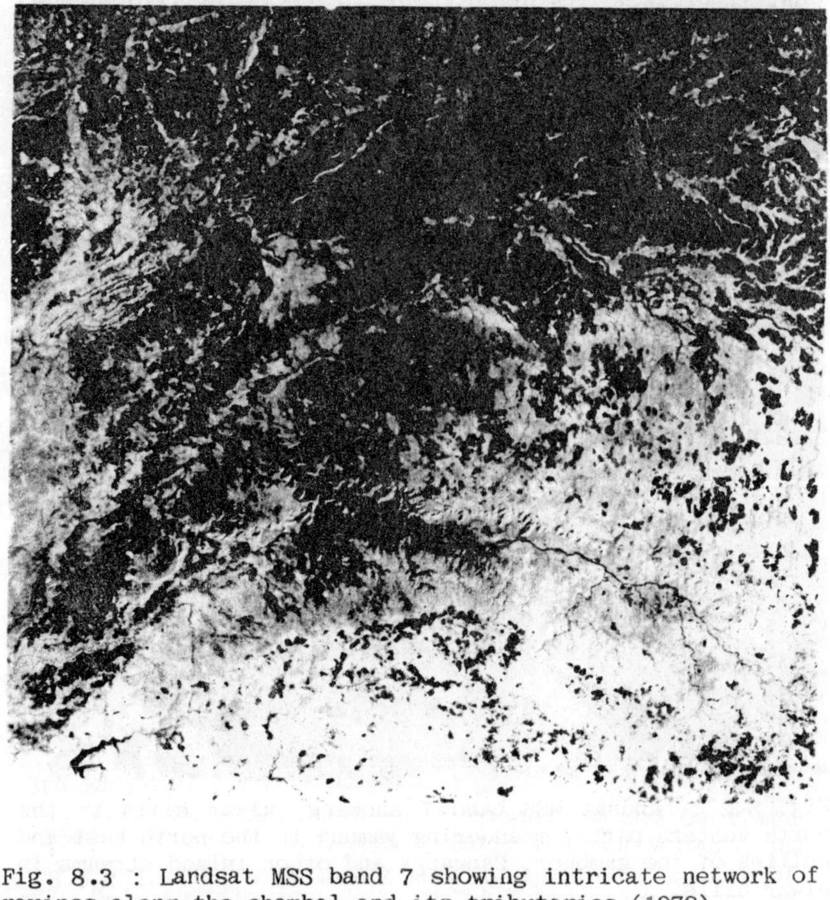

Fig. 8.3 : Landsat MSS band 7 showing intricate network of ravines along the chambal and its tributaries (1972).

In the Chambal region it has been observed that a characteristic feature of ravine on both sides of the chambal is extensive formation of badlands which evidently due to headword erosion along the gullies. But very interestingly the ravines also show a highly dissected features with steep pinnacled remnants. The shales normally form gentle slopes because of its easy erodibility. This can be attributed to the lime content in the sirbu shales and it is this that make this shale very hard due to cementation during the dry season. During the monsoon, when the water level is higher, the ravines get flooded and water enters into the pores of weathered sirbu shales. When the water withdraws, 'blocks' of the shales collapse and thus, there is headword erosion, steep scarp continues and the valley walls of the ravine remain steep. The ravine eats into its sides, due to headward erosion.

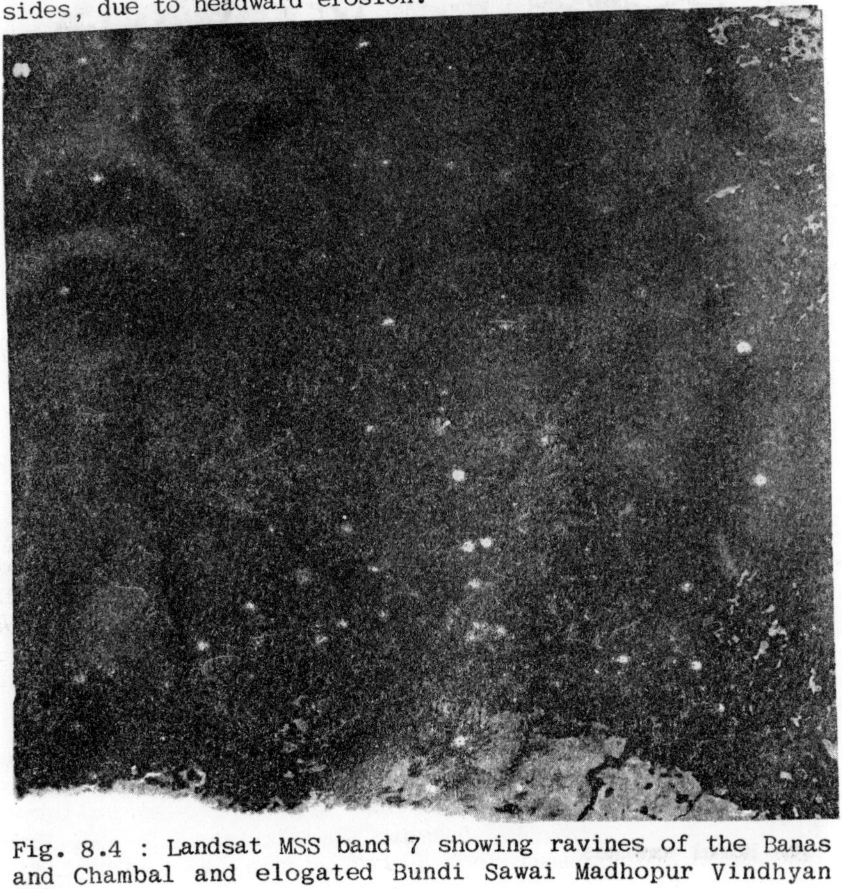

Fig. 8.4 : Landsat MSS band 7 showing ravines of the Banas and Chambal and elogated Bundi Sawai Madhopur Vindhyan hills.

The Mej Ravines

The Mej is a left bank tributary of the Chambal. The middle and lower course of the Mej is also characterized by the presence of ravines. The ravines along this river are evidents in its lower course from Seakoda village upstream. The tributaries like Talera, Mangli, Kural, and Ghorapachad in Bundi district are also attended with network of ravines.

Fig. 8.5 : Landsat MSS band 7 showing ravine lands of middle Chambal and Tripple Scarp and Plateaux (1972)

The Morel Ravines

Among various streams of the Chambal Basin, the Morel river is also characterised by ravines and gullies. Generally the banks of the Morel and its tributaries have been dissected into ravines. Broadly three compact ravine areas may be observed in the Morel Basins. These are:

(i) along the Dhund river between height from 300-360 metres.
(ii) along the Morel and its tributaries between the Bassi-Zatwara hills and Baonli-Lalsot hills at an elevation from 300 - 360 metres.
(iii) along the lower reaches of the Morel river below 180 - 190 metres.

The Dhund is a tributary of the Morel. An intricate net work of ravines is developing along the Dhund between the Achroal hill and its confluence with Amanisah Nallah. The sand dunes of the western foot hills of Jaipur have also been dissected into ravines. Another compact belt of ravines appears along the Morel and its small tributaries. Ravines are most active along the streams that descend from the Lalsot hills. Thereafter, again a compact ravine belt occurs in the lower reaches of the Morel and its tributaries which descend from the scarp of the Vindhyan rocks.

SHAPE AND AREAL PATTERN OF RAVINES

The study of ravines and gullies in the Lower Chambal Valley, Mahi valley, and their tributaries shows that the size, areal pattern and rate of encroachment are mostly related to soil and its thickness. In the Chambal-Yamuna-Mahi zones where the soil is underlain by fine grained clay, which is resistant to rapid erosion, ravines are V-shaped. Such types of ravines have been observed especially along the Chambal and its tributary-the Morel. V-shaped ravines develop where the soil and sub-soils are commonly friable and easily cut by flowing water. Throughout the lower course of the Chambal, Mahi and Yamuna near Etawah and Agra ravines are of this type and shape.

The depth of ravines is dependent on the thickness of the sediments. Sharma (1968) working on the chambal ravines has analysed that there is a close relation between thickness of alluvium and depth of river and ravines. Analysing long and cross profiles of the Chambal alluvium, he studied that the thickness of the alluvium increases as one proceeds from Kota to the confluence with the Yamuna. The bedrock around Kota exposes down 7 to 9 metres, while downstream Dholpur it is found below 45 metres. From this sequence, it may be concluded that before the deposition of the alluvium, the bedrock surface of the valley might has been in the form of a disk or bowl sloping towards north east and so this form has been responsible for greater accummulation of sediments in the lower reaches. Thus the depth of river increases towards lower reaches. The river depth near Kota is 20 metres while beyond Dholpur it increases to 40-50 metres. Similarly, the depth of ravines also increases in the same way. This process indicates the reversel condition in the normal cycle of erosion. Hence, ravines are the result of rejuvenation of the Chambal and its tributaries.

No attempt has so far been made in India on the quantitative expression of ravine pattern. Sharma (1968) has tried to analyse the areal pattern in the Lower Chambal Valley. The study of ravine frequency of this area shows that the frequency varies from Kota to the Yamuna. In Kota area it is minimum and is from 1-5 per unit area. In Mangrol along the Parbati, it is between 5-10 with an exception of some patches where it is from 10-15. From Kota downstream the frequency increases, thereby suggesting the advanced and critical development of ravines.

For the above pattern Sharma (1968) has given two explanations:

1. When one cycle of erosion interrupts by the cymetogenic upwarp, the new cycle generally initiates first on the mouth of the rivers and proceeds slowly upstreams. In the case of the Chambal it may be concluded that after the termination of the tertiary cycle, the Pleistocene cycle was initiated first in the lower reaches and now it is progressing upstream, thereby suggesting the mature and deep ravines in the lower parts and vice versa.
2. Another possible explanation can also be given that the ravines are mainly controlled by the depth of the post-tertiary alluvium. The depth of alluvium increases downstream, hence the ravines get deeper and deeper in the lower reaches.

MECHANISM OF RAVINE GROWTH

Ravine follow a simple pattern of development which is characteristic in all locations of India. Almost invariably they are marked by vertical or nearly vertical headcut and banks which give them a typical rectangular cross section. Once established, the headcut advances, and the channels widen by sapping and undercutting the banks. Abrasion by flowing water, either at the falls or above, is a relatively minor cause of enlargement. The depth or ravine cutting and the gradient of the downstream channel vary widely, apparently dependent on a function of the flow, the character of the eroding sediments, and the slope of the valley floor. Depth of ravines and gullies ranging from few metres to 80 metres have been observed in all ravine infested areas.

Development in this way, ravines may cut for an indefinite distance without any disturbance of the surface, either laterally up upstream, beyond the confines of the channel. In most of the localities ravine cutting generally takes place at depths below the reach of plant roots, the presence of even dense vegetation above headcuts has little influence on the rate of advancement. Besides this, the development of ravines also takes place through swallow holes.

STAGES OF RAVINE GROWTH

Close study reveals that ravines follow a simple pattern of growth on the catchment area. Sharma (1968, 1976) has intensively studied the growth of gullies and ravines in the Lower Chambal Valley and the Morel Basin. He has distinguished four stages of ravine growth. These stages are generally observable in the clayey and clayey loam soils. According to him four stages are as follow (Fig 8.6).

Swallow Hole Stage

In the initial stage parallel swallow holes develop on the flat ground along the river by the mutual action of rain water and the soil particles. The development of swallow holes are more evident on the soil containing calcium carbonate. The kanker nodules in the Chambal Basin, Yamuna Valley and Mahi Valley provide more space to water to percolate into sub-soil.

Piping Stage

The second stage is the formation of pipe like holes. The development of pipes is an important element in the headward extension of many vertical cuts. Near ravine wall pipes often extend into the unravined alluvium for distance of several metres. Even at such distances swallow holes can often be seen.

Piping undoubtedly represents zone where water has passed from the unravined floor of the swallow holes down to the base of the vertical ravine wall along which the water

FIGURE 8.6

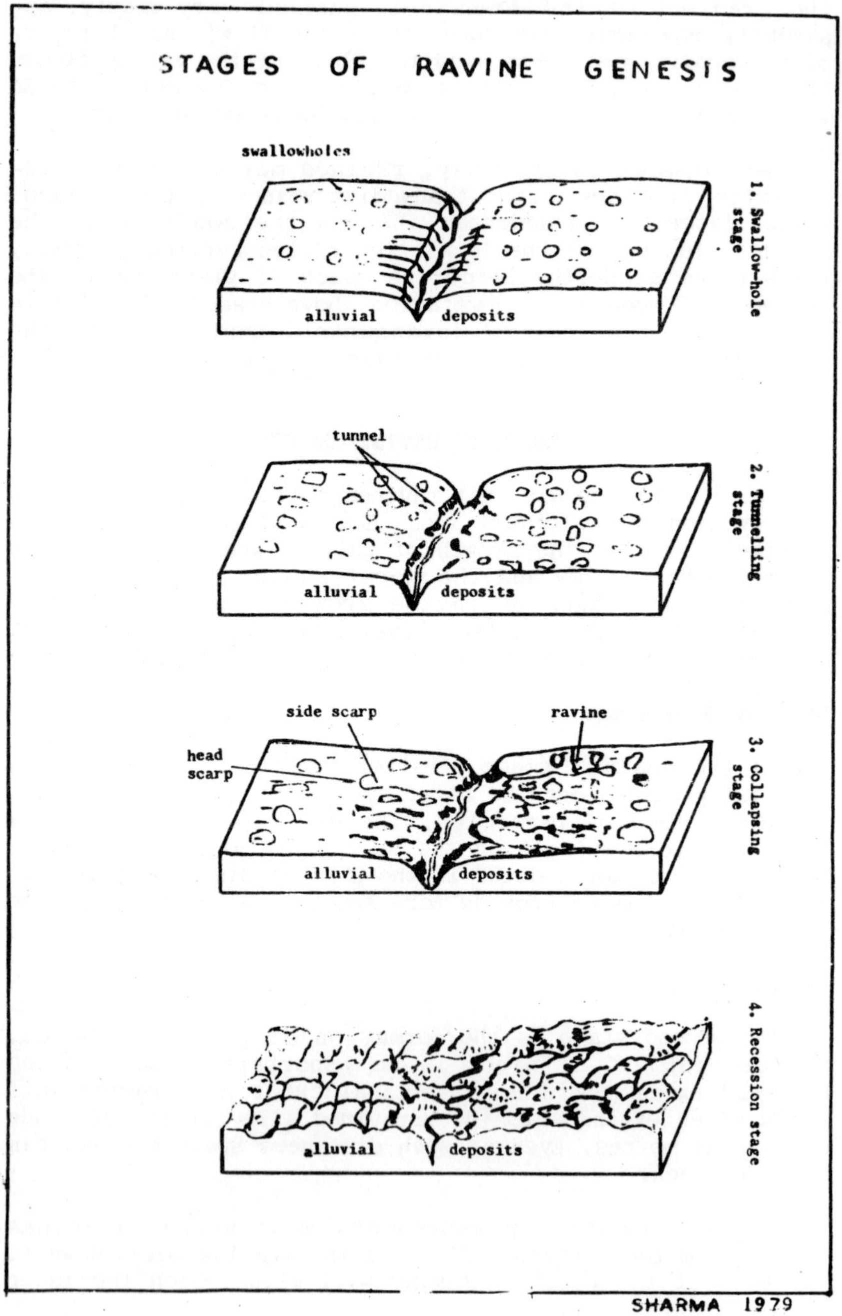

SHARMA 1979

removed not only salts in solution but also fine grained particles of soils.

Collapsing Stage

The movement of water through the 'Kankar Pan' actually transported fine particles and salts away from their original place of deposition and left void. These voids then as a result of concentrated water accelerating the process of soil removal and resulting eventually in the formation of collapsed tunnels from which full fledged ravine develops.

Recession Stage

The collapsed material of the tunnel is subsequently transported by the concentrated flows. The side scarps of the ravines then recede continuously by the process of soil creep. One can observe talus material at the base of the ravine scarps. With the continuous recession of the sides alluvial bluffs are disconnected or isolated. In this stage ravines enlarge their width and size. In subsequently stage vegetation and shrubs establish rapidly on mild slopes and make the ravine stable till the whole cycle is repeated.

Origin of Ravines

In the American south west, the arroyo has attracted scientific attention for more than half century. In the first phase, arroyo tended to be viewed as the result of changes in the vegetal cover brought about by man and his livestock. Later, the discovery of pre-historic gullies led to a shift of attention to possible climatic causes and to elaboration of a widely accepted explanation relating climatic events to erosion-deposition. In the last few years, the study of the pollen record has inspired a new explanation different in important aspects from the older one. The resulting body of literature is immense, but it produced a limited number of explanations for the ravine and gully erosion. Each of the model of ravine genesis is outlined briefly in this paper along with a review of the advantages and disadvantages of each model.

The survey of existing literature on ravine problem brings out the following theories for the genesis of ravines:

1. Human Landuse changes and overgrazing theory
2. Climatic theories:

(i) Climatic change drier to wetter condition
(ii) Climatic change wetter to drier condition
(iii) Change in frequency, intensity or seasonality.
(iv) Climatic change with the grazing 'trigger mechanism'.
(v) Catastrophic floods (extreme meteorological events)

3. Rejuvenation theory,
4. Equilibrium theory, and
5. Integrated theory

Human Landuse Changes and Over-grazing Theory

Man has altered his ecosystem to a great extent. The exponants of this theory associate ravine or arroyo cutting exclusively to human landuse changes and the chief cause being over-grazing. Many early researchers like Dodge (1902, 1910), Hough (1906 a; b), Thornber (1910), Rich (1911), Duce (1918), Leopold (1921), Ross (1923), Miser (1924), Swift (1926), Winn (1926), Bailey et. al (1934), Bailey (1935), Brady (1936), Colton (1937), Cooperider and Hendnicks (1937), Greogry (1938), Knechtel (1938), Cottam and Stewart (1940), Calkins (1941), Thornthwait et. al. (1941, 1942), Bailey et al. (1947), Peterson (1950), Marston (1952), Antevs (1952), Melton (1965), Cooke and Reeves (1976), Womack and Schumm (1977), Cooley, (1979, a), Masannat (1980), Dobyns (1981), Alford (1982), Denevan (1967), Carson and Tan (1977), Zorina et. al (1977) and Bryan et. al. (1982), argued that over-grazing, stripped vegetation from hill sides and valley floors are the chief causes of initiation of ravines and gullies on unconsolidated alluvial materials. The popularity of over-grazing as a major causal factor restored in large part on the temporal coincidence of over-grazing and ravine formation: ravines and gullies seemed to be appear only after cattle were introduced. But as Quaternary researchers have investigated the palaeoclimatic evidences from tropical and sub-tropical areas, asserted that ravines had cut and filled frequently during the thousands of years before the introduction of livestock. Therefore, the grazing hypothesis alone could not explain ravine cutting and although some workers frequently identified grazing as a contributing factor, climatic changes after Quaternary period received increasing attention.

Climatic Theories

Theories to explain ravine genesis by reference to

climatic changes falls into five general categories: erosion of ravine initiated by a shift from drier to wetter conditions; by a shift from wetter to drier conditions; by a change in frequency, intensity or seasonality of rainfall events, climatic change with the grazing a 'Trigger mechanism': and catastrophic meteorological events. The first climatic theory accounts that in times of increasing moisture more runoff would be available for erosion, resulting in ravine development. Dutton (1882), Johnson (1900), Huntington (1914), Bryan (1922), Gregory and Moore (1931), Richardson (1945), Quinn (1957) and Martin (1963) have attributed ravination/arroyo cutting to this factor. Bryan (1922) later on changed his opinion to reverse the basic assumption. He later on postulated that when climate shifts from wetter to drier conditions vegetation cover declines, permitting increasing runoff and resulting in channel erosion (Bryan, 1954). Unlike grazing hypothesis, Bryan's climatic theory was attractive because it easily extended to Quaternary times scales. By the mid 1940's the development towards cumulative generalizations concerning ravine development was obviously at a stand still. Later on Haynes (1968) stated that climate was a controlling factor in ravine formation, but that it was impossible to determine exactly how the connection operates.

After the beginning of recording of climatic data a number of workers discarded the general climatic theory and emphasised on changes in the frequency, intensity, or seasonality of rainfall events. Leopold (1951), Miller and Wendorf (1958), Martin et. al. (1961), Cooley (1962), Schoenwetter and Eddy (1964), Martin (1963), Cooke (1974), Cooke and Reeves (1976), and Malde and Scott (1977) stressed on changes in frequency, intensity or seasonality in ravine genesis. Unfortunately the statistical analysis of rainfall data of only 100 years are available and during this period only few intense storms had occurred therefore adequate testing of hypothesis is difficult to accomplish.

Earth scientists disappointed by the limitation of the human land use changes, grazing and climatic hypotheses proposed an alternative explanation by suggesting that overgrazing acted as a trigger mechanism in disrupting a system. The trigger mechanism hypothesis had the advantage of incorporating much of the available evidence, and could be extended to pre-historic ravines by assuming the action of some other 'trigger mechanism' such as fire (Graft, 1983). Huntington (1914), Gregory (1917), Miser (1924), Richardson (1945), Gregory and Moore (1931), Bryan (1940, (1941),

Leopold (1951, a), Judson (1954), Bryan (1954) and Cooke and Reeves (1976) are some of the geomorphologists who suggested this alternative mechanism.

Catastrophic Meteorological Events

Many research scientists like Reagan (1922, 1924 a, 1924b), Brady (1936), Senter (1937), Wooley and Alter (1938), Wooly (1946), Gregory (1950), Hunt (1953), Hastings (1959), Hasting and Turner (1965), Tuer. (1966), Mosley (1972), Cooke and Reeves (1976), Love (1979), and Sharma and Banerjee (1984) emphasized the occasional large flood or unprecedented rainfall as a solution to the ravine problem. Flood events played a large role in ravine initiation but they provided only partial explanations.

The Equilibrium Theory

The concept of dynamic equilibrium is so attractive that the open system framework also fits well in research on ravine problem. The research of Schumm and Lichty (1963), Schumm (1968a, 1968b), Birkland (1968), Burkham (1981), Baker (1974), Andres (1980), Gerson (1977), Ploy (1977), Pickup (1975), applied successfully the open system approach to ravine development. But the equilibrium concept is not entirely successful in solving the problem of ravine genesis for three reasons:

a. it is difficult to determine how equilibrium would be established between the discontinuous processes and discharge and the continuous form of ravination,
b. the persistence of cyclic evidence in ravine systems has also limited applications of equilibrium concepts which appeared in some cases to be simply old wine in new intellectual bottles (Graft, 1983).
c. Finally the equilibrium approach in ravine research enjoys only limited success because the processes and form might not tend towards equilibrium in the usual sense of the word. Once initiated ravine development seemed to proceed until an entirely new system of processes and form took over the prior conditions (Wertz, 1970).

Integrated Dynamic Theory:

Graf (1977) and Leopold (1978) have called for the development of an integrated dynamic theory for ravine development that encompasses temporal and spatial variation. Graf (1983) stressed that "a successful unified theory must accommodate these opposing situations, and could include the concepts of thresholds, complex responses, equilibrium, rate laws, distance decay functions reversibility, and the comparisions of force and resistance". Based on his experiences in the American south west (1979 c) he proposed that catastrophic theory provided a useful framework for an integrated theory and presented a successful pilot test of the method for ravine/arroyos. Thornes (1980) also arrived at similar conclusions from his experiences with arroyos in Spain. But the major disadvantage of catastrophic theory is that it accounts only for abrupt changes. Lack of sediment data and lack of long term historical records are major problems for evolving a widely accepted well integrated theory for ravine genesis.

The Rejuvenation theory of Ravine Origin

Based on experiences in India Sharma, (1968, 1976, 1981) Ahmed (1968) proposed that the rejuvenation theory provided a useful framework for genesis of ravines. Similar to American south west ravine and gully erosion in India also poses a very difficult problem in fluvial geomorphology. The question why should ravine erosion is mainly concentrated on the marginal areas of the Gangetic Plain? The intriguing aspect of the question is that the intensity of ravine erosion, occurs differently according to geology, slope, vegetative cover etc. Therefore, to look some cause which is independent to geological structure, vegetation, topography, rainfall (changes in frequency intensity, seasonality or catastrophic rainfall events) and human activity if a satisfactory explanation for the force behind remarkable ravine erosion (ranging in depth from 10 m - 100 m and 5 m to 200 m in width) in India.

It will be wrong to say that over-grazing, deforestation, climatic changes and extreme climatic events have not any role in ravine erosion. But there must be some common denominator in different areas of ravine erosion. For example, it is easy to say that recent deforestation is the cause of the severe erosion in the Chambal-Yamuna zone, but is very difficult to substantiate this explanation. First, in the historical past, especially when it was so close to the civilised heart of our country and along the corridor of movement between northern India and western Deccan (Ahmad, 1968). Similarly, the relative abundance of gullies from the more closely occupied country in Dhanbad suggests that the biotic interference is not the main cause of gully erosion. If deforestation were the cause, then Dhanbad would have been most severly gullied. But this is not correct.

Besides this, if aridity along with torrential rain is taken to be the cause, then such rain showers occur in very wide area extending from Punjab to Mysore and Andhra Pradesh and from Gujarat to Bihar. Again the conditions with respect to elevation, forest cover and geological structure in the upper Narmada and Tapti valleys are akin to what they are in the upper Godawari and Krishna Basins, but ravine erosion is much less serious in the latter areas than in the former.

The analysis and careful observation of various hydrological and geomorphic factors in the various parts of ravine affected areas of India leads us to conclude that the most basic cause of marked gully and ravine erosion in ecologically different conditions is the lowering of base level of erosion of streams caused by recent upliftment. The studies of Dunn (1948), Ahmad (1968), Dubey (1968), Pramila Kumar and Rai (1972), Chatterjee and Bagchi (1957), Singh (1967), Varma (1957), Sharma (1968, 1972, 1976), Heron (1938), Radhakrishnan (1967), and West and Chaubey (1964), indicate that the Peninsular Shield should have completely worn out peneplain but the highest elevation in the Chota Nagpur plateau (1160 m), the Bundelkhand (1130 m), the Central Indian Highland (830 m) and the Aravalli Range (1160 m) is definite evidence of recent rejuvenation.

The outermost shore zone between the Mahi and Tapti is marked by shallow marshy conditions pointing out the latest phase of upliftment of land. Bedi (1978) while working in the Mahi Basin of Gujarat has pointed out that the intensive gullying of old lavees, sandy stretches and terraces, which are comparatively of recent age, with some of gullies, having, 'V' to 'U' cross profiles indicate a general uplift. Gullies also exhibit angulate to rectangular patterns which suggest that the fractures in the basement rocks have been reactivated.

Ali (1966) was also inclined to believe that the rigid Peninsular block is still pressing against the Himalaya as a result of this certain areas, particularly where the margin of the shield and Indo-Gangetic depression occurs being elevated. He also believed that the uplift is the maximum in the zone where the distance between the Himalaya and Peninsular upland is the shortest, the area between the Aravallis (the Chambal Valley) and the Rajmahal. This explanation appears fairly convincing and appealing. There are evidences of Pleistocene uplift in the Rajmahal, Chota Nagpur, the

Vindhyan Range, the Chambal-Yamuna Zones and the Tapti and Narmada regions, as side effects of the Himalayan orogenesis.

Neotectonic activity has been reported in number of recent studies by Subramanian and Murleedharan (1985) in Palghat Gap, Valdia (1984) in Balia Valley in Nainital hills in Kumaon Himalayas along the Main Boundary Fault, Ravi Shankar and Dubey (1984) in Tapi-Vaghur confluence (Khan and Banerjee, 1984) and Ahmad (1986) in Rajasthan. Ahmad discussing the neotectonism in Rajasthan has arrived at the similar conclusion about the formation of ravines of the Yamuna-chambal belt. He stressed that it is perhaps not adequately realized that epeirogeny has played a very significant role in the area even in recent times. If this yardstick is applied to the chambal-yamuna ravine area, it would appear that it does not represent a peneplain but an actively upwarping tract in which agents of erosion and denudation are presently active.

The preceding discussion including the observations of various authors in the different part of the Peninsular India is covering enough to accept the suggestion that this part has been rejuvenated in the recent geological past and the present topography has been carved out by accelerated stream action. Thus the cause appears to be the factor that have deepened channels of the Chambal, Yamuna, Mahi, Subernrekha, Koel and other streams and their tributaries. Such deepened channels would induce rapid and forceful flow of water over the bank and selective gullying would develop in the land behind in those parts where vegetation is poorer, slope greater, soil or otherwise erodible.

This it appears very likely that recent and continued uplift is the chief cause of ravine and gully erosion in India. Such a cause has the power of producing results that would be indifferent to all such factors as geology, climate, vegetation, land use pattern etc., (Sharma, 1968). There is no cultural or economic cause which is particularly forceful in India that has caused the development of ravines. In India there are numerous areas of similar aridity as those of the Mahi and the Chambal, similar vegetation or similar soil or similar cattle population or similar human pressure as that of ravine infested areas but there are few areas that are so plagued with gully erosion as that of the ravine areas. Hence, under the present geographical conditions found in India it seems that ravines are one of the features developing due to rejuvenation and not exclusively due to defective land use and cylic changes of climate. Thinking in terms of cycle of erosion ravines may be taken as a feature of early maturity. The study reveals that the first stage for the ravine is, of course, entrenching the master stream itself. Therefore, due to entrenching of the stream the banks are cut and dissected in the form of ravines instead of well defined terraces.

The rejuvenation theory of ravine genesis can further be extended to explain arroyo cutting in other parts of the world like U.S.A., Spain, U.S.S.R., Britain. The Plate Tectonic results of American south west reveal that this plate still pressing westward and seismologically falls into major to moderate earthquake zone. San Andreas Fault and creation of Cascade range and Pacific coast ranges during Pleistocene to Holocene period led us to conclude that the trigger force behind initiation of arroyo might be neotectonism and the other events like deforestation, over-grazing and climatic changes are contributing factors. However, in order to confirm, detailed geomorphological studies from this point of view are essential.

Russian geomorphologists like Yakushova (1986) and Kostenko (1975) have also sighted examples of ravine and gully erosion from the uplands of the Russian Plain. According to the data provided by Kostenko (1975) within the boundaries of Moscow region in the areas of continental glaciation indicate that a complex network of young gullies inserted into the gullies of more ancient generations is being developed. What has a marked effect on the complex system of gullies according Yakushova (1986) are the neo-tectonic movements, difference in the composition of rocks and the original irregularities of the relief. In the central Russian upland the most recent tectonic uplifts caused on entrenched dendrite erosion dissection. The cutting down of trees and an overall ploughing up of the territories promoted the further intensive growth of the gullies (Doskach, 1975).

Based on direct geologic evidence, rates of recent tectonic uplift in various parts of the world are on the order of 4 to 12 m per 1000 years, using an average rate of 8 m / 1000 years earth surface might rise from sea level to an elevation of 80 to 90 m in about 10000 years (Holocene period) giving rise to ravines upto a depth of 80 m to 90 m in the Chambal Valley. The planation surface gained potential energy from neotectonism. Potential energy changes to kinetic energy when fluvial erosion initiate. This type of episode development of ravines can be visualised as occuring

FIGURE 8.7

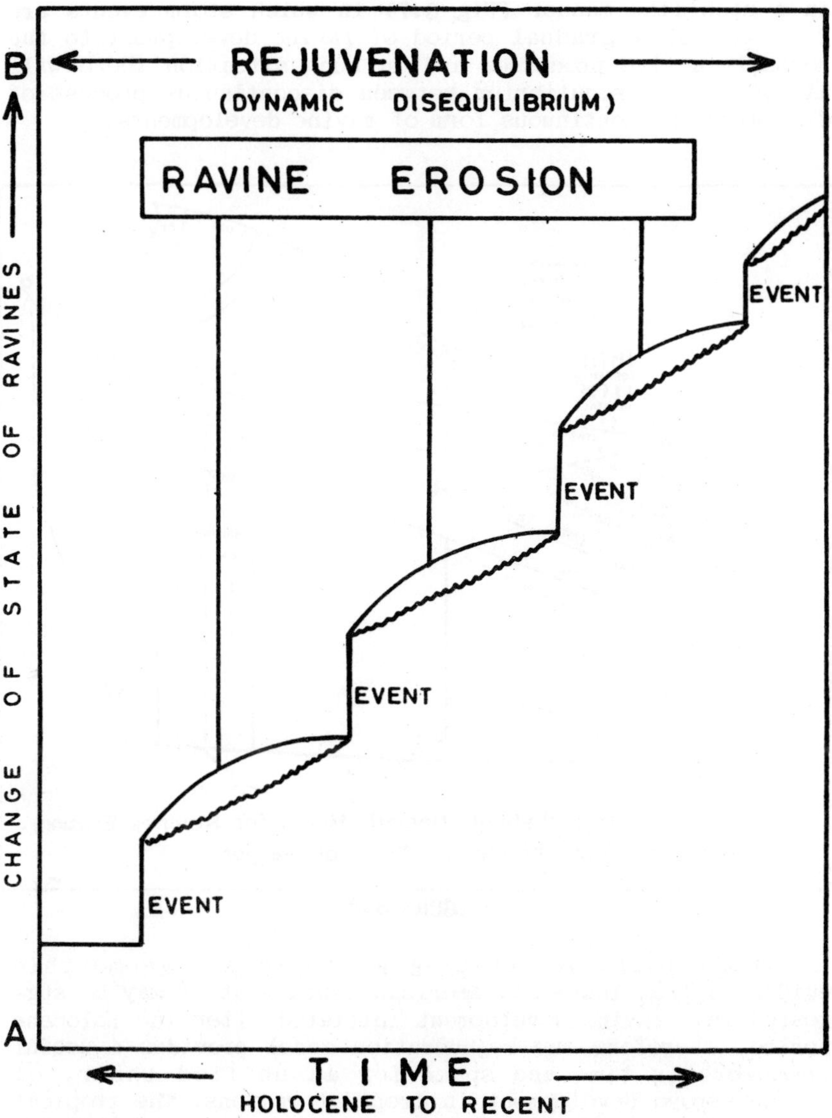

DISEQUILIBRIUM REJUVENATION MODEL OF RAVINE ORIGIN

in a step like manner (Fig 8.7) in which storm events are followed by a gradual period of ravine development to the normally active processes and in such situation there will always be a disequlibrium between discontinuous process of discharge and continuous form of ravine developments.

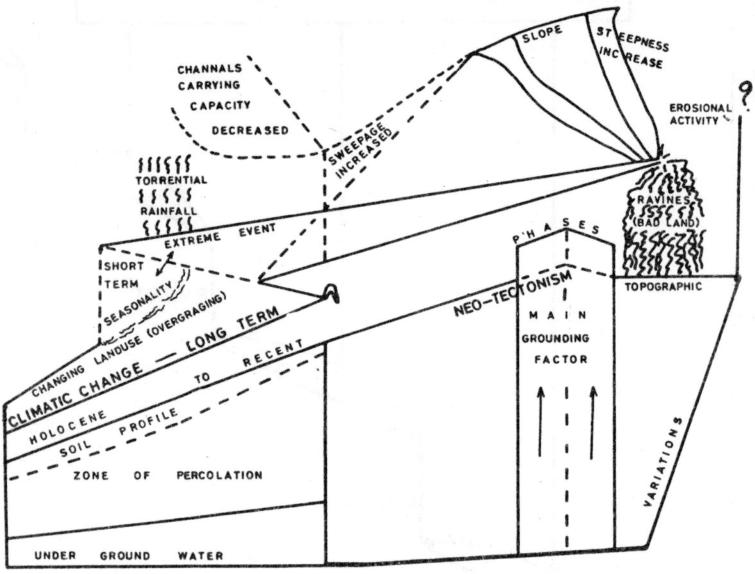

Diagramatic Hypothetical Nested Model for Ravines Erosion in varying Time-space in Tropical Region.

FIGURE 8.8

On the basis of existing geologic and geomorphic evidences from India and American south west it may be suggested that ravine development initiated after the Holocene period, therefore the rejuvenation model provides a useful framework in time and space for an unified theory of ravine/arroyo development in tropical regions. The tropical regions bears incomparable evidences of neotectonic activity and climatic changes from Holocene to recent period. Thus, it can be suggested that the neotectonic activity provided a grounding force behind incision of river courses and development of ravines. The climatic changes and the man's role might have promoted or contributed the ravine erosion by cutting down trees, unwise use of land and overgrazing.

The hypothetical nested model for ravine erosion in varying time-space in tropical region is presented in a diagramatic manner in Fig. 8.8.

The ravine erosion is a serious problem of agriculture in Rajasthan. In the foregoing analysis a brief review of the problem and existing theories of ravine development has been discussed. It is clear from the overall study of research work pertaining to origin of ravines that a widely accepted unified theory has not yet formulated. However, the neotectonic theory provides a useful framework for development of ravines. Deforestation, misuse of land and climatic changes during quaternary period promoted the acceleration of ravines in tropical areas.

NINE

Planer Surfaces and Inselbergs

The concept of erosion surfaces has had rather a controversial history and more has been written for and against this concept than on any other geomorphic idea. Between 1950 and 1986 geologists and geographers recognized many erosion surfaces in different parts of the world. Some early workers attributed regional planation to aeolian action, but being inconsistent both with general theory and field evidence, the idea has never really enjoyed wide acceptance. Others have interpreted ancient planate remnants as owing to marine planation, a possibility that exercised the minds of geologists and geomorphologists for many years. Now a days most of workers now agree that planation surface is owing to running water. On this hypothesis, two major models of landscape evolution have developed over the years. One is of Davis (1909) peneplanation concept and the other, the pediplanation concept of King (1962).

A third alternative model has also been expounded by Hack (1960). This model considers erosion as non-cyclic where ensuing the forms are independent of time. In this concept of dynamic equilibrium the erosion system is considered analogous to thermodynamic system of which the different parts are in a state of dynamic equilibrium.

German workers such as Budel (1935, 1948) have emphasized that climatic changes in the geological past caused a change in landscape character without crustal movement being in any way involved. According to them climatic morphological factors are all the more important in producing planation surfaces of faint relief.

On the basis of above review it may now safely be concluded that by and large planation surfaces are surfaces of erosion, the agent more often being fluvial and less marine and aeolian. These surfaces are generally associated with mountains and plateaux. It is now realized that some of the surfaces are the result of deposition and perhaps other agents as well, hence a non-genetic term planer surface is used now a days for these surfaces, though a majority of them are caused by erosion.

The most commendable work on geology and geological history of the Aravallis is that of Heron (1923, 1925, 1938, 1953). His work still remains the best account so far providing a wealth of background information on the area. Although a great deal of advancement in the field of regional litho-stratigraphic sequence and tectonic history of Rajasthan has recently been made, Heron's contribution to the physiographic history of Rajasthan is monumental. Sen (1972) also has studied the geomorphology of the Aravalli range with special reference to the residual erosional surfaces. Her observations supplement Heron's study. Sen has also applied the concept of dynamic equilibrium.

In Rajasthan especially in the Aravalli region five geological cycles have been established by the Geological Survey of India. These are the :

a. Bhilwara cycle ;
b. Aravalli cycle ;
c. Delhi cycle ;
d. Vindhyan cycle ; and the
e. Deccan trap cycle.

Radiometric dating of the first three cycles have also been made, as stated earlier in Chapter 1. These various cycles are separated by well-marked unconformities. The remains of ancient cycles are lost in obscurity and the earliest of peneplanes preserved in the relief of Rajasthan is of pre-cretaceous period. In the absence of any fossils of the pre-Cambrian period, it has become very difficult to date the various planer surfaces. However, in the present study the following techniques and evidences have been considered in order to analyse the complex nature of various planation surfaces :

1. The accordance of summit levels which represent the uniform skyline of the ranges.
2. Flat summits of varying areal extent over mountains and plateaus.
3. Truncation of rocks of varying resistance : In the Aravallis the rocks of different resistance stand at

equal height.

4. Hard formations such as quartzites, conglomerates, limestones and sandstones stand as ranges of hills while phyllites and schists form the lowland also dotted by isolated inselbergs and hogbacks.
5. Numerous examples of superimposed drainage in the Udaipur, Chappan hills, along the Chambal and Mej in the Kota plateau, Bundi hills and triple scarps.
6. Erosional scarps facing the Chambal and Kunu rivers.
7. Antecedent and transverse courses of the Banas at Rajmahal and Kali Sindh at Gagrun also indicate rejuvenation and upliftment of the Aravalli planer surface.
8. The incised meanders identified along the Chambal and its tributaries and the Morel (Sharma, 1972, 1976, 1979).
9. The ravine formation of the Chambal and its tributaries: It is generally stated that ravines developed owing to defective land uses and climatic changes. But a study by Sharma (1968) in the lower Chambal valley reveals that they are formed owing to regional rejuvenation of the Chambal resulted into entrenching of the stream course into its own deposits.
10. Knick points along the Chambal at an elevation of 375 m, 275 m and 225 m along the Bhimlet in Kota plateau and along the Menali river east of Mandalgarh and numerous waterfall sites in Chittorgarh hills and Kanjarda plateau.
11. Laterite and bauxite deposits over the Deccan Trap in south and eastern Kota district and Chittor district. This recent reporting of laterite deposit has added a new chapter to the denudation cycle and paleoclimatic sequence in Rajasthan.

MORPHOMETRIC ANALYSIS

For the quantitative expression of various planer surfaces, several morphometric techniques such as superimposed profiles, projected profiles and serial relief profiles, and altimetric frequency histograms have been used.

The superimposed profiles which were drawn from east to west across the Aravalli range exhibit the morphological unity if any and the existence of certain levels of erosional or structural surfaces. The superimposed (Fig.9.1) profile of the whole region shows the existence of three planer surfaces between the heights of 550 m to 900 m and above, 400 m, 550m and 180 m to 300 m. Between 350 m and

400 m one more surface is also evident. The projected profile (Fig. 9.2) and superimposed profiles both indicate that the first and second surfaces merge at a height of 500 m. Serial profiles were also drawn across the Aravallis from north-west to south-east (Fig. 9.3). These profiles show the undulation of the surface morphology the Aravalli. The altimetric frequency histogram of the whole of Rajasthan also brings out the concentration of frequencies from 250 m to 400 m, 450 m to 550 m, and 600 m to 900 m and above (Fig. 9.4).

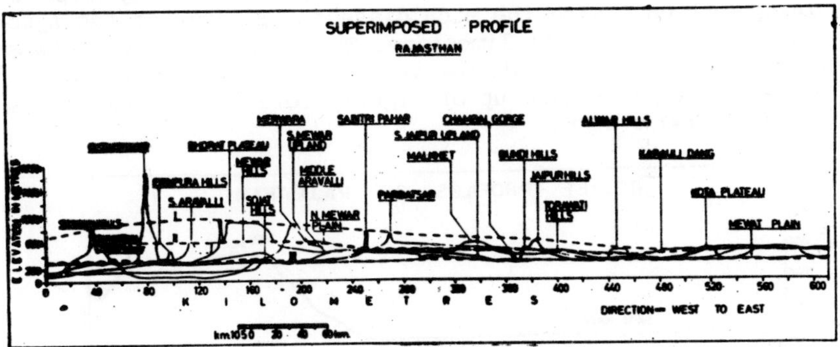

FIGURE 9.1

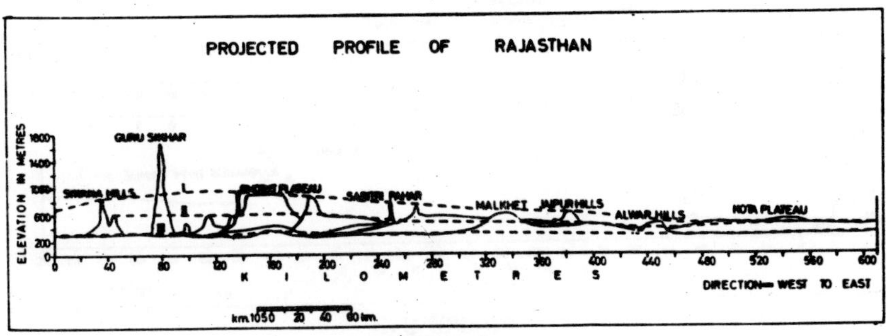

FIGURE 9.2

THE ARAVALLI RANGE

The most conspicuous geomorphic feature in Rajasthan is the Aravalli range which dominates the entire landscape. This range is considered as one of the oldest mountain ranges in the world. It trends diagonally from north-east south-west for about 530 km. In plan the range appears somewhat like two fans, joined handle to handle. Both the northern and southern sections open out like fan the former embracing the hills of Alwar and Torawati and the latter those of Mewar. The northern fan is highly complicated, consisting of many isoclines developed in the hills of Jaipur and Alwar. It also consists of flat topped monoclinal ridges of hard Alwar quartzites and strike valleys carved out of the soft Ajabgarh phyllites. The southern fan on the other hand consists of folded hills of quartzites of the Aravalli super group alternating with deep valleys of soft Ajabgarh mica schist, giving rise to a ridge and valley type of topography enclosing the much dissected plateau of the Ajabgarh calc-schists and limestone. The Bhorat plateau between Kumbhalgarh and Gogunda is very conspicuous. Tne middle Aravalli range is made up of tightly folded synclinal hills 600 m to 900 m high.

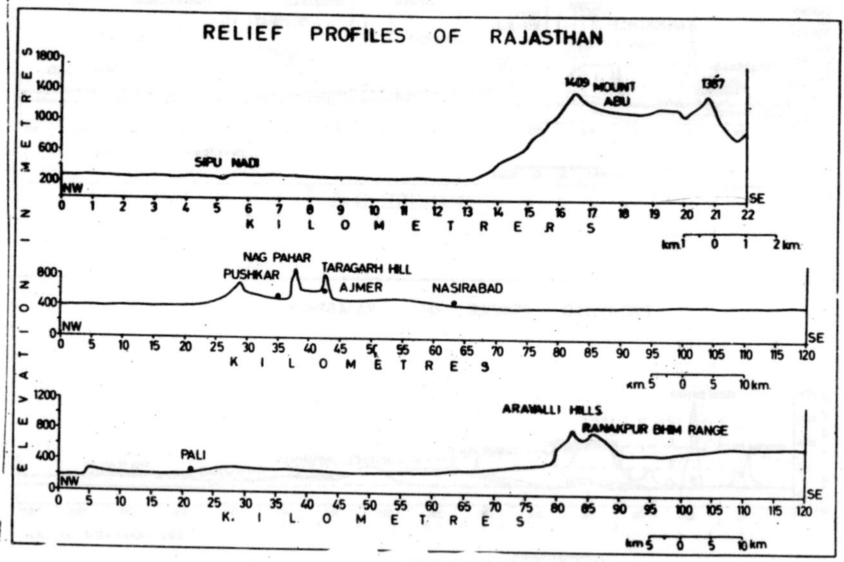

FIGURE 9.3

The highest portion of the Aravalli range lies between the fort of Kumbhalgarh and Gongunda a plateau locally called the Bhorat lying at a level of nearly 900 m, from it bold strike ridges of the Alwar quartzites, with almost level tops, project to about 1000 m (Fig. 9.5). The highest summit in Rajasthan is on the great Abu batholith of Erinpura granite, separated from the main Aravalli range by a narrow valley.

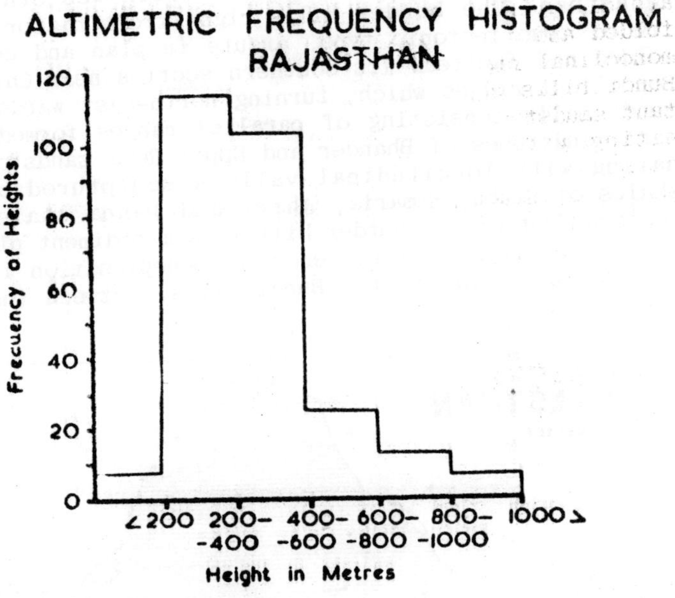

FIGURE 9.4

The relief of the Aravalli range is very intimately dependent upon the lithological character of the rocks of which it is made. The Aravallis are principally phyllites. The intercalated quartzites are usually thin and inconstant, but remarkably hard and resistant to erosion and denudation. When sufficiently thick and continuous, as in the south and south-west of Udaipur city, they take part with the Alwar quartzites in the formation of the range. These phyllites are very easily eroded and form monotonously rolling country intercepted by ramifying shallow valleys, with little difference in elevation between the valley bottom and the smoothly rounded knolls. There is little variation in average level over the area occupied by the Aravalli phyllites.

On the either side of the Aravalli range lies the Vindhyan country from Chittor in the south-west to Dholpur in the north-east. The Vindhyans are separated by the Great

Boundary fault from the pre-Aravalli and the Aravalli ranges. This is the north-western continuation of the Vindhyan basin. The whole of the Vindhyan basin is by no means flat but got folded in the north-western region of Rajasthan. The border hills right from Chittor to Lakhari are of true tectonic type, arcute in plan and considerably folded and faulted. Its southern section near Chittor forms monoclinal ridges which, turning north-east wards, join the Bundi hills consisting of parallel ranges formed of resistant sandstones of Bhander and Upper Rewa sandstones alternating with longitudinal valleys sculptured out of soft shales of Sirbu, Simaria, Jhiri and Panna shales. At the western foot of the border hill lies a pediment plane formed of pre-Aravalli. The evidence of peneplanation is afforded by many wind gaps in the Bundi hills. Between these border ....

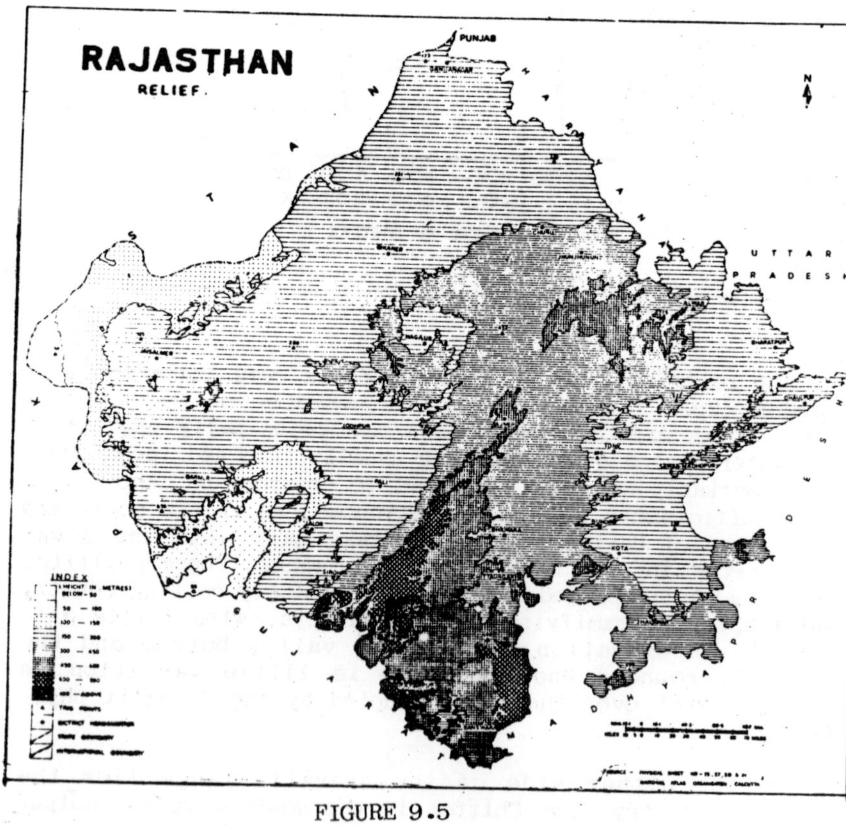

FIGURE 9.5

hills of the Vindhyan and the Aravalli ranges lies an extensive alluvial plain, underlain by meta-sediments of Bhilwara and Aravalli super groups, the entire surface being dotted with low monadnocks and inselbergs. Its peneplain character is evident in a traverse from the Aravalli to the Banas through Shahpura where the rocks of varying hardness gneiss, granite, phyllites and Schists have been truncated by a peneplane now lying between 450 m and 525 m above sea level. The higher altitude of the peneplain on the south is owing to warping. At the Chappan hills on the south, the peneplain terminates and is considerably dissected to give a hilly character to the terrain.

The purpose of describing the general character of the Aravalli and the adjoining areas is to emphasise the fact that there is a variation in the morphogenesis of different parts which has ultimately resulted in different surface levels.

REGIONAL DISTRIBUTION OF PLANER SURFACES

On the basis of morphometric techniques, study of landsat imageries and aerial photographs, substantiated with extensive field observations, the following four planer surfaces can be identified (Fig. 9.6)

a. planer surface above 600 m upto 900 m (I) ;
b. planer surface 400 m to 550 m (ii) ;
c. planer surface 300 m to 400 m ; and
d. planer surface 180 m to 300 m (IV).

Planer Surface I : Pre-Cretaceous

The pre-cretaceous planer surface has developed across rocks of varying resistance, steeply dipping Alwar quartzites and Aravalli quarzites. This is most extensively preserved and relicts are found on Siwana hill, Bhorat plateau, Sabetri pahar, Jaipur hills, Alwar hills, Samoda hills, Achrol hills in Delhi quartize and Chittor hills, Bundi Sawai Madhopur hills, Karauli Dang, Kota plateau, triple scarps and around Jodhpur in the Vindhyan rocks. When looked at from a distance, these groups of ridges are striking for their wall-like appearance and their long level tops. These features were remarked on first by Fraser in 1824. The bevelled surfaces of the Aravallis and Delhi Super group are somewhat different from those of the Vindhyans. The rocks are usually steeply dipping or even vertical and their tops have been denuded and are remarkably flat especially at Bhorat plateau, Jaipur hills, Samoda hills, Newai hills, Alwar hills and Torawati hills. The Vindhyan surfaces in most of the cases are horizontal except in the border hills of Bundi-Lakheri which characterized by bevelled surfaces. The Vindhyan plateau such as characterized by bevelled surfaces. The Vindhyan plateau such as Kota, triple scarps, Dangland of Ķarauli, the monocline hills of Chittorgarh are bounded by steep scarps. In the south-eastern part of Rajasthan near Jhalawar, the Vindhyans are covered by the Deccan trap. At places the Deccan trap has been denuded and the pre-Deccan Trap topography (Vindhyan topography) has been resurrected.

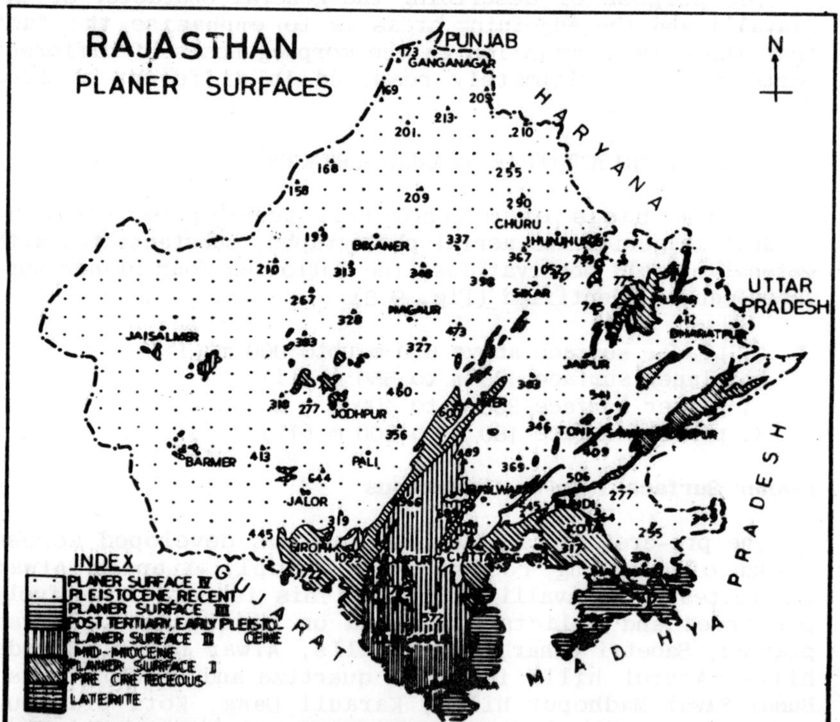

FIGURE 9.6

<u>Age of planer surface I</u> : In the absence of any datable sediments over this surface, it is very difficult to determine its age. However, Heron (1938) on the basis of structural evidence has pointed out that Aravalli range appears to have been peneplaned during the Mesozoic era, at sometime

not very much earlier than the Cretaceous period. This may well have been the first peneplanation since the range arose in pre-Vindhyan times. The peneplain has since been warped in such a way that its central position near udaipur city is now some 1300 m above sea level, its north-eastern end near Delhi about 330 m and its south-western end in Gujarat about 300 m above sea level. It has been pointed out by geomorphologists that land cannot be worn out completely down to sea level for its base level of erosion must always be determined by the final gradient of the streams eroding it, below which they cease to operate. This may well be true to estimate for upliftment of the Aravalli range, by considering the present level of the plains, which will give us values close enough for our purpose, as the plains of India have very nearly reached the stage at which their rivers can no longer east them away. We may take this as about 220 m at Delhi, 200 m to 300 m in western Rajasthan west of the centre of the ridge, and 300 m on the Gujarat margin. On these assumptions the uplift of the range since peneplanation, has been some 100 m to 125 m at its ends, and a maximum of 1150 m in the centre (Heron, 1938).

On the basis of these assumptions Heron (1938) has pointed out that the rejuvenation of the Aravallis and the formation of the Great Boundary Fault may be closely connected and contemporaneous. Thus, the Aravalli range was peneplaned in the Mesozoic, re-elevated before the beginning of the Tertiary, and the southern prolongation of it was let down in the Tertiary, after the accumulation of the Deccan trap.

Planer Surface II: Mid Miocene

The second planer surface is also very extensive in the south-eastern part of the Aravalli range. It is truncated by a pre-cretaceous peneplane surface in the border hills of Bundi-Lakheri. It has developed over rocks of varying hardness gneisses, granites, phyllites and schists. It is best developed on the plains to the east of the Aravalli range, but forms the floor of the strike valleys within the range. Rising above it we have not only the great even-topped ridges of Alwar quartzites, truncated by the Mesozoic peneplane, but many residual resistant masses owing to the harder epidiorites and granites, quartz reefs, minor quartzites, limestones and such less easily eroded rocks. For considerable areas this planer surface is covered with surface deposits of not very great depth, through which the streams have cut down to the rocks. A study of toposheets and geological formations reveals that it is not so much a continuous plane as a series of planes, at slightly differing levels, falling in steps from the centre of the range to the plain bounding it. The heights of these planes are determined partly by the character of the rocks of which they are composed, and partly by the extent to which erosion has been retarded, in certain of the higher valleys, by rock-barriers impeding the drainage and hindering the deepening of the streams. As an example of the effect of the nature of the rocks on the levels of adjacent stepped planes, one might take the Ajabgarhs, above the strong Alwar quartzites, which consist of three types of rock exposed in the heart of the range :

a. limestones and calc-gneisses with normally few intrusives ;
b. flaggy calc-schists with numerous small pegmatites ; and
c. mica schists riddled with small and large pegmatites.

These three types are intrinsically different in their lithological character and, in consequence, in their behaviour under denudation : the calc-gneisses and limestones being the most resistant and mica schists the least ; in addition the type of rocks are weekened by the pegmatite intrusions in them. As a result, a set of three stepped planes is developed from the rocks, that on the calc-gneisses and limestones being the highest, forming long narrow plateau along the heart of the range, flanked by shelves cut out in the calc-schists, and bounded by lower belts along the mica schists. Roughly, there may be differences of some 20 m to 40 m between the levels of the three planes taken at any cross-section, and existing streams have deeply incised them all. The contour lines of 300 m, 450 m and 600 m roughly demarcate the boundary of the three planes. Heron (1938) analysing the nature of the rocks and the three planes, pointed out that if it is not a fortuitous effect resulting from the even erosion of foliated rocks, which looked at broadly are mostly homogeneous, it may perhaps represent the Tertiary peneplane which has been detected by various workers in Central India and Chota Nagpur. The rejuvenation of the surface is deduced by the incised courses of streams of the Chambal and its tributaries in the upper reaches.

Planer Surface III

Planer surface III is developed on the Deccan Trap in

the south-eastern part of Chittorgarh, Banswara, Kota and Jhalawar districts. The Deccan Trap here forms an extensive flat topped upland bounded by gentle concave scarps. The trap basalt once covered a very extensive mature Vindhyan topography in Chittorgarh and Jhalawar-Shahbad area. In Banswara district in Mahi Erau Doab the Deccan Trap overlies the Aravallis and at places the Aravallis have been resurrected. In Rajasthan 6,7,8,9 (four) flows of basalt have been identified.

Recent investigations by the Geological Survey of India and field observations in Kota and Jhalawar area by the author reveal, the existence of bauxite and laterite over the Deccan Trap surface. The Deccan Trap flows are laterized. This condition doubtless marks the conditions of weathering after cessation of the eruptions. Examples of this may be seen in the caps of the small peaks at around 400 m at Harnaoda east of Aklera, at Manoharthana in Jhalawar district (Fig. 9.7), at around 530 m near Dhaman Dungari, Rampura Doli, and at around 520 m near Bhanej, Chotti, Ambali in Chittorgarh district. The thickness of the capping in Jhalawar district is 14 m while in Chittorgarh area 2 m to 10 m Bauxite has been found near Baselio, Majola, Mamoni, and Sherol Khera in Shahbad tahsil of Kota district in the form of blanket cappings. The average thickness of the bauxite zones is 3m, but the maximum thickness is more than 15 m at some places.

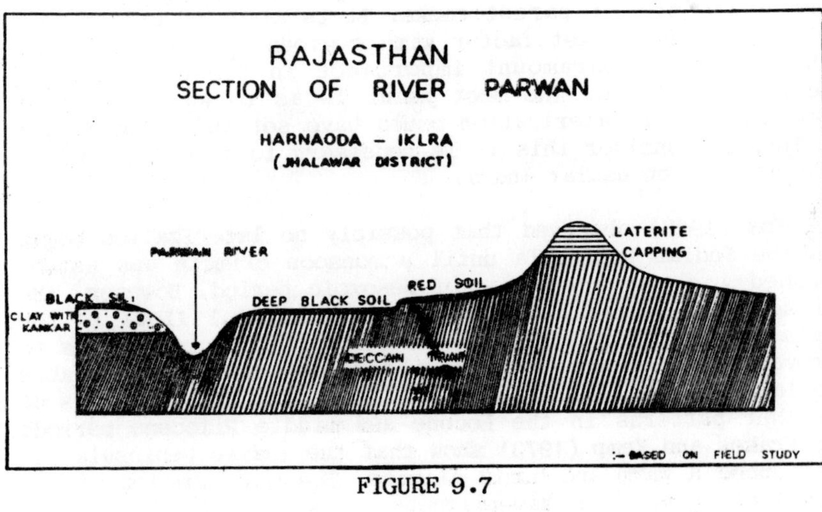

FIGURE 9.7

Composition of laterite :

Laterite of this area is highly ferruginous, vesicular and pesolitic to boundary in form. It is of variegated colours such as grey, red, black, yellow, blackish and brown. It is the result of insitu chemical weathering of basalt aided by organic activity under a tropical climate. Bauxite is also a product of insitu alteration of the Deccan Trap rocks.

Laterite and Palaeoclimatic Inferences

Laterite and bauxite capping over the Deccan Trap in Rajasthan constitutes an integral part of this planer surface and shows the influence of climate in the shaping of landforms, especially in post Trap times. It is generally agreed that formation of laterites and bauxites in landforms, implies the preponderance of chemical weathering over mechanical weathering. Such a condition may be expected to prevail towards and last stages of planation by erosion, when the intensity of mechanical weathering tends to wane owing to the decay of relief features, and when chemical weathering may gain ground over mechanical weathering subject to the prevalence of climatic conditions favourable towards laterization. Under a stable tectonic set up sustained and deep chemical weathering may well be expected to lead to the formation of laterites. It is well known that laterization is governed by several factors including climate, topography, ground water quality and the movement and character of parent rocks. It is also recognized that climate is a blanket factor with regional implications and that it is of paramount importance in the evolution of residual profiles. The moot point is as to when a climate favourable for laterization could have set in in Peninsular India. To consider this it is necessary to trace the palaeo climate of Peninsular India.

Fox (1923) observed that possibly no laterization began in the Indian Peninsula until a monsoon climate was established i.e. at the close of Mesozoic period. However, the presence of prolific tropical and subtropical flora in the Palaeocene sediments of the subcontinent bears witness to prevalance of a tropical to subtropical climate favourable to laterization (Sahni and Kumar 1974). The global models of weather patterns in the Eocene and middle Pliocene periods by Frakes and Kemp (1973) show that the Indian Peninsula experienced a warm and humid climate. The continuation of the tropical climate in Mio-pliocene times is indicated by the occurence of several species of evergreen moisture loving fossil flora in Mio-pliocene sandstones exposed near Pondicherry in the southern part of Peninsular India (Ramanujam, 1968). The occurence of laterite and bauxite capping over Mio-pliocene sandstones traced along the east coast (Subramaniam and Mani, 1978, 1978) and over the upper Jurassic sediments in Palni, Anamalai and the Nilgiri hills in south India (Vaidyanadhan, 1967), reveal that climate conducive towards laterization has been prevalent in the Peninsula since the beginning of Tertiary era. It is now reasonable to suggest that in Rajasthan the warm and humid phase started from the Tertiary period and it is logical to say that soon after the cessation of the basalt flow, the processes of laterization began, giving rise to residual laterite and bauxite capping over the Deccan Trap. Hence this planer surface may be attributed to the post-Tertiary or early Pleistocene period. The radiometric data which will provide the exact date of this surface are still awaited.

Laterite formation over the Deccan Trap has also been reported by Choubey (1967) in the central Indian highlands.

Planer Surface IV : Pleistocene Surface

The fourth planer surface is the most extensive of all the surfaces. It is developed between 180 m to 250 m on either side of the Aravallis. To the south in the central Rajasthan, this surface meets the planer surface of the Tertiary age where it rises to about 450 m ; on the western side of the Aravalli range it lies at about 330 m. Near the mountains the level of alluvium rises some what rapidly, but this is owing to the heaping up of wind blown sand in a ramp. This surface is covered with sandy hills of different shapes. The Bap upland and Jaisalmer area are covered with sandy hills of different shapes, and also gravels, boulders and cobbles. In eastern Rajasthan along the Chambal and its tributaries this alluvial surface has been dissected into the form of ravines and gullies, indicating the initiation of a new cycle of erosion in recent time (Sharma, 1979). The age of this surface can be assigned to the Pleistocene or sub-recent period.

The existence of palaeochannels, buried pediments, interdunal flats, playas and aggraded older alluvial plains reveals that since the Quaternary era this planer surface has witnessed alternate arid and humid phases of climate.

INSELBERGS

The study of evolution of inselbergs has long been of particular interest of geomorphologists in various parts of the world. Considerable work on inselberg formation has been done by Falconer (1911), King (1948), Handley (1952), (Mabbutt, 1952, 1966), Linton (1955), Budel (1957), Ollier (1960, 1965, 1981), McCraw (1965), Thomas (1965), Palmer, et. al. (1962), Caine (1967), Browne (1964), Demek (1964), Boye and Fritsch (1973), Clayton (1956), Bradley (1963), and Twidale (1978, 1981, 1981).

The study of evolution of inselberg and related microform has received a passing reference by geomorphologists of Rajasthan. However, large parts of Rajasthan, specially south-western and south-eastern parts are dominated by inselbergs of various size and form. The granitic, gneissic, schists and rhyolitic landscapes are characterized by the presence of sharply defined residual hills that rise abruptly from the adjacent plains and pediplains. In eastern part of Rajasthan these inselbergs, residual hills and low lying tors break the monotony of the planated surface. The significance of inselberg landscape lies with the fact that their geomorphic history is associated with the evolution of the Aravallis. Hence it deserves special discussion as a morphogenetic feature.

NATURE OF GRANITIC ROCKS

The intrusive granites and rhyolites occur mostly in the south western part of the Aravallis. Few patches of granites are found along the eastern side of the Aravallis in Berach Basin, Bhilwara area, Tonk area and in north eastern Rajasthan at Ajitgarh, Seoli, Bairat and Harsora. Banded gneisses are well exposed in eastern Mewar plain (Fig. 1.1)

The granites of Rajasthan are of different age. The radio-metric dating of various granitic bodies are as follow:

1. Erinpura granites and gneisses	C 740 million years
2. Sendra Ambaji granites and gneiss	C 800 million years
3. Godhra granites	C 955 million years
4. Kishangarh synite	C 1490-1910 million years
5. Synoorogenic granites and gneiss	C 2275 million years
6. Berach granites	2585 million years

7. Untold and gingala granites — 2950 million years (+ 150 million years)

Typical granite is made up of quartz, feldspar and mica. It is fine to coarse grained, well jointed and without bedding. Berach granite occurs over a crescent shaped area of about 100 Km in length and 30 Km in width in central Rajasthan in the vicinity of Chittorgarh (Fig. 1.1). Berach granite is essentially non-porphyritic with no traces of banding, medium to uniform in grain and varying in colour from reddish pink to greenish yellow. The feldspars orthoclase and acid plagioclase are much decomposed to kaolin, sericite, calcite and epidote. The ferromagnesion element characteristically scanty comprises with biotite and green hornblende which occurs as a rule as streaks of brownish and greenish decomposition products, mostly chlorite and epidote. In contrast Erinpura granite occurs as bosses and sills of all size. The outcrops are found extending from Ajmer and Beawar in the north east to the main belt of the type area of Erinpura and Mount Abu in the south west. In eastern Rajasthan, it shows well developed gneissic structure especially at the contact of the mica schist. This granite varies in texture and grain size, but has a constant chemical composition with biotite as the principal ferromagnesian mineral. The Malani suits of igneous rocks are represented by the Siwana and Jalor granites. The rhyolites are porphyritic and well jointed.

Structure of Granites

The granites are characterized by sheet structure. They posses well defined vertical and horizontal joints. Diagonal joints are also distinct. These structural joints in granites are the dominant factor for developments of inselbergs.

About 1500 lineaments have been deciphered in various parts of Rajasthan (Fig. 1.3). The majority of the lineaments show evidence of faulting and major fracture systems. These lineaments are mostly seen controlling the distribution and the ground plan of the granitic inselbergs. In particular there is a strong correlation between lineament and inselberg and tor alignment.

DISTRIBUTION OF INSELBERGS

Inselbergs in Rajasthan occur over a wide range of climatic conditions - arid, semi-arid and sub-humid. The granitic and rhyolitic inselbergs of Jalor, Siwana occur in arid and savanna climatic conditions. Mount Abu and Sirohi rock tower inselbergs and associated with high rainfall from 1000 m to 1650 mm. Isolated inselbergs are also found in semi arid tracts of north eastern and central Rajasthan.

Siwana, Jalore, Sendra, Desuri, Falna, Mount Abu, Sirohi, Bhilwara, Malpura-Diggi, Tonk, Ajitgarh, Ajmer, Bairat, Chittaurgarh, Udaipur and Kishangarh are some localities where different forms of inselbergs are found.

HEIGHT RELATIONSHIPS

On the basis of height inselbergs can be divided into:

1. high level inselbergs, and 2. low level inselbergs.

High Level Inselbergs

The huge batholithic inselbergs of Mount Abu and Sirohi area ranges in height from 1000 m to 1722 m. The main inselbergs are Vashisht Ashram (1083 m), Ram Kund Ka Pahar (1327 m), Kura Ka Pahar (1401 m), Deoli (1352 m), Abu (1409 m) and Gurushikhar (1722 m).

Low Level Inselbergs

The low level inselbergs are below the height of 1000 m and most of them are above 600 m. The inselbergs of Jalor (736 m), Israna (839 m), Machhel (664), Palasia (835 m), Kundal (834 m), Sela (681 m) and Jinpur (634 m) are above 600 m. Same is the height of Ajmer and Udaipur inselbergs. In the Berach basin around Chittaurgarh most of the domical inselbergs are at heights of ranging from 487 m to 531 m. In Malpura and Tonk area these are between 450 m and 500 m.

WEATHERING PROCESSES

Deep weathering is common in granites, gneisses and rhyolites. Some great depths have been reported in various parts of the world are tabulated in Table 9.1 :

TABLE 9.1 **Depths of weathering in different parts of world.**

Author	Area	Depth in Metres
Thomas, 1965	Nigeria	100
Ollier, 1960	Uganda	100
Ollier, 1965	Queensland	45
Ollier, 1965	South Wales	300
Ollier, 1965	Victoria	80
Ollier, 1965	Western Australia	40
Razumova and Kheraskov, 1963	Russian Platform	1000 to 1500
Demek, 1965	Czechoslovakia	30

The field investigations in eastern Rajasthan around Tonk, Malpura, Jahajpur, Bhilwara and Shahpura indicate that the recent mantle of weathering consists of 1.1 m to 3 m thick saprolite zone (schist and gneisses decomposed to various degree but more or less retaining the original rock structure) grading upward into a 50 Cm to 2.50 m thick fine to medium grained grity, structureless zone comprising of hydromicas, clays, angular pieces of quartz, altered felspars and nodulas of Kankar which ultimately passes towards the surface into a 0.50 m to 1.50 m thick soil horizon comprising brownish grey to greenish black silty loam.

Relicts of the former deep profile of weathering can be surmised by the extensive planation, presence of inselbergs, pediment surfaces and patches of older regolith blow the oldest fluvial sediments of the Banas and its tributories. The existence of laterite and bauxite over Deccan trap surfaces of Kota, Jhalawar and Chittaurgarh areas at about 400 to 550 m height undoubtedly marks the condition of weathering after cessation of the Deccan trap eruption. The thickness of laterite cover ranges from 2 m to 15 m. The pediment slopes of Aravalli and Quartzitic hills of Newai, Bichun, Alwar and Jaipur are covered by deep profile weathered regolith. This is well supported by a 450 m deep oxidation of the sulphide deposits of Rajpur Dariba and Zawar deposits of Udaipur district, Deri and Iswal deposits (Sirohi district), Lohakkan and Sawar deposits of Ajmer district and Chauth Ka Barwara of Sawai Madhopur district. On the basis of these evidences it has been inferred that deep weathering in Rajasthan perhaps required long and persistent period of warm and moist climates and may be of Tertiary or even Mesozoic periods when such climates were prevailed in the world. Browne (1964) also concludes that during a long period of Cretaceous planation the granite was deeply weathered. In fact, the Cretaceous was a period of extensive planation and the climate over much of the world was suitable for extensive chemical weathering.

Exfoliation by spheroidal weathering, unloading, flaking and granular disintegration are the most important processes of mechanical weathering operating along joints, cracks, and sheet structures. Chemical weathering is most prominent in the Mount Abu area which is responsible for formation of cavities and weathering hollows. Granular disintegration of granites is wide spread in Jalor, Siwana, Sendra and in eastern Rajasthan unloading and flaking results in peeling of thick curved slabs on the domal surfaces of granites and gneisses.

FORMS OF INSELBERGS

Three basic forms of inselbergs are found in Rajasthan. They are :

1. Domical inselbergs (Bornhardts) steep sided, bald, domical hill with convex side slopes. Kunat Mata ki Dungari, Kadila inselbergs of Malpura Diggi are classic examples of domical inselbergs. These inselbergs are 90-100 m in height from surrounding plain and are composed of fine to medium grained granites. Their profile is convex. They are attended by horizontal diagonal and vertical joints along which intensive mechanical weathering in the form of granular disintegration is quite intense. Their tops are covered with tors. The rhyolitic hills of Jalor Siwana area are also rugged and domical type of inselbergs.
2. Castle Koppies (rock towers, stacks) known as tors are also steep sides but they are angular and blocky, though their blocks are insit and display an orderly arrangement determined by orthogonal joint sets. Such forms are best developed in Mount Abu Inselberg complex.
3. Nubbins are more or less conical hills that seemingly consist of a chaotic mass of large blocks and boulders. A profile (Fig. 9.8) near Kuraria shows the development of knick point, pediment surface and sheet erosion.

Micro Forms

Common micro forms on granitic inselbergs of Rajasthan are mostly weathering hollows, flared slopes, and caverneous weathering (tafoni) features are best developed in Mount Abu, Sendra, Malpura, Ajmer, Jalor, Sirohi and Siwana areas.

Flared slopes represent as a particular form of weathering front. They are no way related to flat lying or sheeting joints, which they transact. Such slopes represent the former hill plain junction, where scarp foot weathering occurred, but it appears that they have lateron be modified and abraided by wind action in Sendra area.

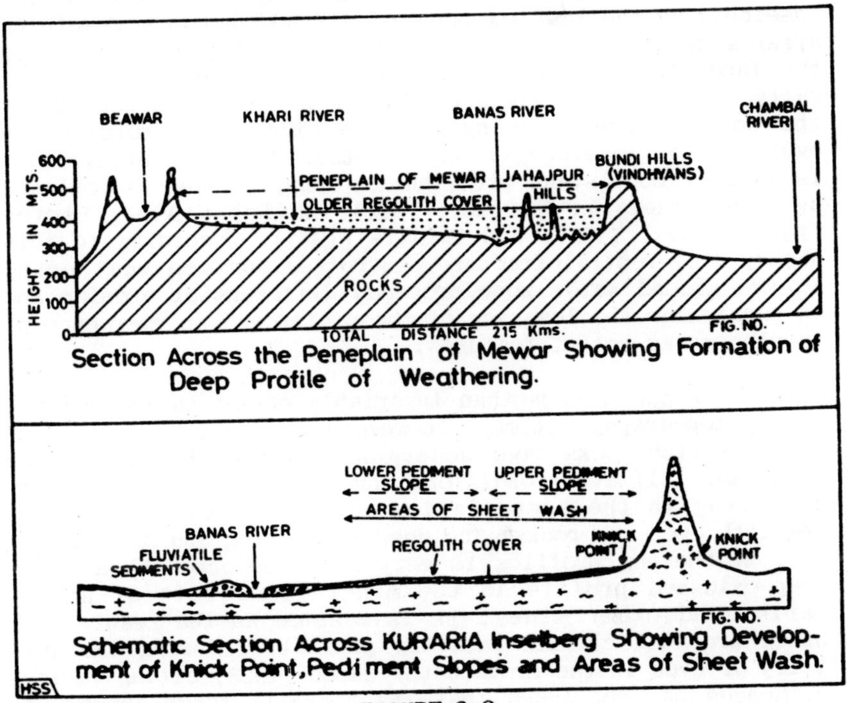

Section Across the Peneplain of Mewar Showing Formation of Deep Profile of Weathering.

Schematic Section Across KURARIA Inselberg Showing Development of Knick Point, Pediment Slopes and Areas of Sheet Wash.

FIGURE 9.8

Hollows on granitic inselbergs and boulders include shallow pans and deep cylinderical pits on gentle slopes, armechair shaped hollows on steeper slopes, caverneous weathering beneath overhangs or at a joint intersections on rock walls. It seems likely that many such hollows develop from irregularities on the rock surface resulting from the effects of sub surface weathering on weaker parts of rock fabric. On exposed surfaces such irregularities retain water in contact with the rock and permit the growth of algae and lichens throughout a wide climatic range. A more rapid rate of weathering results as it becomes deepened. The micro climate of the rock surface is important and the weathering stage of the exposed surface may influence the rate of sub aerial modification.

ORIGIN OF INSELBERGS

No agreement has been reached as to the origin of inselbergs. Two concepts have long been dominated geomorphological thinking. Many workers like King (1953) have interpreted inselberg as the last remnants of circumdenudation remaining after a cycle of scarp retreat and pedimentation has reduced the landscape to low relief. Though the model is theoretically feasible, the field evidences on the whole contradict the scarp retreat having played a significant role in the evolution of inselbergs. On the other hand, field evidence is largely consistant with the two stage concept formulated by such workers as Falconer (1911), Hadley (1952), Linton (1955), Budel (1957) and Twidale (1981). According to this theory differential fracture controlled surface weathering in the ground water zone proceeds differential erosion and in particular the lowering of the plains and an emergence of inselbergs as relief features.

Inselbergs of Rajasthan invariably occur in the multicyclic topography. It may, therefore, be conceived that the feldspar rich rocks have undergone deep weathering under favourable climatic conditions (humid tropical) of the past resulting in the accumulation of a very thick cover of regolith over the rocks. The deep weathering and presence of deep weathering profiles is very well supported by a 450 m deep oxidized profile in the mineralized belt of Rajpur Dariba and other mines. It is assumed as the result of a humid tropical climate possibly prevailing during the Tertiary or even in the Cretaceous. Heron (1938) citing various evidences has concluded that the Aravalli range appears to have been peneplained during the Mesozoic, at sometime not very much earlier than the Cretaceous. The peneplain has since been upwarped in such a way that its central portion near Udaipur and Mount Abu is from 800 m to 1722 m above sea level and its north eastern and near Delhi is about 250 m and its south western and in Gujarat is also about 250-300 m above sea level. If we accept this thesis, it can safely be concluded that during a long period of Cretaceous planation of the Aravallis, granites, gneisses and schists were deeply weathered. The upwarping of the Aravallis revived erosion and partial stripping produced high level inselbergs of Mount Abu and surrounding areas. Lesser period of extensive weathering during post Eocene period as evidenced by laterite deposits on Deccan trap of Kota, Jhalawar and Chittaurgarh regions. Renewed erosion during Tertiary period began to make new erosion surfaces of the Mewar plateau. The formation of this surface was relatively easy because of the deep weathered mantle was ready to be stripped. The highly resistant rocks like Delhi quartzites escaped deep weathering and appear to have lowered by erosion, mass wasting and rectilinear recession of slopes. Some of the blocks of unweathered rock might have remained protruding above the regolith cover. The protruding portions above the surface of regolith cover appear to have formed the present strike ridges. Formation of lower inselbergs and tors perhaps took place by simple stripping and exhumation of the unweathered masses. Exhumation since post Eocene period followed and has continued to the present.

A section between the Aravalli hills in the west and the Vindhyan plateau in east illustrates the deep weathering profile formed over the feldspar rich rocks which has been stripped off to give rise to the peneplained tract of the Mewar (Fig. 8.8). Rajasthan witnessed alternate humid and arid phases of climate during Quaternary period. The alternate arid phases have modified the forms of inselbergs which can be seen in abrasion of tafoni and other micro forms.

CONCLUSION

Inselbergs of Rajasthan invariably occur in multicyclic topography. The hypothesis of a two cycle origin of inselberg by deep weathering followed by exhumation by stripping of weathered material has got support with certain modification in the case of Rajasthan. detailed field investigations are still required to confirm this hypothesis.

NOTE

1. personal communication by Rajendra Shukla of Geological Survey of India.

PART THREE
MORPHOGENETIC REGIONS

TEN

Morphogenetic Rgions and Land Systems

The foregoing detailed analysis of morpho-climatic systems provides the basis for the morphogenetic regionalization of Rajasthan. It has been felt that the delimitation of the morphogenetic regions of Rajasthan should be based on both present day phenomena and past morphogenetic features. Further, the classification cannot solely be based on climatic data because climate affects morphogenetic processes and forms in both a direct and indirect manner. An analysis of the morphometric properties of drainage basins and spatial pattern of drainage density, stream frequency, slope, dissection index, etc. indicates that drainage density and stream frequency are spatially most coherently distributed and form the most manageable basis for regionalization. Gregory and Gardiner (1975) state that because of its dependence on climate and basin form, drainage density provides a suitable basis for delimiting geomorphological regions in relation to the morpho-climatic system.

Climate indirectly affects soil and vegetation pattern. Therefore, soil and vegetation in association with climate also forms a basis of morphogenetic regionalization.

Thus on the basis of climatic, pedogenic, vegetation, drainage density and stream frequency, the following five morphogenetic regions have been delineated (Fig. 10.1).

a. arid morphogenetic region;

b. transitional (between arid and semi-arid) morphogenetic region;

c. semi-arid morphogenetic region;

d. sub-humid morphogenetic region; and

e. humid morphogenetic region.

The Aravallis forms the boundary between the transitional zone and semi arid morphogenetic region. The 300 m isohyte marks the boundary between the arid and transitional regions. The boundary between the semi-arid and subhumid regions is based on 650 mm isohyte. The 750 mm isohyte or the + 93 moisture index (krishnan and Thanvi 1970) forms the boundary between the subhumid and humid zones. The boundaries of soils, drainage density and vegetation roughly coincide with the boundaries mentioned. Hence it can be said that the morphogenetic regionalization procedure is a combined effect of various sets of criteria. The disparities in the morphogenetic systems and processes are clearly marked from one region to another (Table 10.1).

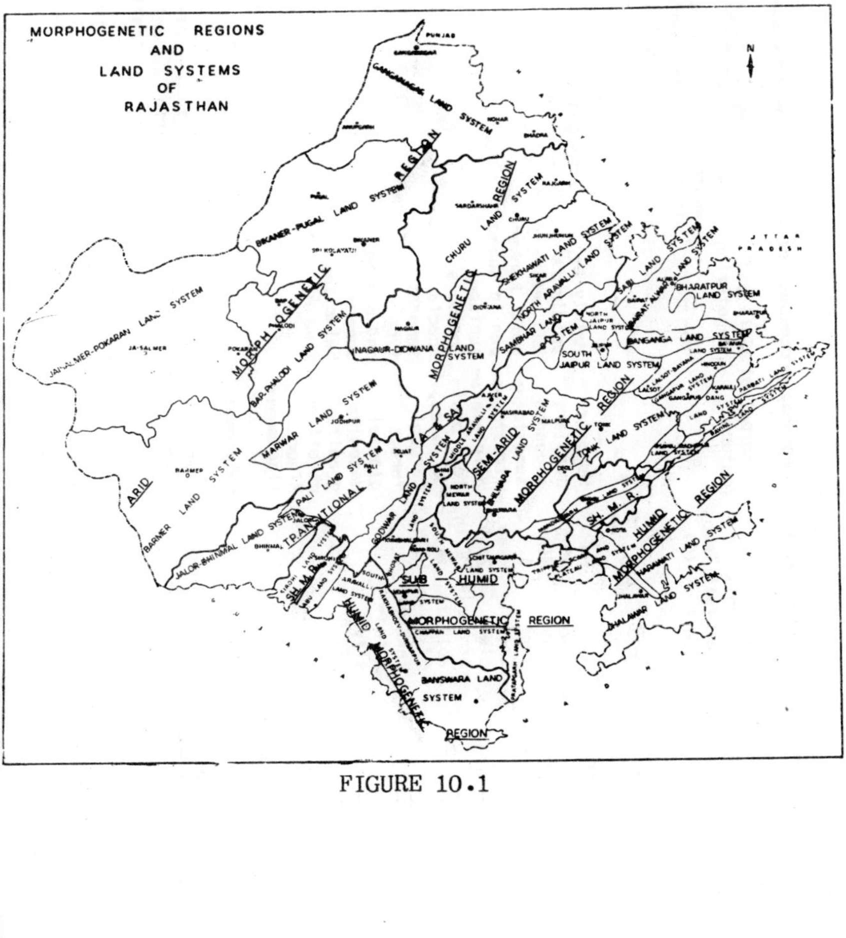

FIGURE 10.1

TABLE 10.1. **Morphogenetic regions; processes, climate and landform association of Rajasthan**

Morphogenetic Region	Climatic elements: Rainfall (mm)	Climatic elements: Temp. (C)	Lithological features	Soil and vegetation	Physical weathering	Morphogenetic processes: Chemical weathering	Morphogenetic processes: Mass movement	Morphogenetic processes: wind action	Morphogenetic processes: Fluvial	Morphological features
1	2	3	4	5	6	7	8	9	10	11
Arid morphogenetic region	100 300 even less Coefficient of variability 67.67 Annual Intensity 16.66	Max. 46 Min. 3 Mean 24.5 PE in mm 1878.2	Mesozic tertiary sequence sandstones limestones comglamerates, boulder beds, Carboniferous glacial deposits, granites, rhyolites Minor lineaments	Mainly aridisols and entisols, kanker pan in subsurface, Prosopis acacia senegal type and acacia senegal balamites type	Strong	Very weak	Slight	Strong wind speed km/ph 10.7	Fluvial slight Sheet erosion rill erosion surface drainage absent runoff nil	Sand dunes of various forms and types, parabolic, transverse, longitudinal, scrub, copple barchans. etc. pediments gravelly/rocky flats buried, interdunal flat pans, older aggraded alluvial plains, playa saline (ranns), glacial boulders, fossilized dunes, ventifacts, wind abraided grooved surfaces scarps, buttee, mesa, inselbergs, etc.

Morphometric properties

Basin slope in degrees	Drainage density	Stream frequencies	Relief ratio	Enlongation ratio	Texture ratio	Circularity ratio
12	13	14	15	16	17	18
19.80	2.34	3.57	0.086	0.81	1.88	0.56
33.97	2.28	2.28	0.098	0.86	2.71	0.53

TABLE 10.1 Contd.

1	2	3	4	5	6	7	8	9	10	11
Transitional (A & SA) morphogenetic region	300-500 Co-efficient of variability annual 45.30 Annual rainfall intensity 15.67	Max. 42 Min. 1 Mean 21.5 Evapotranspiration (mm) 1503.2	Malani rhyolites, Erinpura granites, Vindhyan sandstones limestones Eocene formation. major and minor lineaments	Aridisols and entisols mainly camborthids, calciorthids, torripsammants, paleorthids and inceptisols Kankar pan (calcrete) Prosopis, Acacia senegal type and Acacia senegal balanites type	Moderate	Weak	Moderate	Moderate Mean wind speed km/ph 7.3	Strong to weak	Sand dunes of various types, interdunal flats or pans, sandy undulating buried aggraded alluvial plain, rills and gullies, ranns, and playa lakes pediments, gravelly/rocks, scree material, bajades, etc. Azonal features - Jayal gravel, ridge, fossilized dunes, rhyolitic and granitic inselbergs, tors bornhardts, castle, kopples nubbins, etc.

12	13	14	15	16	17	18
33.87 34.93	2.32 2.29	3.14 3.10	0.218 0.246	0.75 0.84	1.66 3.17	0.64 0.58

TABLE 10.1 Contd.

1	2	3	4	5	6	7	8	9	10		11
Semi-arid morpho-genetic region	450 750 Coeffi-cient of vari-ability 40.20 Rainfall Inten-sity 17.5	Max. 44.0 Min. 2.0 Mean 23.0 Eva-tran-spir-ation in mm 1555.7	Gneisses, granites, Alwar quar-tzites, Phyllites, Aravalli quartzites Basic and ultra basic rocks Major and minor lineaments, Great bou-ndary faults Syncline and anticlines	Entisols (Torripsa-mments) ustiflu-vents alfisols (haplu-stalfs) Inceptisols (ustochr-epts) Kankar pan in subsoil horizon Vegetation-Anogeissus pendula type Acacia senegal	Moderate		Weak	Str-ong	Str-ong mean wind speed km/ph 9.9	Mode-ate to strong	Gullies and rills, scree ratematerial (talus cones), tokankarized surface strong exposed along stream surfaces, older, aggraded alluvial plains, younger alluvial plains,planation surfaces, flat summits, granite, inselbergs pediments, bornhardts, quarzitic tors, inland beheaded, streams, str-eam migration, obstacle dunes, fossilized dunes. hogback ridges.

12	13	14	15	16	17	18
24.97	1.96	2.70	0.075	0.81	1.82	0.58
39.58	1.49	1.86	0.058	0.91	3.47	0.89

TABLE 10.1 Contd.

1	2	3	4	5	6	7	8	9	10	11
Sub-Humid morpho-genetic region	500 750 Coeffi-cient varia-bility 40.27 Annual rain-fall inten-sity 25.70	Max. 42.0 Min. 2.0 Mean 22,0 Eva-po-trans pira-tion in mm 1321.2	Alwar quart zites, gneis-ses, ultra-basic rocks of Aravalli super group, phyllites Vindhyans, sand stones shales, lime stones, Dec-can Trap Major and minor linea-ments, sync-linorium, anticlino-rium	Alfisols haplustalfs vertisols, (chromus-terts) pellusterts Kankar pan in sub-soil santalum type north-ern dry deciduous forest Anogeissus	Moderate		Str-ong	Mode-rate to str-ong	Weak Mean wind speed km/ph 5.7 Strong	Denudational hills structural ⌐lateaux, pediplain, inselbergs, tors, bornhardts, castle kopples, nubbins etc. Integrated net work of drainage ways, gullies, sheet erosion, alus cones, older flood plains, stream migration, laterites, ferricretes, pediments, elongated rlges of quartzites and Aravalli super group, meandering streams, braided streams, Transverse and antecedent stream valleys, water-falls.

12	13	14	15	16	17	18
18.87	1.18	1.63	0.046	0.71	1.49	0.59
29.71	1.47	2.16	0.104	0.82	8.33	0.63

TABLE 10.1 Contd.

1	2	3	4	5	6	7	8	9	10	11
Humid morpho-genetic region	750 above up to 1600 Coeffi-cient of vari-ability 38.12 Rainfall inten-sity 21.51	Max. 46.0 Min. 4.0 Mean 25.0 Eva-po-tran-pira-tion in mm 1540.3	Alwar quartzites ultrabasic rocks, gra-nites (Erin-pura) phyll-ites Deccan trap, Vindhy-ans, sand-stones shales and lime-stones major and minor lineaments	Alfisols, vertisols inceptisols, and their combinat-ions. Kankar pan below 1 m Teak and Mitragya-Anogeissus accuminata type, sub-tropical evergreen forests southern tropical dry, decid-uous tecto-nagrandis	Weak	Strong		Mode-rate	Weak Mean wind speed km/ph 5.3 Strong	Dissected plateaux igneous plateau, mesa, buta structural, sedi-mentory plateau, infilled, river valleys, dissected pediments, flat topped ridges, planation surfaces, inselberge rock towers (batholithic mass of Mt Abu - stepped topography, river terr-aces, natural lavees, gullies and ravines incised meanders, transverse and super-imposed valleys, water-falls, escarpment slopes, potholes, rapids, etc. domal structures.

12	13	14	15	16	17	18
30.66 27.40	3.39 2.84	5.36 4.30	0.299 0.123	0.79 0.86	2.83 5.48	0.49 0.56

According to Budal (1963) Rajasthan forms a part of the tropical zone of planation surface formation including subzones such as:

a. tropical sheet wash zone; and
b. tropical hot deserts and marginal deserts.

While according to Tricart and Cailleux (1965) Rajasthan falls within the limit of:

a. deserts and degraded steppes without severe winters
b. savannas;
c. semi-desert steppes; and
d. inter-tropical forests.

The boundaries of the morphogenetic regions of Rajasthan roughly coincide with Tricart and Cailleux's zones, with some modifications. The five morphogenetic regions of Rajasthan have further been divided into 45 land systems and land units based on a recurring pattern of landforms, soils, vegetation, rocks, erosional pecularities and other morphological characteristics.

The morphogenetic regions and land systems are as follows:

1. The arid morphogenetic region which includes the following land systems:

   Jaisalmer-Pokaran;
   Barmer;
   Bap-Phalodi;
   Bikaner-Pugal;
   Marwar;
   Ganganagar.

2. The transitional morphogenetic region which includes the following land systems:

   Jalor-Bhinmal;
   Pali;
   Godwar;
   Nagaur-Didwana;
   Churu;
   Shekhawati;
   North Aravalli;
   Sambhar.

3. The semi-arid morphogenetic region, which includes the following land systems:

   North Mewar;
   Middle Aravalli;
   Bhilwara;
   Tonk;
   Sawai Madhopur-Ranthambhor;
   Ravine Belt;
   Karauli Dang;
   Parbati;
   Gangapur-Hindaun;
   Lalsot-Bayana;
   Banganga;
   South Jaipur;
   North Jaipur;
   Bharatpur;
   Bairat-Alwar;
   Sabi.

4. The subhumid morphogenetic region, which includes the following land systems:

   Bundi;
   Chittorgarh;
   Pratapgarh;
   Triple scarp and plateau;
   Chappan;
   South Mewar;
   Udaipur;
   Bhorat;
   Sirohi.

5. The humid morphogenetic region, which includes the following land systems:

   Mount Abu;
   South Aravalli;
   Rekhabdev-Dungarpur;
   Banawara;
   Jhalawar;
   Harawati.

Land systems have been designated either after the name of a major town or the historical-traditional name of the region.

ARID MORPHOGENETIC REGION

This morphogenetic region extends over the area having less than 300 mm annual rainfall. The coefficient of variability of rainfall is 67.67 while the average annual intensity of rainfall has been worked out as 16.66 mm. The mean annual temperature is recorded as 24.5°C. The potential evapotranspiration of the region is 1878.2 mm.

The entire region is covered with sand-dunes and sand-sheets of various types and forms. The rocky outcrops disrupt the monotony of this vast sandy desert. The area is formed of Mesozoic Tertiary rocks. The undulating rocks are largely sandstones, shales and limestones, which form a series of wind abraided, low, flat-topped ridges. The granitic and rhyolitic rocks are also well exposed and form isolated inselbergs, castle koppies, tors and tafonis. The region is also characterized by minor lineaments.

The soils of this region are characteristically different from those of other morphogenetic regions. Aridisols and entisols are well distributed and are characterized by hard kankar pan in the subsurface.

Prosopis acacia senegel type and acacia sengel balanites types are predominant over the entire area.

Physical weathering is dominant throughout the region. Chemical weathering is generally absent owing to lack of humidity. Physical weathering occurs through thermal expansion. The evidence of physical weathering can be seen in the form of scree material, talus cones and rock fragments in the vicinity of rocky surfaces. Mass movement processes are slight. Wind action is very strong and is evident in the wind-abraided surfaces, grooved surfaces, striation marks on rocky surfaces, and in the orientation of dunes of various types. The wind speed of the region is 1-7 kilometres per hour which is very high as compared to that of the other regions. The wind direction is generally from south-west to north-east and is responsible for the formation and orientation of the longitudinal dunes.

Fluvial action and processes are negligible and occur only in the form of sheet erosion during the rainy season. Surface drainage is absent. Only inland drainage of the centripetal type in the vicinity of ranns is perceptible. The area is free from surface runoff.

The average morphometric properties such as drainage density, stream frequency, relief ratio, elongation ratio, texture ratio and circularity ratio have been worked out and are different from those of other regions (Table 10.1).

The climato-morphological features which distinguish this morphogenetic region and others are: sand-dunes of various forms and types such as parabolic, transverse, longitudinal scrub coppice and barchans, interdunal flats or pans, gravelly/rocky pediments, old aggraded alluvial plain, saline depressions and ranns, glacial boulders, fossilized dunes, gravel ridges, ventifacts, wind-grooved surfaces, scarps, mesas, and inselbergs, etc. The former stream courses and aggraded plains indicate the humid phase and kankar pan and other features indicate the arid phase.

This morphogenetic region on the basis of lithology, surface configuration, soil morphology and erosional peculiarities has been divided into six land systems.

Jaisalmer-Pokaran Land system

This is a biggest land system unit in Rajasthan. It is also called the Chattani land system.

Rocky plains, isolated flat-topped hills, and vast sandy hills diversify the topography of this region.

Geology:

Geologically, this land system consists of Sandstones, phasphatic sandstones, limestones, clay, coal, felshic sandstones, quartzitic sandstones, ferruginous sandstones calcarious sandstones, limestones, conglomerate, silt stones and shales of the Jurassic-Cretaceous-Eocene period.

Geomorphology

The Jaisalmer-Pokaran land system lies in the south-western part of the Rajasthan desert. For some distance both east and west of Jaisalmer city there are virtually no sand accumulations. The undulating rocks which form a series of wind-abraided, low, flat-topped ridges with a general north-east to south-west trend, are part of a rocky plateau which slopes away very gradually to the Indus plains on the north-west. Further to the west there are dune fields which marks the transition from the rock plateau to the alluvial plains of the Indus, but there the undulating surface can be seen free from both sand and alluvial silts. The land north of Jaisalmer town is known as Toba land and to the south-west Tibba land, a complex dunal system. Longitudinal sand-dunes or siefs predominate in this part. Flat-floored depressions or pats occur sandwiched between two sief dunes—sites for verdant oases. Sand-dunes appear to be moving all the times, though slowly at about 5 m per year.

The landscape east of Jaisalmer continues much like that around Jaisalmer, the predominant impression being one of gently undulating troughs and ridges with occasional slightly higher hill rising above the general level. The area around Pokaran is composed of boulders of the Gondwana age. Small knolls of granite, undercut and polished by aeolian erosion, are very characteristic of the region.

The more important ranns in the area are Mitha, Kharia, Kanodwala, Kharowala and Pokaran Ranns, which receive the drainage of Kakaniwadi, Sukriwadi and Lathiwadi, forming a centripetal drainage pattern. The relief is up to 271 m.

Vegetation

The area is covered by clumps of grasses up to 1 m high. The landscape is mainly open scrub with scattered trees. There is a certain amount of vegetation in the trough between ridges, including small acacia trees.

Soils

Soils are mainly aridisols and entisols of which torripsamments, calciorthids and paleorthids are the most common. Quartzipsamments are also found in south western part.

Land units

On the basis of genetic character and erosional pecularities, the Jaisalmer-Pokaran land system can further be divided into the following land units:

a. rocky/gravelly pediments;
b. flat buried pediments;
c. sandy undulating buried pediments;
d. older aggraded alluvial flats;
e. dunal fields;
f. sandy undulating interdunal flats;
g. saline flats or ranns; and
h. hills and plateaux.

Barmer Land System

The Barmer land system is also called Malani land system. Rocky plains, isolated flat-topped hills and vast sand-dunes diversify the topography as in the case of the Jaisalmer land system. The difference is that this land system has some streams of the Luni which maintain their water courses, and that the sand-dunes are of a stable nature.

Geology

The terrain is made up of mainly sandstone, silt stone, conglomerate, shales, Badhaura sandstone of the Jurassic period, Malani rhyolites, Sewana granites and Quaternary aeolian sand.

Geomorphology

The Malani land system has one typical stony surface near Barmer, surrounded by more or less fixed sand-dunes of the barchan type. Near Chohtan and Siwana there are granitic inselbergs and knolls, sculptured characteristically by the wind. The height of these inselberg ranges from 391 m to 834 m near Siwana. The effect of the daily range of temperature on the rock masses resulting in their splitting and peeling, are evident. Chemical weathering is minimum and mechanical weathering is dominant. The tops of the granitic hills are dome shaped, while the tops of the rhyolitic hills are pointed and their sides are steep. The drainage radiates from these hills in all directions. severe water erosion has created rills and gullies from 70 cm to 1 m depth and 2 m to 3 m depth respectively. East of Barmer, the river terraces of the Luni are evident. The terraces are made up of river silt and are now dotted with sanddunes. The level difference between the terraces is sometimes as much as 50 m. The interdunal space varies from 130 m to 150 m. The dunes run through aeolian and fluvial phases of sandbars are evident on the recent floodplain of the Luni. Saline depression, like Thob are also numerous. These saline depressions appear to be the relicts of past integrated drainage. The depth of the saline depressions is from 0.25 m to 3.5 m. They are inundated during the rainy season.

Fluvial action appears to have stripped off the Vindhyan sandstone overlying the Malani rhyolites and as such remnants of sandstones are now visible only here and there. The granites and rhyolites have also been stripped off and dissected, leading to the formation of extensive alluvial plains. Kankar pan is also evident below 90 cm, mainly in the piedmont zones.

Soils

Aridisols and entisoles, or combination of both, form most of the surface soil of this land system. The area is mainly covered by torripsamments, calciorthids and paleorthids of mixed aridisols and entisols.

Vegetation

Surface vegetation consisting of scanty bushes, scrub and acacia, dominate the landscape.

Land units

The following land units have been identified in the land system:

a. hills or inselbergs of granite and rhyolites
b. rocky /gravelly pediments;
c. flat buried pediments;
d. sandy undulating buried pediments;
e. older aggraded alluvial flats;
f. sandy undulating older alluvial plains;
g. dunal fields;
h. sandy undulating interdunal plains;
i. saline depression; and
j. recent floodplains.

Marwar Land system

The Marwar land system is characteristically different from other described land systems in the sense that it is mainly rocky and stony desert. Dunes occur in the extreme north-western part of this land system. Sandstone plateau and domes dominate the landscape.

Geology

Horizontal sandstones, shales and limestones of the trans-Aravalli Vindhyans; rhyolites and phyllites of the Delhi super group are found in the Marwar land system.

Geomorphology

The land system forms the eastern limit of the Rajasthan desert. The steep bare sandstone hills, domes and plateau dominate the entire landscape. The summits are flat. The plateau is bounded by a steep free face or scarp face with detritus and pediment below. Phyllites also form hills and inselbergs. The plateau falls away steeply to the south and east, and rather more gradually to the north and west. The area is drained by the right bank tributaries of the Luni namely, the Jogri and Mitri. The Jogri once was the part of the Luni and probably used to join the Luni river near Pachpadra salt lake. The streams are ephemeral and intermittent in nature. The general slope is from north to south. Sandy plains and dunes occur in the north-west and north-east near Osian bhakar and Jodhpur bhakar. The southern portion is stipped off in the form of an alluvial plain. Parabolic, longitudinal, transverse barchans and scrub coppice dominate the landscape in dunal country. The relief is up to 475 m.

Physical disintegration in the form of block separation and granular disintegration is common in sandstones and rhyolites, respectively. Chemical weathering is absent.

Soils

The surface soils are mainly aridisols including camborthides and calciorthids. Inceptisols also occur in the north-eastern part. Around Jodhpur both aridosols and entisols are found in a mixed form.

Vegetation

Vegetation is of the scrub type. Calligonum polygonoids, Crototaria buried and Capparis decidua are the main shrubs while Panicum turgidum, Lasiurus sindicuss, Cenchrus ciliaris are the main grasses.

Land units

The following land units have been identified in this land system:

a. rocky/gravelly pediments;
b. flat pediments;
c. hills and plateau;
d. sand dunes;
e. interdunal plains;

f. saline depressions; and
g. old aggraded alluvial plains.

Bap-Phalodi Land System

This land system consists of an upland of stony character.

Geology

Sandstones, limestones and conglomerates of the Jurassic period form the surface. Bap boulder beds of the permo-carboniferous glacial deposits and quaternary silt and aeolian sand constitute the Bap-Phalodi land system.

Geomorphology

The Bap-Phalodi upland is a peneplain littered with pebbles representing the coarse products of rock disintegration, the fine ones being winowed away by wind. They are heaps up to a height of about 50 m above general ground level, forming stony wastes. The size of boulders and gravels is very large,ranging in diameter from 1 m to 2 m. with faceted and grooved surfaces. The northern part is a sandy plain while the south-eastern region from Shergarh to Ghantiyadi is dominated by dunes of various size and shapes. The relief is up to 383 m.

Surface drainage is absent. Only some inland drainage in the interdunal plains is present. Strong wind action is discernible.

Soils

Aridisols sometimes mixed with entisols are the main soil orders. Torrisammants, calciorthids and paleorthids are main soil groups associated with this land system.

Vegetation

Scrub and grasses of 1 m height dominate the surface. Acacia arabica, khejra and akera are also found on the way from Phalodi to Bap, and east of Phalodi.

Land units

The following land units have been identified:
a. rocky / gravelly pediments;

b. flat buried pediments;
c. sandy undulating plains; and
f. dunal fields.

Bikaner-Pugal Land System.

Dunal and sandy plains make up this land system.

Geology

Geologically this system is made up of limestone, clay and quaternary aeolian sand and silt.

Geomorphology

The Bikaner land system is a part of most dreary part of the Rajasthan desert. The surface is dominated by dunes of various forms and sizes known as thal and sandy plains at a relatively lower level in the form of a depression called thali. The area between Bikaner and Kolayat is almost flat country with some saucer-shaped depressions. The surface soil is admixed with rock fragments and gravel. Around Kolayat the surface is more undulating and has been sculptured by inland fluvial agencies. Evidence of paleochannels is also found in the form of gravel and quartz sand of fluvial origin. The north-western and north-eastern region contain sand-dunes of the barchan type with a steep concave northern face which rise to a maximum height of 50 m. A number of salt soaked playa basins such as that of Lunkarnsar diversify the north-eastern topography. The relief is up to 238 m.

Soils

The soils are mainly entisols or mixed aridosols and entisols. The surface colour of the soil around Bikaner is red. Near Kolayat, torrifluvents of the entisol order are widespread.

Vegetation

A mixed xeromorphic thorn forest type of vegetation is prevalent here. Psammophytic scrub desert vegetation dominates the north-western part of this land system, owing to very low annual rainfall.

Land unit

The Bikaner-Pugal land system has further been divided

into the following land units:

a. rocky / gravelly pediments;
b. flat buried pediments;
c. flat interdunal plains;
d. sandy undulating interdunal plains; and
e. sand dunes.

Ganganagar Land System

The surface character of this land system is quite different from that of the other land systems described earlier under the arid morphogenetic region. It forms the outer fringe of the Rajasthan desert and contains more sand free land.

Geology

Quaternary aeolian sand and old alluvium constitute this land system.

Geomorphology

Being on the outer fringe of the Rajasthan desert, this land system is a dune free country, mainly made up of older aggraded alluvial material. The area is also known as Rajasthan bagar. Saucer-shaped depressions are common as in Rawatsar Rohi and Nohar Rohi (rohi means green field). Except for the wide dry bed of the Ghaggar (the Saraswati) and one or two other beds joining it (the Drishadwati, etc.), there is no indication of fluvial erosion in the area in the present time. Transverse dunes are common south of the dry bed of the Ghaggar in the area between Suratgarh and Rawatsar.

Soils

The surface soil consists of clay and sand of aeolian origin. Torrisammants and torrifluvents are common throughout the land system.

Vegetation

Steppe vegetation prevails in this region.

Land units

The following land units have been delineated in this land system:

a. older aggraded alluvial flat;
b. younger alluvial plain;
c. sandy undulating plain; and
d. dunal fields.

TRANSITIONAL MORPHOGENETIC REGION

Between 300 isohyte and the Aravalli range the vast elongated stretch of land represents a true transitional morphological character both of arid and semi-arid. On the eastern margin of the arid region the action of running water is evident. Rills and gullies replace the wind abraded polished surfaces of the arid region. In the southern part of this morphogenetic region the great volume of running water flowing is clearly apparent from the large number of water channels of the Luni which descend from the western slopes of the Aravalli range. In the rainy season the streams are filled with rushing water. The combined water power erodes away the ground at the base of the Aravallis. The northern region is morphologically different from the southern, in the sense that there are very few stream courses discernible on the surface. Only a few streams from the western slopes of the central and northern Aravalli descend and lose themselves in sandy country. The savanna character of the southern part is replaced by complex dunal region with intervening flat alluvial plains.

The mean annual rainfall in this region is from 300 mm to 500 m. The coefficient of variability is 45.30 and annual intensity of rainfall 15-67 mm. The mean annual temperature is 21.5°C. Evapotranspiration is 1503.2 mm, which is less than the arid region.

This morphogenetic region is formed of trans-Vindhyan sandstones, limestones and shales, Malani rhyolites, Erinpura granites and Ecocene formations. Major and minor lineaments are common. The Luni-Sukri lineament is a major one which controls the stream courses of the Luni and Sukri.

Prosopis acacia senegal and Acacia senegal balanites types dominate the landscape.

The surface soils are of the aridisol, entisol and inceptisol orders. Kankar pan is common in all the soil profiles.

Physical weathering is moderate and chemical weathering

is weak. Mass movements and wind actions are also moderate. The wind speed is about 7.3 km per, and strong to weak action of running water is observed.

The morphometric properties of this region are characteristically similar to those of the arid region, except for the basin slope which is higher in this zone (Table 10.1).

The morphological features are: sand-dunes of various forms and types, interdunal flats (pans), sandy undulating buried aggraded older alluvial plains, rills, gullies, playa lakes, scree material, bajadas in the piedmont zone, fossil dunes, fluvial soils, gravel ridges, inselberg, tors, castle koppies, nubbins and younger alluvial plains.

This morphogenetic region has been divided into eight land system.

Jalor-Bhinmal Land System

This land system consists of gently undulating plain dotted with granitic and rhyolitic inselbergs and hills.

Geology

Eninpura granites, Jalore granites, Malani rhyolites, alluvium and aeolian sand in the foothills constitute the geological feature of this land systems.

Geomorphology

In the Jalor-Bhinmal land system inselbergs and low hills are eroded below the cretaceous Aravalli surface. Consequent upon the upliftment of the Aravalli range a second cycle of erosion was started when the low-lying areas of the eroded surface of the Aravallis and the exposed Erinpura granite suites were deeply weathered. The uplands, however, were unaffected. The exposed rocks were worn down by the subaerial denudation. The Jalor and Malani rhyolites are younger than the Erinpura granites. The rhyolite with its vertical and horizontal joints is more resistant to weathering and may be found as rugged inselbergs and domed inselbergs. The Jalor hills are flanked at their bases by piedmont plains and are composed of thick colluvial debris derived from the adjoining hills. The obstacle dunes and wholebacks are formed on the colluvial sediments of the piedmont plains. The area is drained by the Jawai, Khari, Bandi and Sagi rivers of the Luni system. The drainage pattern is dendritic to subparallel. The land system is affected by severe wind and by slight to severe water hazards.

Soils

Aridisols are the main surface soils. Aridisols and entisols in mixed form also cover the plain area. Inceptisols cover the north-eastern part of the land system.

Vegetation

The Acacia senegal community dominates the entire surface. Prosopis cineria associated with the younger alluvial plains and older alluvial plains are significant.

Land units

The following land units have been identified in this land system:

a. hills and inselbergs;
b. rocky/gravelly pediments;
c. flat buried pediments;
d. old aggraded alluvial flats;
e. pediplain; and
f. younger alluvial plain.

Godwar Land System

Piedmont plain constitutes this land system.

Geology

Erinpura granites, Ajabgarh phyllites and trans-Vindhyan sandstones, shales and limestones, and Quaternary alluvium constitute this land system.

Geomorphology

The Godwar land system consists of a long rolling piedmont plain bordering the western slope of the Aravalli range. Inselbergs dominate the landscape around Jawai Bandh, Sanderao, Desuri and Falna. Bajada and colluvial material form the typical character of the foothills, which extends up to Sendra in Beawar. The morphological features of this region are dominated by inselbergs, tors, tafonis and nubbins. Flared slopes are common. Domed inselbergs of Erinpura granite also form a conspicuous feature of the landscape.

The area is drained by the headward tributaries of the Jawai and other streams of the Luni system.

Soils

The soils belong to the aridisols and inceptisols orders.

Vegetation

The foothills and hill slopes of the Aravalli range are covered with thick vegetation. However, most of the Aravalli slopes and inselbergs are dotted with the cactus variety of the flora.

Land units

The following land units are found here:

a. hills and inselbergs;
b. flat buried pediments; and
c. pediplain.

Pali Land System

Rolling plains dotted with isolated hills and inselbergs constitute the Pali land system:

Geology

Geologically, this land system consists of biotite schists, granitic gneisses, Jalor and Erinpura granites, limestone, phyllites and quaternary alluvium.

Geomorphology

This is a bolson territory. Residual low hills or inselbergs rise above the floor of the bolson near Pali, Chanod, Kalia and Sojat. The area is drained by the left bank tributaries of the Luni River: Sukri, Guhiya, Bandi and Mirti, giving rise to a perfect plain. the rivers are seasonal and their courses are shallow and brainded. The relief is up to 644 m.

Soils

The surface soils are mainly aridisols with camborthids and calciorthids groups. Kankar is found below 1 m and even less.

Vegetation

The vegetation consists of mainly of the members of the Acacia senegal community; also of Capparis decidua, Prosopis cinereria etc.

Land units

The following land units have been identified in the Pali land system:

a. hills and inselbergs;
b. rocky/gravelly pediments;
c. flat buried pediments;
d. saline flat aggraded older alluvial plains; and
e. younger alluvial plain.

Nagaur-Didwana Land System

Flat plains and sandy dunal plains make up this land system.

Geology

Granites, trans-Vindhyan sandstones, shales, limestones, phyllites, quaternary aeolian sand and alluvium are the main geological features of this area.

Geomorphology:

Geomorphologically this land system is very interesting in the sense that it contains a gravel ridge at Jayal_a fossil feature in the present desertic condition. Sandy plain and dunal field form conspicuous features of the landscape. The Didwana salt lake is a saline depression. Surface drainage is very meagre. Longitudinal dunes from Merta road to Didwana are a characteristic feature of the area. The dunal complex merges into the Churu land system. The relief is up to 473 m.

Soils

The soils consist of aridisols, entisols and in some patches, torripsammants of entisols.

Vegetation

Mainly grasses and bushes of exophytic plants of the Acacia senegal community and akara catch the eye of observer in this region.

Land units

The main land units in the Nagaur-Didwana land system are :

a. flat buried pediments;
b. the Jayal gravel ridge;
c. dunal fields;
d. older aggraded alluvial flats; and
e. saline flats.

Churu Land System

Complex dunal fields undulating sandy plain constitute this land system.

Geology

The region is mainly covered with Quaternary aeolian sand and alluvium.

Geomorphology

Sand dunes form a conspicuous geomorphic feature of this land system. The area is characterized by numerous inter-dunal flat which distinguishes this land unit from the others. The pans are circular or sub-circular to ovate in shape. The relief is up to 398 m.

Soils

The surface soil consists of camborthids, caliorthids, torripsammants and paleorthids of aridisols and entisols soil orders. Kankar pan is present in the subsoil.

Vegetation

The vegetation consists chiefly of the Acacia sengal type. Bushes and xerophytic plants dominate the landscape.

Land units

The following land units are found in the Churu land

system:

a. dunal fields;
b. sandy undulating interdunal plains;
c. sandy undulating aggraded older alluvial plains; and
d. flat interdunal plains.

Shekhawati Land System

This land system consists of sandy plains with green fields.

Geology

Quartzites occur in the vicinity of Jhunjhunu. Some exposures of granites, but mainly quaternary aeolian sand and alluvium are found.

Geomorphology

This is an eastern continuation of the Churu land system. But the sand thins out towards Shekhawati. Because of its closeness to the Aravalli range, it receives water courses flowing down the western flank of the range. This region is also called Rohi land, meaning green fields and prosperous villages. The dunes are of the stable type. However, barchans also prevails around Sikar and Jhunjhunu.

Soils

The soils are mainly toripsammants, calciorthids, paleorthids and comborthids of the aridisol and entisol orders.

Land units

The main land units are:

a. sandy aggraded older alluvial plains;
b. flat interdunal plains;
c. sand dune fields;
d. hills; and
e. sandy undulating interdunal plains.

North Aravalli Land System

This is mountainous country characterized by quartzite ridges and phyllitic valleys of the folded Delhi super group.

Geology

Mainly quartzites and phyllites of the Delhi super group, granites and pegmatites of post-Delhi intrusion, quaternary aeolian sand and alluvium occur in this region.

Geomorphology

This land system is made up of part of the dissected Aravalli peneplain of the Mesozoic period. The summit plains are well preserved. The quartzites form ridges and the phyllites invariably have given rise to valleys. Ridges are flat-topped monoclinals. The fluvial courses are mainly alligned to the synclinal structures and zones of softer phyllites. The topography is of the hogback type. This region is characterized by high relief and continuous ridges of quartzites. Abundant scree material is found at the base of these hills. The prevalence of conspicuous dip slopes instead of scarps is owing to the high degree of rock dip. The defined drainage radiates from every cluster of hills. These rivers dies out in the piedmont zone. The Katli, Sota, Dohan and Sabi originate from the slopes of this part of the Aravallis.

Soils

The soils are mainly mixed with rock pieces, boulders and gravel. Toripsamments of the entisol order dominate the surface soil.

Vegetation

As mentioned before there is a gradual increase in the amount and variety of the vegetation from west to east with increasing rainfall. The growth of vegetation is more luxuriant on the side slopes of the hills than near their tops. The Udaipurwati hills support fairly dense forest.

Land units

The following land units have been identified:

a. hogback ridges;
b. intermontane valleys; and
c. pediplain.

Sambhar Land System

The Sambhar land system falls both in the transitional zone as well as in the semi-arid region. Saline flats with centripetal drainage patterns are its salient features.

Geology

Weathered mica schists, felspathic quartzites, pegmatites and quaternary aeolian sand, silty clay with fresh water fossils, calc congomerates, etc. occur in this region.

Geomorphology

The lake is a bolson feature in a semi-arid and transitional climatic zone. Both aggradational and degradational features have been studied. The structural hill ranges, formed of felspathic quartzites, are scattered sporadically in the area, and their intervening areas are either blanketed by aeolian cover forming buried pediments, or they form pediplane surfaces. Most of the area around Sambhar is covered with sand dunes and sand sheets. This gives rise to a sandy undulating topography of this region: Recent and reactivated dunes are mostly seen in the northern part of the lake area near Kajpura. The recent dunes are mostly obstacle dunes. The relief is up to 742 m.

Soils

The surface soils are saline of mixed aridisols and entisols.

Vegetation

The vegetation of the interdunal flats consists mainly of khejara and acacia.

Land units

The following land units are evident in this land system:

a. dunes / interdunal flats;
b. playas / bolson;
c. Pediments / buried pediments;
d. structural hills;
e. inselbergs; and
f. fluvial terraces along the Mendha river.

SEMI-ARID MORPHOGENETIC REGION

The intensity of fluvial processes and other mass movement processes east of the Aravalli range increases with increasing rainfall. The character of the topographic and surface configuration is different from the other regions. Dunal fields are replaced by sand free and aggraded alluvial plains. The rills and gullies on the hills slopes and along the stream courses become evident. The sand sheets, owing to the availability of more moisture, tend to stabilize and are greatly oxidized. The action of mass movement in the form of scree material is apparent along the foothill zones.

This region receives a rainfall of between 450 mm and 650 mm and sometimes up to 750 m. The annual rainfall intensity is worked out as 17.5 mm. The coefficient of rainfall variability is 40.20 per cent. The average annual temperature is 23°C. The PE is 1555.7 mm which is lower than that of the arid region.

Alwar quartzites, granites and gneisses, phyllites basic and ultra basic rock and Vindhyans constitute major part of this morphogenetic region. Major and minor lineaments are common. The Great Boundary Fault divides the older Aravallis from the Vindhyans. Synclines and anticlines also dominate the folded structures.

The soils of this region are of the entisol, alfisol and inceptisol great orders. Kankar pan is found in the subsoil and is exposed in the vicinity of streams.

The vegetation is mainly of the Anogeissus pendula and Acacia senegal type.

The geomorphic processes and physical weathering processes are moderate while the chemical weathering is still weak owing to the lack of moisture in the air. Mass movement processes are strong. Wind action is still strong even with the increase in the amount of rainfall. The mean wind speed is 9.9 km. per hour. Fluvial action is moderate to strong.

The morphometric properties of this morphogenetic region are presented in Table 10.1. There is a decrease in drainage density even with the increase in rainfall and network of drainage.

The mean morphological features of this region are: gullies and rills, scree material in the form of talus cones, older aggraded alluvial plain, younger alluvial plains, flat-topped monoclinal structural hills, granitic inselbergs both dome shaped or tors, pediplains, inland beheaded stream's, stream migration, obstacle dunes, fossilized dunes, hogback ridges and valleys, fluvial terraces and buried pediments.

This morphogenetic region has been divided into 16 land systems, as:-

North Mewar Land system

This land system consists of rolling high plains.

Geology

The North Mewar land system includes rocks of the Bhilwara super group comprising metavolcanics, composite gneisses, schist, granites, hornblende schist, biotite schist and gneiss.

Geomorphology

This land system is a transitional zone between the Bhorat and Bhilwara land systems. It has been dissected below the Mesozoic Aravalli surface. The true form of the peneplain, is evident in a traverse from Kankeroli to Asind and Bhilwara where the rocks of varying hardness gneisses, granites, phyllites and schists have been truncated by a peneplain now lying between 450 m and 525 m in the South. The presence of large and small reservoirs with a basin like form is characteristic of the entire area. Stream migration in the Banas river near Uncha is evident. Boulders and cobbles are widespread. The Banas and its tributaries drain the land system.

Transverse valleys are also evident. Braided river courses and rock islands are also observable along the Banas river. Biological and chemical weathering can be seen in the hilly regions. The relief is up to 689 m.

Soils

The soils of the land system are heavy textured, and are grey to black in colour. Haplustalfs of the alfisol order and ustochrepts of the inceptisol order are dominant.

Vegetation

Subsidiary edaphic type of vegetation such as Boswellia serrata Anogeissus pendula and Acacia senegal in this land system is commonly found.

Land units

The region can be further divided into the following land units:

a. hills;
b. pediplain;
c. younger alluvial plain;
d. older floodplains; and
e. dissected pediments.

Middle Aravalli Land system

This land system is also called the Merwara land system. It extends from the Sambhar lake in the north to Bhorat plateau in the south.

Geology

Geologically mafic and ultra-mafic associated with migmatization, Delhi quartzites, phyllites, granites, etc. make up this region.

Geomorphology

The Middle Aravalli land system is a tightly folded synclinal hills of 600 m to 900 m high. It is traversed by a number of wind gaps, which indicates that the ancient fluvial courses across the Aravalli range vanished after the mesozoic diastrophic activity. The western slopes are steeper than the eastern. The Sagarmati, Saraswati, Luni, Sukri and tributaries of the Luni and Sukri rivers drain the western slopes of this region; while the Khari and Die, tributaries of the Banas, drain the eastern slopes. Trellis drainage pattern is evident. Bajada slopes are common in the foothill zones. High lands are conspicuous between two parallel fold ridges.

Soils

The soils of the Middle Aravalli land system are hilly.

Vegetation

Thor is the dominant species on the Aravallis. Acacia senegal types of plants are also common.

Land units

- flat-topped hills;
- intermontane valleys;
- intermontane highlands; and
- bajada zones.

Bhilwara Land System

This is the biggest of all the land systems of the semi-arid morphogenetic region. It is a peneplain surface dotted with monadnocks and domed inselbergs.

Geology

Rocks of the Bhilwara super group consist of granitic gneiss, schist, Jahajpur granites and mata complex.

Geomorphology

This land system is a north-eastern continuation of the north Mewar peneplain passes through Shahpura and truncated with the lower peneplain surface of the quaternary period in the north-east near the Banas plain. Antecedent stream courses are evident. For example, near Raj Mahal the Banas has breached the quartzitic ridges. Inselbergs and structural hills with rolling pediments and pediplains dominate the entire landscape of this land system. The hill tops are rounded, with concave side slopes. Deoli, Nasirabad, Malpura and Hindoli are part of this system which is characterized by isolated hills, domed inselbergs and extremely flat rocky terrain. The relief is up to 584 m.

Soils

The soils are heavy textured mature and grey to black in colour. Alfisols and inceptisols are most common orders found. In some pockets torripsamments of the entisol order can be seen.

Vegetation

The chief vegetation is of the subsidiary edaphic type.

Land units

The following land units have been identified:

a. hills and inselbergs;
b. older aggraded alluvial plains;
c. pediplains;
d. younger alluvial plains; and
e. pediments.

Tonk Land System

Rolling plains dotted with isolated hogbacks and structural hills constitute this land system.

Geology:

The area is composed mainly rocks of the Aravalli and Delhi super groups comprising quartzites, phyllites, schist, gneisses and alluvium.

Geomorphology

The character of the Tertiary peneplain of the Bhilwara land system is maintained in this region, which is also a perfect rolling plain. Hogback ridges,monoclinal ridges with buried pediments are characteristic of this land form. The area is traversed by the Banas and its tributaries. Sand bars and river terraces are found along the Banas. Escarpment slopes near Boanli and Niwai, and pointed elongated ridges around Tonk are common. Weathered material in the form of scree is abundant. Stream courses are senile and shallow. Gullies along the Banas and its tributaries are also observable. The relief is up to 541 m.

Soils:

In this land system the soils are mainly of the alfisol order.

Vegetation:

The vegetation of this region is of the subsidiary edaphic type.

Land units:

The major land units are as follows:

a. structural hills;
b. older aggraded alluvial plains;
c. younger alluvial plains;
d. pediments;
e. fluvial terraces;
f. obstacle dunes; and
g. saline flats where the kankarized surface is exposed by erosion.

Gangapur Land System

Plain country with ridges constitutes the Gangapur land system.

Geology:

The rocks of the Aravalli and Delhi super group are dominant ridge makers. Alluvium constitute the plain.

Geomorphology:

The enclosed basin between karauli and Gangapur is a very conspicuous feature of this land system. Gullies and ravines on the foothills and along the river courses are common. The Karauli river, a tributary of the Morel drains the whole region. The relief is up to 250 m.

soils:

Soils of this land system are of alfisol order. Kankars are common in the subsoil.

Vegetation:

The vegetation consists mainly of subsidiary edaphic type.

Land units:

The following land units are identified:

a. hills;
b. ravines and gullies;
c. older aggraded alluvial plains;
d. buried pediments; and
e. younger alluvial plains.

Lalsot Bayana Land System

Hilly region with intervening valleys make up this land

system.

Geology:

This land system is made up of rocks of the Aravalli and Delhi super groups arkose orthoquartzites, quartz schist and conglomerates.

Geomorphology

Continuous and elongated ridges separated by longitudinal valleys dominate the landscape. The hills reflect the trend of the thin persistent bands of quartzites. The steep barren slopes with serrate edges and craggy outlines are breached at intervals by wind gaps which represent the former fluvial courses. The Churhao is the main river draining this hilly land (Fig. 10.2).

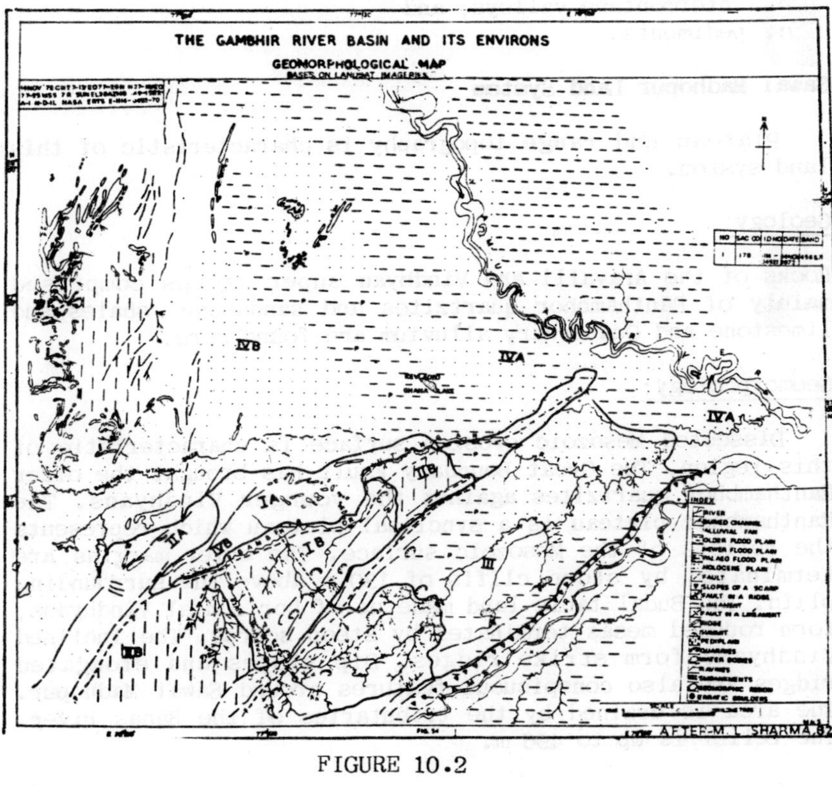

FIGURE 10.2

Steep scree-covered slopes are common. The paralle drainage pattern dominates.

Soils

Soils of the inceptisol order are found in the valleys the Lalsot Bayana land system.

Vegetation:

The vegetation consists mainly of the subsidiary edaphi type.

Land units:

The following land units are identified.

a. hills;
b. intermontane valleys; and
c. pediments.

Sawai Madhopur Land System

Plateau and range topography is characteristic of this land system.

Geology:

Rocks of the Aravalli and Vindhyan super groups comprised mainly of Ranthambhor quartzites and sandstone, shales and limestone and Quaternary alluvium are found here.

Geomorphology:

Dissected Mesozoic erosion surface is characteristic of this region. The Great Boundary Fault has brought the older Ranthambhor quartzites against the younger Vindhyans. The Ranthambhor plateau is a synclinal plateau which represents the relicts of the Mesozoic surface. The outer margins are terminated by steep cliffs of 180 m above the surrounding plain. The Budal table land made up of horizontal Vindhyans, form rounded mesas terminated by steep scarps. The inclined Vindhyans form strike ridges. Dip slopes and elongated ridges are also conspicuous features around Sawai Madhopur. The area is drained by the tributaries of the Banas river. The relief is up to 496 m.

Soils:

The soils are mainly hilly.

Vegetation:

Dense forests cover the Ranthambhor plateau Subsidiary edaphic type of vegetations cover the Sawai Madhopur land system.

Land units:

The region can be further divided into the following land units:

hills and plateau;
strike valleys; and
enclosed basins.

The Karauli Dang Land System

The plateau of the Vindhyan sedimentaries comprises this land system.

Geology:

The area is composed of rocks of the Vindhyans, consisting of Sandstones, shales and limestone. Older Aravallis are found west of the plateau.

Geomorphology

The range character of the Vindhyans becomes less prominent to the north of the Banas water gap and here the dang of Karauli is met with, which is dissected into flat topped hills. In the south it is bounded by the Chambal scarp, which has also been dissected to give rise to five terraces along the Chambal. The area is drained by the head streams of the Gambhir.

Soils:

Thin layers of soil mainly hilly, cover the Karauli 'Dang' land system.

Vegetation

The vegetation is mainly of the subsidiary edaphic type.

Land units:

The main land units are as:

a. plateaux;
b. pediplains; and
c. narrow valleys.

Ravine Belt

The Chambal between Etawah and its confluence with the Yamuna forms a belt of intricate network ravines and gullies.

Geology

This land system consists of shales and limestones of the Vindhyans super group and alluvium.

Geomorphology

The main geomorphic feature is the ravines. From the Banas water gap to the Yamuna, the noticeable features are floodplain bluffs and alluvial knolls. Incised meanders of the Chambal indicate slight upliftment. Sharma (1968) has attributed the genesis of gullies and ravines to the lowering of the local base level of erosion caused by cymatogenic upliftment. Four stages of ravine growth have been studied (Fig. 10.3). At places, headward erosion is also responsible for the evolution of ravines where the subsurface soil is soft and easily worm away. The mud flow also gives a helping hand to the lateral encroachment of ravines. This process is common near the lower reaches of the Chambal around Dholpur, Barhi and Ater during the rainy season. The belt of ravines is 2-3 km wide on both the sides of the Chambal. The local relief is up to 210 m.

Soil:

The soil of the ravine belt made up of loamy sand to clay loam developed from the Vindhyan shales and brought by the river. Soils belongs to the ustifluvents group.

Vegetation:

The area is devoid of vegetation. Recently bamboo has been introduced to check the growth of ravines. Vegetation is of subsidiary edaphic type.

Land units:

The following land units are found in the ravine belt;

a. ravines;
b. inter ravine flat lands;
c. flat marginal lands; and
d. terraces at the base of the Chambal scarp.

Parbati Land System

This land system consists of a flat riverine plain.

Geology:

This land system is made up of sandstone, limestone and shales of the Vindhyan group.

Geomorphology:

The Parbati is the major tributary of the Gambhir. It forms a basin in the eastern part of Rajasthan. It is a structurally controlled river which flows parallel and diagonal to the Great Boundary Fault. Steep scarps, linear ridges and intervening valleys are conspicuous geomorphic features. Ravines and gullies also form significant geomorphic features in the upper and middle reaches of the river (Fig. 10.3).

Soils:

In this land system clay loam to loamy sand of ustifluvent group is found.

Vegetation:

The vegetation of this region is the Edaphic type.

Land units:

The main land units found in the Parbati land system are:

a. dissected plateaux;
b. riverine plains;
c. ravines and gullies; and

d. escarpment slopes.

Fig 10.3

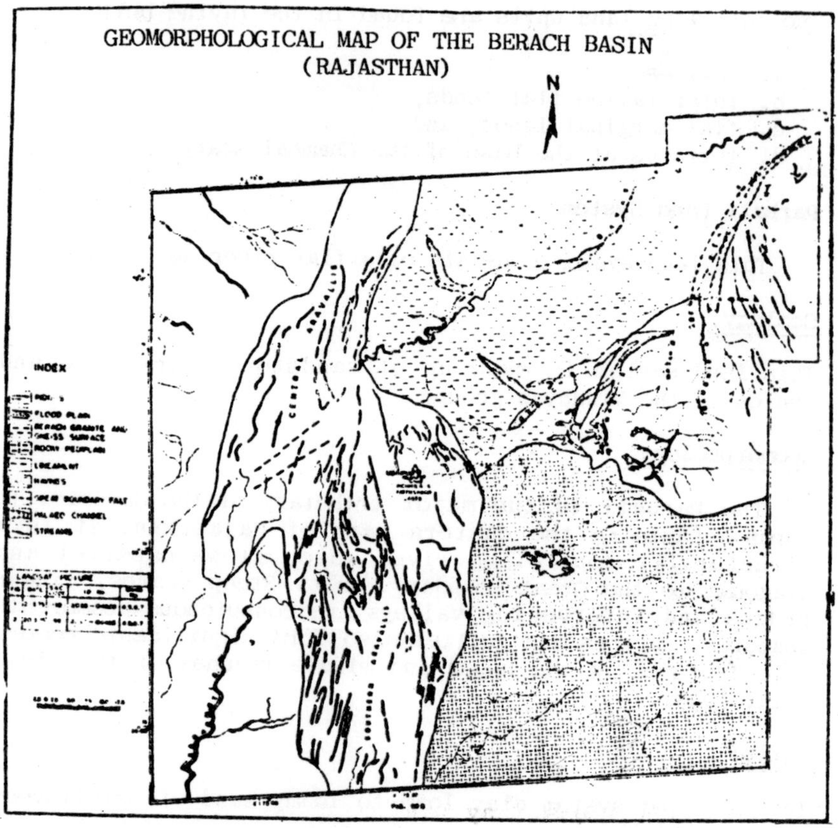

The Banganga Land System

Flat fluvio-aeolian plain constitutes this land system:

Geology:

The area is composed of quartzites of Delhi super group, phyllites and limestone in the subzone Quaternary alluvium is widespread.

Geomorphology:

The greater part of this land system is an undulating

allurial and sandy plain with a gentle easterly slope. The Banganga is an ephemeral and beheaded stream (Sharma, 1983). It disappears near Pana after branching out into several channels. The presence of aeolian and lacustrine sediments indicates alternate phases of climatic changes. A transverse valley in the upper course is evident. The relief is up to 500 m.

Soils:

In this land system soils are mainly aeolian-fluvial and lacustrine.

Vegetation:

The vegetation of this region is anogeissus pendula type.

Land units:

The following land units have been identified:

a. hills of quartzites;
b. mounds of various sizes;
c. lacustrine plains;
d. fluvio-aeolian sandy plains; and
e. terraces.

Bharatpur Land System

Flat riverine plain constitute this land system.

Geology:

The Bharatpur land system is composed of Quaternary sediments.

Geomorphology:

The Bharatpur land system consists of a flat riverine plain. Patches of aeolian sand in the form of loam are also found. The area exhibits remnants of old oxbow lakes, meanders, etc. Clay mounds and levees are common features. The relief is up to 177 m.

Soils:

Ustifluvents soils are found in the land surface of the Bharatpur land system.

Vegetation:

The vegetation of this area is mainly anogeissus pendula type.

Land units:

The region contains the following land units:

a. palaeo-floodplains;
b. mounds;
c. Ghana lake;
d. meander belt areas; and
e. lavees.

North Jaipur Land System

This land system is characterized by upland and hilly topography

Geology:

The Aravallis and Delhi super groups comprising quartzites, phyllites, schist, gneiss, granites, Quaternary sand and alluvium constitute this region.

Geomorphology:

This is a triangular land system at a height of 450 m. The eastern part is characterized by elongated hills of quartiztes. The valleys have been carved out in the areas of phyllites. The quartzitic hills are broad, flat-topped and wall like. The area provides an excellent example of aeolian and fluvial processes and resultant land forms. The area is drained by the upper streams of the Mendha, Sabi and Dhund rivers. Inland drainage is common in the Jaipur hills. Buried pediments, planer surfaces, hogback ridges, gullies, obstacle dunes, sand sheets and scree slopes are the main geomorphic features. The relief is up to 745 m.

Soils

The soils of the north Jaipur land system belong to entisol and alfisol orders.

Vegetation:

The vegetation consists mainly of the anogeissus pendula

type.

Land Units

The main land units found in this land system are:

a. aeolian landform : sheets, dunes;
b. fluvial land forms : plains, younger, gullies, terraces, etc.
c. structural hills;
d. denudational hills and inselbergs; and
e. buried pediments.

South Jaipur Land System

Flat fluvio-aeolian plain constitute the south Jaipur land system.

Geology:

Quartzites and gneisses, quarternary aeolian and alluvium deposits are main rocks of this land system.

Geomorphology:

Fluvio-aeolian landforms dominate the entire land surface. This land system is lower than the north Jaipur land system. Gullies are the chief features along the course of the Dhund and Bandi rivers and on the obstacle dunes. Buried pediments are common occur in the areas where the gneisses have been exposed on the surface around Phagi. Sand-dunes and sand sheets are common. Older aggraded alluvial and recent aluvial plains are also found. The relief is up to 456 m.

Soils:

The soils of ustisols and alfisols order are dominant in this region.

Vegetation:

The vegetation of this land system is of anogeissus pendula type.

Land units:

The following land units have been identified:

a. structural hills;
b. older aggraded alluvial plains;
c. young alluvial plains;
d. buried pediments; and
e. sand dunes and sand sheets.

Sabi Land System

Sandy undulating riverine plain constitutes this land system.

Geology:

The Sabi Land System is composed of Delhi super group mainly quartzites and phyllites, Quaternary aeolian sand and alluvium.

Geomorphology:

The sabi, an inland drainage, forms a flat basin dotted with isolated structural hills of quartzites. The river is famous for flooding during rainy season. Aeolian and fluvial processes and landforms are dominant. The stream courses are dissected into gullies and ravines. This land system is bounded on the west by the Torawati hills and south-east by the Alwar hills. The basin opens out towards the north-east. The relief is up to 340 m.

Soils:

The soils of this land system are of the torrisammant group of the entisol order.

Vegetation:

The vegetation of this land system is of prosopis acacia senegal and anogeissus pendula type.

Land units:

The following land units have been identified :

a. hills;
b. sand dunes and sand sheets;
c. older aggraded plains; and
d. recent floodplains.

Bairat-Alwar Land System

Region of hills and mountains, constitute this land system.

Geology:

The main rock types of this land system are Kushalgarh limestone, phyllites, quartzites of Delhi super group; schists marbles, phyllites and mica schist of the Aravalli super group. The folds are consistently overturned to the east with axial planes dipping steeply 60° to 80° west wards.

Geomorphology:

The structure and composition of the rock types have controlled the sculpturing of the topography of the hill ranges. The Alwar quartzites being harder than Raialo rocks, extend as main ridges. Ridges such as those around Sohna, Alwar, Kushalgarh, Bairat, Dariba and Partabgarh are determined by fold plunges. Local termination of ridges may be the result of sedimentary thinning of the quartzites. The Raialo limestone and quartzites, forming steeply dipping alternate layers, often form low hills with a ribbed appearance because of the differential weathering of quartzites and limestone under a semi-arid climate. Strike valleys are a notable feature of the region. The hills are flat topped and characteristically represent former base levels of erosion, upriser and now being redenuded to a new lower level. The Sabi and Barah rivers, the tributaries of Banganga drain this region. The deep incision of the present rivers through the ancient base level down to the existing one is illustrated by the flat-bottomed valley excavated through the hills by the Barah river at Kushalgarh.

The role of a semi-arid climate is very well shown in the form of weathering and resultant micro features along the hill slopes, namely talus cones and scree material. This area represents the best example of mass movement processes and weathering under semi-arid climatic conditions. The relief is up to 689 m.

Soils:

The soils of this land system are hilly, with ustifluvents occurring in the valleys.

Vegetation:

Anogeissus pendula type of vegetation dominates this land system.

Land units:

The following land units have been identified:

a. flat topped hills;
b. valleys between ranges of hills;
c. piedmont zones; and
d. alluvial plains.

SUBHUMID MORPHOGENETIC REGION

The region bounded by 500 and 750 mm isohytes is a subhumid morphogenetic region. It occurs in two parts; southeast of the Aravalli range covering Udaipur, Chittorgarh and Bundi district, and south-west of Aravallis including part of Sirohi district. These regions are centered on the highland of Mewar, the tripple scarp of Kota, Bundi, Chittorgarh and Sirohi. Much of the land in this morphogenetic region is mountainous, with steep slopes and a sentalum, Mitragya Anogeissus acuminate type of vegetation. Runoff and biochemical processes are indicated. The drainage density indicates that the runoff from these areas is high.

The rainfall ranges from 500 mm to 750 mm and the coefficient of variability is 40.27. The annual rainfall intensity is higher than that in the semi-arid region, that is 25.70 mm. The mean annual temperature is 22°C, and PE rate is 1321.2 mm.

The region is composed of a wide range of geological formations ranging from the Bhilwara super group to the Deccan trap and other intrusive rocks. The Vindhyans are also exposed in the eastern part of this region. Major and minor lineaments are present.

The soils of this region are haplustelfs of the aridisol order. Chromusterts and pellusterts of the vertisol order are also widespread. Hilly soils are very thin on the mountain tops. Kankar pan of moderate thickness is also exposed along the stream courses.

The morphogenetic processes have changed with the pattern of rainfall and temperature. Physical weathering and mass movement processes are moderate in intensity. Chemical weathering and the action of running water are dominant.The intensity of chemical weathering sets this region apart from the semi arid region as a morphogenetic unit. Wind action with a lower mean wind speed of 5.7 km per hour is very weak. The sculpturing is a result of fluvial processes in action since early geological times.

The morphometric properties are presented in Table 10.1.

The main morphological features are plateau, pediplains, inselbergs and tors, Nubbins and Castle koppies, integrated drainage, gullies, talus cones, concave and rounded slopes, older floodplains, laterites, ferricrits,pediments, meandering streams, braided stream courses, waterfalls and rapids, superimposed stream valleys and infilled valleys. The sheet erosion is prominent.

This morphogenetic region has been divided into nine land systems.

Mandalgarh-Bundi Land System

The Mandalgarh-Bundi land system is characterized by hilly topography.

Geology:

Sandstones, limestones and shales of the Vindhyan super group, separated by the Great Boundary Fault from the old Aravallis. Granites, gneiss, dolomites and phyllites are main rock types of this land system.

Geomorphology:

This land system is a region of elongated sandstone ranges of hills from Mandalgarh in the south-west to Lakheri in the north-east. From Mandalgarh north of Kushia Dungar the hills are in a detached form with very steep slopes or free face. North-east of it there is a continuous chain of three elongated ranges of Kaimur, Rewa and Bhander sandstones. Dip slopes are common. The tops of these hills have called bevelled surfaces, with steep scarp slopes towards the south-east. The longitudinal valleys have developed into shales and limestones. The Mej river forms a transverse gorge at Khatkar. The border hills on the Vindhyans highland are true tectonic in plan and considerably folded and faulted. At the western foothills lies a piedmont plain formed of Gwalior shales. Wind gaps are common in these hills. The relief is up to 545 m.

Soils:

Mainly hilly residual soils are found in this land system.

Vegetation:

Subsidiary edaphic type of vegetation dominates this entire land system.

Land units:

The following land units are identified.

a. ranges of sandstone hills;
b. inter-montane valleys;
c. piedmont plain; and
d. pediments.

Chittorgarh Land System

Hilly and upland country constitute this land system.

Geology:

Berach granites and gneiss, phyllites, slate, dolomite, marble, quartzites and conglomerate of the Bhilwara super group or pre-Aravallis, sandstone, shales, limestone and conglomerate of the Vindhyan super group and Deccan trap are main rock types of this land system.

Geomorphology:

The southern section of the border Vindhyan hills near chittorgarh forms a monoclinal ridge. The area is characterized by linear ridges and valleys. The stream course have been deeply incised on the Vindhyan plateau section south of Chittorgarh. The Menali and Gambhir rivers drain this region and form waterfalls along their tributaries. The pattern of drainage in this part of the land system is trellis and parallel. The south-western and southern part is a level plain made up of the pre-Aravalli rocks such as Berach granites. This broad pediplain has developed on granitic gneisses (Fig. 10.4) The extreme southern portion is a highland which is a eastern continuation of the south Mewar land system. The steep slopes and relatively undulating surface make the eastern plateau section very significant from geomorphic point of view. The Deccan trap topography is undulating, forming flat topped, rounded hills and valleys. At places the mature topography of the Vindhyans have been exhumed owing to the removal of the Deccan trap. Lateritic deposits are found over the Deccan Trap. The relief is up to 611 m.

FIGURE 10.4

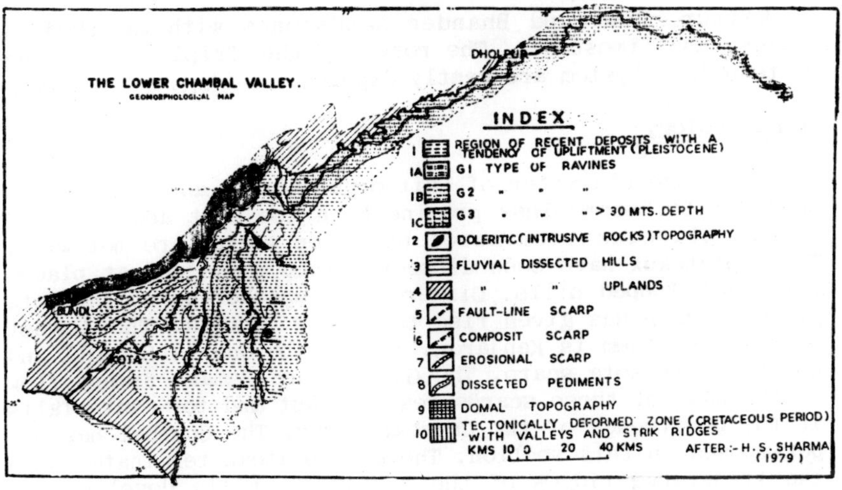

Soils

Alfisols and vertisols mixed and alfisols alone are main soils of this land system.

Vegetation

The vegetation of this land system is sentalum type and anogeissus pendula type.

Land units:

The following land units have been identified:

a. hills and plateau;
b. pediplains;
c. upland;

d. younger alluvial plains.

Triple Scarp and Plateau:

This land system is characterized by Plateau and scarp country.

Geology:

This land system is formed of:

Kaimur, Rewa and Bhander sandstones with interbedded shales, and limestones. The rocks of the Triple Scarp and Plateau land system are gently dipping.

Geomorphology:

The range character of Chittorgarh hills and Mukandwara-Bundi hills become less prominent in the east and here the triple plateaux terminated by steep scarps are met with. These plateaux have been dissected into mesas and at places into flat Topped hills. Differential weathering on the Kota plateau scarp has given rise to double scarps. The lowest scarp among them is Kanjarda. The middle one is Singoli and uppermost is Kota scarp. The Gunjali and Bamani rivers flow at the base of these scarps west to east and form waterfalls at the confluence of the Chambal river. The Chambal has superimposed in this section. The Kota plateau terminates with the transverse ridges of the Mukandwara hills developed on gently dipping hard strata. Two gorges are conspicuous in this section_Gandhisagar and Kota Bairage. The structure is very simple. The relief is up to 614 m.

Soils:

The soils of this land system are of vertisol order.

Vegetation:

Sentalum type of vegetation dominate this land system.

Land units:

The land system is further divided into the following land units:

a. plateau;
b. scarplands;

c. strike valleys; and
d. hilly sections.

Pratapgarh Land System

Rolling upland country constitute this land system.

Geology:

This land system is composed of Meta Volcanics, phyllites, mica schist of pre-Aravalli super group and Deccan trap.

Geomorphology:

The isolated flat topped Deccan trap hills and plateaux terminated by steep scarps on the west are the main geomorphic features. Erosion had succeeded, during Tertiary times, in making an extensive level surface across the pile of lava over which flow the headwaters of the Retam, Sivana Sai (Tributaries of the Chambal river) and Erau of the Mahi river separated by gently undulating divides. The relief is up to 526 m.

Soils:

The soils of this land system are of Vertisols type.

Vegetation:

Teak type of vegetation dominate this land system.

Land units:

This land system consists of the following land units;

a. plateaux;
b. scarp slopes; and
c. pediments.

Chappan Land System

The Chappan land system is characterized by dissected upland.

Geology:

Meta volcanics, phyllites, mica schist, marble dolomite,

carbon phyllites and Deccan trap are main rocks of this land system.

Geomorphology:

The region south of the Mewar land system is not a plain as it is to the north, but is deeply and intricately dissected. It is characterized by hillocky topography. The Jakham, Somi and Erau rivers drain this region. The northern portion is drained by the Gomati. This is the south-eastern part of the Tertiary planer surface dissected below the Mesozoic Aravalli surface. Deep incised courses of rivers indicate rejuvenation and examples of superimposition are numerous in this section. The relief is up to 696 m.

Soils:

The soils of this land system are of Alfisols and Vertisols.

Vegetation:

Teak type of vegetation is widespread in this land system.

Land units:

The main land units of this land system are:

a. hills;
b. valleys; and
c. lowlands.

Udaipur Land System:

The hilly and basin topography constitutes this land system.

Geology:

The main geological formation of this system are Meta volcanics, phyllites, mica schist, quartzites, ultra basic, carbon, phyllites of Aravalli super group and post Delhi intrusive granites.

Geomorphology:

The planer surface of the Mesozoic age is best developed

on the hill tops of the Udaipur land system between height 900-1100 m . The region is characterized by folded hills of Aravalli quartzites alternating with deep valleys of soft Ajabgarh mica schist, giving rise to a ridge and valley type of topography, enclosing much dissected plateaux of Ajabgarh calc schist, phyllites and limestones. The phyllitic valleys with rounded knolls are characteristic feature of this land system. The linear quartzitic ridges persist up to Iklingji, Nathdwara and near Kankorali it merges into the undulating highland of Mewar. The relief is up to 1100 m.

Soils:

Alfisols and vertisols are main soil types in this land system.

Vegetation:

The vegetation of this land system is of Anogeissus pendula types.

Land units:

The following land units are identified in this land system:

a. ranges of hills;
b. basins;
c. valleys; and
d. pediments.

South Mewar Land System:

The south Mewar land system is characterized by undulating highland.

Geology:

Schist, gneiss, ambhibolite,migmatite, phyllites, slate of pre-Aravalli and Berach granites and post-Aravalli granites (intrusive), quartzites and conglomerates of Aravalli super group are main rock types found in this land system.

Geomorphology:

This rolling upland falls north of the Udaipur and west of the Chittorgarh land systems. In the west the Bhorat land system of plateaux forms the boundary. Geomorphologically, it is a rolling undulating rocky plateau. It has been dissected by a dendritic and rectangular drainage pattern. Elongated hills with rounded summits break the monotony of this vast plateau. This region exhibits numerous depressions or water bodies. The area is drained by head streams of the Berach river. It has a planer surface dotted with monad rock developed on pre-Aravalli gneiss, schist and migmatites. The relief is up to 637 m.

Soils:

The soils are of alfisols, sometimes mixed with vertisols.

Vegetation:

The vegetation of this land system is Anogeissus pendula type.

Land units:

The following land units have been identified:

a. hills;
b. pediplains; and
c. pediments/rocky pediments.

Bhorat Land System

Plateaux and hilly country constitute this land system.

Geology

This land system is composed of calc schist, quartzites and conglomerates of Kumbhalgarh and Gogunda group of the Delhi Super group, phyllites and mica schist.

Geomorphology:

The land above 670 m largely plateau and hilly, is known as the Bhorat-Kumbhalgarh-Gogunda plateau. The western edge of the plateau is steeper and forms the western boundary of Mewar with Marwar, whereas the eastern slope is gentler and gradually fades into the high Mewar land system. The slope along the Rajnagar-Kelwara-Ranakpur line is greater than that east of kelwara. The tract south of Kelwara-Kumbhalgarh is significant for its parallel ridges, conical peaks, narrow valleys and innumerable swift flowing streams and rivulets. The Banas originates from this section south of Kumbhalgarh fort. Compared to its southern counterpart, the tract north, north-east of Kumbhalgarh has a gentler slope; narrow and low ridges encircle fairly large valleys.

Soils:

The soils of this land system are hilly.

Vegetation

Dense forest with subsidary edaphic type Boswellia Serrata. Type of vegetation dominate this land system.

Land units:

This land system has further been divided into the following land units:

a. hills;
b. plateaux; and
c. narrow valleys.

Sirohi Land System

The Sirohi Land System is of hilly and inselberg type of topography.

Geology:

Erinpura granites, mica schist, conglomerates, dolomites, etc. are main geological formation of this land system.

Geomorphology:

The Sirohi land system extends from the south of Mount Abu to Sirohi in the north. The area is characterized by granitic inselbergs. There are inter-montane basins which are filled with residual soils admixed with quartz grains. Towards the Mount Abu mass the topography is stepped and dotted with inselbergs. The slopes of the hills are cut by rills which run parallel and divide these hills into rounded inselbergs. Granular disintegration and decomposition is a characteristic feature of this region.

Soils

Residual soils admixed with quartz grains are confined to the low basins. Inceptisols are also found in this land system.

Vegetation

Mixed vegetation Acacia senegal balanites type and Mitragya Anogeissus acuminata type are dominant.

Land units:

The following land units are identified in this land system:

a. inselbergs and hills;
b. inter-montane basins; and
c. pediments

HUMID MORPHOGENETIC REGION

With the increase in the amount of rainfall in the south south-eastern part of Rajasthan the morphological features also change. This morphogenetic region is confined to the Mount Abu mass, Dungarpur, Banswara and Jhalawar districts.

The rainfall is more than 750 mm and generally between 1000mm to 1600 mm. The rainfall intensity is 21.51 mm and the coefficient of variability is 38.12. The mean annual temperature is 25°C and PE 1503 mm.

The geological formations which are confined to this zone are Alwar quartzites, ultrabasic rocks, granites, phyllites, Vindhyan sandstone, shales, limestones and Deccan trap. Numerous major and minor lineaments are also found which control the orientation of the streams.

Soils of alfisols, vertisols, inceptisols and their combinations are found throughout this region expect for hilly soils which occur in the mountainous tracts.

The vegetation pattern changes from the subsidiary adaphic type of a semi-arid climate to teak, Mitragya anogeissus accuminate and the subtropical evergreen forest of a humid climate.

The geomorphic processes such as physical weathering and

wind action are weak. Chemical weathering and landslide activities are predominant. The action of running water is maximum.

The morphometric properties of this morphogenetic region are shown in Table 10.1. The drainage density and stream frequency are higher than other morphogenetic regions.

The morphological features which dominate the landscape are dissected plateaux on igneous and sedimentary rocks, mesas, buttecs, infilled river valleys, dissected pediments,flat-topped ridges, inselbergs, river terraces, stepped topography, gullies,ravines, transverse and superimpose driver valleys, waterfalls, scarps, rapids, domal structures, incised meanders , exhumed topography etc.

On the basis of degree of erosion and lithological variation the region can further be divided into the following six land systems:

Mount Abu Land System

Batholithic mass stacks and inselberg topography constitute this land system.

Geology:

The whole area is composed of Erinpura granites. It is structurally a batholith.

Geomorphology:

The south western part of the Aravalli mountain is characterised by a topographic form sculptured on Erinpura granite by fluvial agencies and chemical weathering processes. The Mount Abu mass is separated from the main Aravalli range by a valley carved out by the western Banas river. Its highest peak, Gurushikher, is 1722 m above sea level. The slopes of this mass are extremely precipitous especially on the western and northern sides. Continued chemical weathering has given rise to several caverns and holes over this surface, the most phenomenal being the lake Nakki tal. The cavernous features are elongated, oval and circular with cross and vertical joints, which make chemical weathering easy. The whole mass has been dissected into deep valleys of 300 m to 400 m. giving rise to stepped topography. Three to four such steps can be observed. Morphologically, stacks are developed on granites. The rolling surface of this mass stands as a contrast to the general ridge and valley topography of the southern Aravalli range.

Soils:

The soils of this land system are hilly.

Vegetation:

Subtropical evergreen type vegetation and forests, dominate this land system.

Land units:

Only two types of land units are found in this land system:

a. the batholith mass; and
b. valleys and basins.

South Aravalli Land System

Ridge and valley topography constitutes this land system.

Geology:

Pre-Aravalli super groups and Delhi super groups - Alwar quartzites and phyllites are dominant types of rock found in this land system.

Geomorphology:

The plateau and range character of the Aravalli range changes south of the Bhorat Plateau, and the morphology is dominated by ridges and valleys. The quartzites which form the ridges and valleys have been carved out into areas of phyllites and limestones. The Wakel and tributaries of the western Banas and Sabarmati drain this region. The hill slopes are very steep. Valleys are broad and rounded knolls of phyllites are invariably found in the basins. The superimposed courses of streams are evidence of upliftment of this peneplain surface; maximum upliftment is exhibited here.

Soils:

Hilly soils are found in this land system.

Vegetation:

The vegetation of this land system is sub-tropical evergreen forest - Mitragya Anogeissus acuminata type.

Land units:

The region may be divided into the following land units:

a. ridges and hills;
b. valleys and basins;
c. pediments; and
d. infilled valleys.

Rakhabdev-Dungarpur Land System

Dissected upland constitute this land system.

Geology:

Rakhabdev ultrabasic suits, rocks of Aravalli super group, quartzites and phyllites of Delhi super groups are main rock types of this land system.

Geomorphology:

The area to the south of the Udaipur land system and the Chappan land system is also a dissected upland. The ridge character of the Aravalli range continues but the summits are at lower levels. Most of the area is drained by the Som and its tributaries. The river courses are very sinuous and are controlled by lineaments. Surface runoff is accelerated. Slopes are bare and summits are rounded. Around Dungarpur broad valleys and rounded summit plains are exhibited. The relief is up to 696 m.

Soils:

Around Dungarpur soils are hilly. Alfisols and vertisol orders dominate.

Vegetation:

The vegetation of this land system is of Teak forest. type.

Land units:

The following land units are identified:

a. hogback ridges;
b. broad flat plateau;
c. valleys; and
d. pediments.

Banswara Land System

Upland with ranges of hills and valleys constitute this land system.

Geology:

Meta volcanics, carbon phyllites, marbles,dolomites and Deccan trap are main geological formations.

Geomorphology:

This land system exhibits dissected uplands and long narrow with intermittent valleys carved out in areas of phyllites. The eastern half of this land system is an upland with ridges of steep precipitous slopes. The western part is a lowland developed on softer Aravalli formation. An uninterrupted dissected carp of Deccan trap runs from Banswara up to Kushalgarh. This part appears to be a fossil landscape and has been exhumed where older Aravalli formations are found just a position with the Deccan Trap, indicating that the trap was poured out upon a very uneven surface of the pre-Cretaceous age. The younger trap hills invariably are lower in elevation because of rapid spheroidal weathering of basaltic lava. They are flat topped, blackish in colour and are covered with natural vegetation. Whereas the resurrected hills of Aravalli formation rise higher, are craggy and steep sided and often are of the hogback type. The western part is a broad undulating plain carved out by the Mahi river and its tributary-Erau. The course of the Mahi and its tributaries are controlled by fracture patterns. Stream courses are meandering. The relief is up to 493 m.

Soils:

Mainly of vertisols and alfisols orders of soils are found in this land system.

Vegetation:

Teak type - subtropical evergreen type of vegetation is found in this land system.

Land units:

The major land units of this land system are:

a. hogback hills;
b. plateau;
c. pediplains; and
d. narrow valleys filled with rocky fragments and boulders

Jhalawar Land System

The Jhalawar Land System is like broad sedimentary plateau with gently dipping hill character.

Geology:

Sandstones, shales and limestone of Vindhyan super group and Deccan Trap are main rock types of this land system.

Geomorphology:

This region is a rolling plateau of sandstones which owes its origin mainly to the upliftment of an earlier original bedding plane (structural platform) or to a fossil peneplain. The whole of the flat rocky terrain is of horizontal nature but the rocks towards the north become inclined and now form the parallel sandstone ranges of the Mukandwara hills which extend from the Chambal gorge to the east of Gagraun. The south-eastern part of this land system is a plateau developed on Deccan trap. The laterite has also been developed on Deccan trap indicating the tropical condition of the climate during post-Eocene times.

Soils:

The soils are of Vertisols, inceptisols, alfisols types.

Vegetation:

The vegetation of this land system is anogeissus pendula and teak type.

Land units:

This land system has further been divided into the following land units:

a. elongated hills;
b. sandstone plateau;
c. deccan trap plateau; and
d. pediplains.

Harawati Land System

Riverine plain constitute this land system.

<u>Geology:</u>

Shales and limestones of the Vindhyans form the subsurface. Thin alluvium also cover the entire surface.

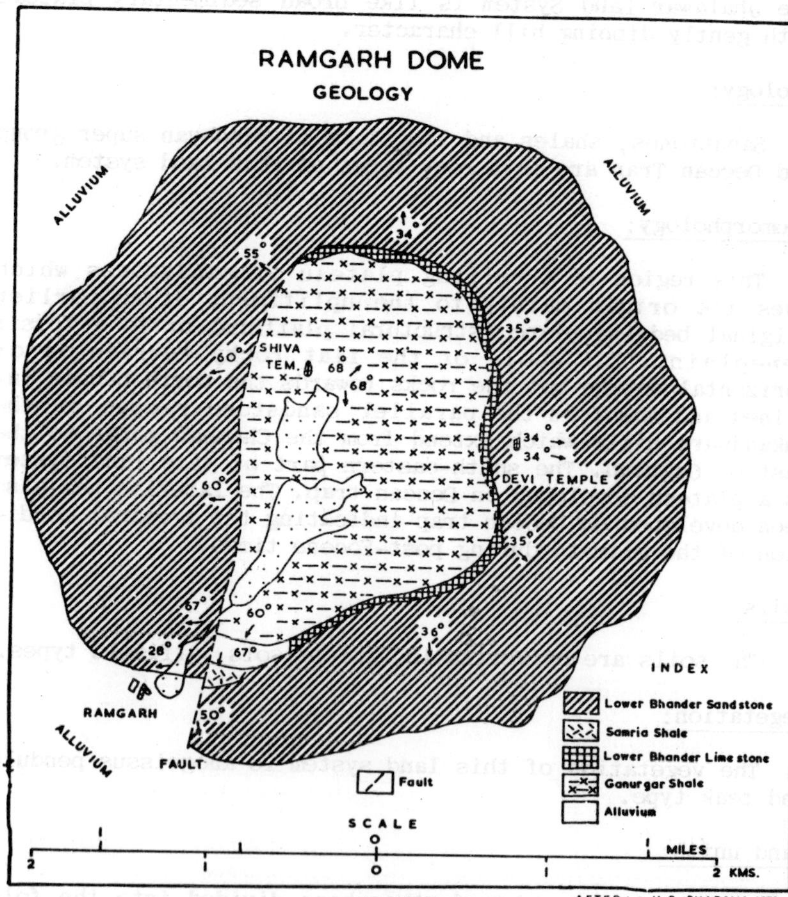

AFTER :— H.S. SHARAMA, 1973 .

FIGURE 10.5

Geomorphology:

This region is a remarkably flat riverine plain developed on Vindhyan shales (Fig. 10.5). The Kali Sindh, Chambal and Parbati rivers unitedly form a triangular drainage basin. The plain is characterized by ravines and gullies, incised river courses, and domal structures. The incised meanders indicate slight upliftment. It is a pleistocene surface. The Ramgarh structural dome is topographically expressed as a circular basin. It is composed of Bhander sandstones underlain by Ganurgarh shales. The relief is up to 269 m.

Soils:

The soils are of vertisols.

Vegetation:

The vegetation of this land system is Anogeissus pendula type.

Land units:

This land system is divided into following land units:

a. flat riverine plains;
b. ravine belts;
c. older flood plains; and
d. domal structure.

ELEVEN

Summary and Conclusion

In this morpho-climatic study of Rajasthan an attempt has been made to analyse the morphological form and processes. The study reveals that the area has been affected by the epeirogenic and cymatogenic forces rather than the orogenic movements since the Mesozoic period. The landforms and planer surfaces of different ages have evolved through long periods of subaerial degradation and aggradation interrupted by short periods of cymatogenic uplifts and climatic changes. The major existing landforms have resulted owing to the climato-morphogenetic processes operating during post-Tertiary period on various litho-units under different climatic phases. The geomorphic evidences like buried channels, old aggraded alluvial plains, fossil dunes, interdunal flats, pans, kankar pan (calcrete), playas, inselbergs, boulders and gravels, ventifacts sand pediments, palynological and archaeological evidences make it evident that this part of the country has experienced two humid and two arid phases of climate during Quaternary period. The study thus upholds the findings of scientists of central Arid Zone Research Institute, Jodhpur regarding climato-genetic history of Rajasthan during Quaternary period.

Morphogenetic Environment - Geological Structure

Geological structure forms the foundation of the landforms of an area. Geologically, Rajasthan is made up of diverse rock types ranging from the oldest Archaean metamorphites to Sub-recent to Recent alluvium and aeolian sand. Three-fourths of the state is covered by aeolian sand and alluvium, the remaining area exposes a wide variety of hard rocks comprising crystalline and sedimentary rocks. The crystalline rocks include various types of metamorphic schists, quartzites, marbles and gneisses of pre-Cambrian age with associated acid and basic intrusive rocks. The sedimentaries include the rocks of Upper pre-Cambrian Vindhyan Supergroup and of Permo-Carboniferous to Jurassic, Cretaceous and Tertiary ages. The south-eastern extremity of the state is occupied by a pile of basaltic flows of Deccan Traps of Eocene age.

Lineaments

Rajasthan exhibits a spectacular array of lineaments of varied dimensions and directions which control the morphogeny of Rajasthan to a great extent.

A regional analysis of the spectral signatures obtained by landsat imagery has revealed numerous lineaments of varying length, intensity, and azimuth. About 1,500 lineaments have been deciphered in various parts of Rajasthan. These have been divided into major, intermediate and minor-lineaments based on their spatial extent of more than 300 kilometers/100 kilometers and less than 100 kilometers respectively. Four major lineaments of more than 300 kilometers length identified in Rajasthan are named as Major, Intermediate, and Minor Lineaments.

Lineaments in the south central Rajasthan in Udaipur-Banswara signifies the presence of neo-technonics as indicated by tilting of Quaternary river terraces, deep seated nature of the few linears associated with the implacement of ultramafics and carbonatite intrusive bodies which have also been confirmed by the geo-physical profiles run across the Trans-Aravalli Lineament near Palanpur.

The Nature of Quaternary Deposits-Soils

Soils are the products of biochemical and physical process operating upon earth materials under various topographic and climatic conditions. They reflect as much as do landforms the climatic and geomorphic history of the region in which they are involved. Paleosols and polygenetic soils may furnish information from which it is possible to reconstruct past climatic and topographic conditions. It has been found in the present study that paleosols are generally associated with buried landscapes, with exhumed portions of ancient landscapes and with relict features of previous geomorphic cycles. 'Kankar Pans' or Calchi layers in the soil of Rajasthan have been interpreted as having been formed under periods of relative aridity. So the detailed study of the soil profiles and soil series along with geomorphic study has given a fairly complete record of the climatic changes which have taken place in Rajasthan since early Quaternary time down to the present.

The study of soils of Rajasthan has been done from two points of view:

- as an indicator of past climatic history, and
- as a denominator of Morphogenetic regions of Rajasthan.

Present and Palaeoclimatic Studies

Rajasthan lies within a region which according to Koppen (1931), Thornthwaite (1931), and Meigs (1953) classifications may be described as having arid (Ed), semi-arid (Dd), transitional between Ed and Dd, and sub-humid (Cw). The area under study receives limited rainfall, the rainfall variability is high and evaporation losses are very high. Almost 90 per cent of the rainfall occurs between June and September. During the remainder months of the year rainfall is light and sporadic and several consecutive rainless months and years also are not uncommon in arid part of Rajasthan. Annual and daily ranges of temperature are high. Marked seasonal contrasts in humidity, clouds, wind speed and velocity and radiation are significant.

Rajasthan bears incomparable geomorphological, archaeological, palynological and pedological evidences of climatic change. The climatic phases, have been used to construct the chronological sequences of the climatic changes in different parts of the world during the Quaternary period. In Rajasthan especially in desert part several climato-morphogenetic studies have been conducted by Stein (1942), Wadia (1960), Ghose (1964 and 1965), Abhichandani and Roy (1966), Bryson and Barrlis (1967), Raikes (1968), Ahmad (1969), Whilhelmy (1969), Verstappan (1970), Singh, G (1971), Roy and Pandey (1971), Meher-Homiji (1973), Singh, et. al. (1974), Mishra and Rajaguru (1975 and 1980), Vishnu Mittre (1976), Singh (1977), Singh and Ghose (1977), Ghose et. al. (1977), Sood et. al. (1978), Allchin, Goudie and Hegde (1978), Yash Pal et. al. (1980) and Agrawal, et. al. (1978, 1979, and 1980). In this study particular emphasis has been made on geomorphological, archaeological, palynological evidences as follows:

Geomorphic Evidences

- Paleo-Drainage channels

- Kankar Pan-calcret as an indicator of climatic changes
- Buried Pediments
- Sand dunes-fossil dunes and reddened dune sands
- Interdunal flats (pans)
- Boulders and gravels
- Laterite and Bauxite
- The recent reporting of laterites and bauxites in south-eastern Rajasthan over the Deccan Trap aiding new chapter in the post-Tertiary climatic history.

Archaeological Evidences

Palaeo-botanical Evidences

The present study refute the origin of faceted gravels and boulders in Rajasthan desert as a product of Pleistocene glaciation as suggested by some workers. Evidences show that nowhere did the Pleistocene glaciation come down to lower than 40° N latitude except in mountaneous region. Their origin has been attributed to fluvial agencies accompanied by wind abrasion.

Morphogenetic Forms and Processes -
Studies of Aeolian Forms and Processes

It has been studied that Rajasthan had experienced two major humid phases with intervening arid phases during the later Quaternary period. During intervening arid phases dune building processes took place in western Rajasthan and also towards east of the Aravalli range upto Jaipur-Bharatpur. The presence of stabilized dunes east of the Aravalli range indicated that formerly the desertic conditions were more wide spread and it subsequently strand upto the present desert boundary. Within the desert belt, there are also two distinct zones of aeolian and fluvial processes. The 300 mm isohyte form the boundary between aeolian and fluvial activities. West of 300 mm isohyte, the aeolian processes are more intensive which have re-activated and degraded the stabilized dunes and created new actives dunes. To the east of 300 mm isohyte, the fluvial processes are dominant which have dissected the old existing landforms through rills and gullies. The dunal forms of eastern Rajasthan have also been dissected into the form of rills and gullies especially on wind ward obstacle dunes.

The 100 years data of rainfall and temperature of Rajasthan do not give an indication of any departure from the normal, but the presence of dunes in eastern Rajasthan has created lots of confusion amongst geomorphologists. Indeed, sanddunes and sandsheets are found in the eastern Rajasthan at many localities. A particularly large sand field occurs in the Sambhar area, west of Jaipur and extends almost to Jaipur. The dunes extend a considerable distance east and north of Jaipur, however, therefore a large number of impressive obstacle dunes are found at Banada, Jhir, Dausa, Lalsot, Newai, Bhairana, Achroal and Gangapur. This would appear to mark the eastern most extension of the sand features in Rajasthan.

The existence of sand features in eastern Rajasthan have been misinterpreted as a product of present aeolian process. The geomorphological, archaeological, pedological and botanical evidences suggest that sanddunes of eastern Rajasthan are fossile features and of relict state of nearly 10,000 years. These dunes have well developed soil profile, well Kankarised, oxydised, and decalcified. The percentage of clay and silt ranges between 17 to 20 per cent and have a subdued form with sign of reactivation in areas of particularly strong biotic interference. The sanddunes of Sambhar area do not indicate any sand drift except local movement. The Jaipur city itself was planned on the stable dune about 255 years ago. The dunes of Jhalana and Shastri Nagar area were once well vegetated and stabilised. But because of rapid urganisation and expansion of city during last 16 years, these dunes have been reactivised and thus the loose sand transfers locally and seasonally.

Therefore, the thesis of desert expansion towards east through various gaps of the Aravallis in the form of a sand movement appears inadequate. However, over exploitation of resources, overgrazing, large scale deforestation in the Aravalli range by man and animals has accentuated the reactivation of sanddunes and accelerated soil erosion in eastern Rajasthan, causing serious ecological imbalance. The continued denudation of the Aravalli forests would have serious repurcussions and jeopardize the future of the agriculture of eastern Rajasthan.

In the present study special attention has also been paid to classify dunes to show relationship between wind regime and dune orientation, processes of sand movement and temporal changes in dunal forms. In this area six types of dunes have been identified:

Old system:
- Obstacle dunes

- Parabolic dunes
- Longitudinal dunes
- Transverse dunes

New system:

- Barchan dunes
- Scrub coppice dunes.

The detailed study of interdunal pans leads us to conclude that they must be considered as virtually fossil landforms, which originated in post-Holocene time as a result of blowouts. The blowouts become the site of gathering ground of seasonal water from adjoining dunes in which clays and sandy clays were deposited. The desiccation of pans is confirmed by the occurrence of thick layer (1-3 meters of Kanker Pan ($CaCo_3$-11-15 per cent).

Studies of Fluvial Morphology

East of 300 mm isohyte fluvial processes in the form of sheet, gully and surface wash are predominant and aeolian process become less effective. East of the Aravalli Range there is a region of predominantly fluvial activities except in the north east where both fluvial and aeolian activities transform the surface form morphogenesis.

The evolution of landforms and planation surfaces makes it evident that the existing morphogenetic features in Rajasthan are the outcome of a long and continuous morphogenetic processes which began after the upwarping of the Aravallis and the Central Vindhyan Plateau in the Cretaceous period. In the subsequent periods the fluvial cycles were interrupted twice due to the intermittent upheavals of the Himalayas. After the last phase of the Himalayan Orogeny in the Pleistocene period, the present cycle of erosion started as a result of which most of the rivers of Rajasthan (mainly the Chambal and its tributaries) were rejuvenated and thus they are actively engaged in down cutting their courses. The ravinus are thus the result of rejuvenation of streams caused by neo-tectonic activity.

Broadly, the following drainage systems have been identified in Rajasthan:

- **The disorganized Ghaggar Drainage System**
- **The Chambal-Banas Drainage System**
- **The Mahi Drainage System**

- The Luni-Jawai-Sukri Drainage System
- The Sabarmati Drainage System
- The Western Banas-Sipu Drainage System
- The Gambhir Drainage System
- Inland Drainage System
  - . The Banganga System
  - . The Katli-Dohan-Sabi and Barah system
  - . The Sambhar Didwana-Kuchaman Drainage System
  - . Inland Drainage of Jaisalmer-Barmer-Lik River and Ghugari System.

The study of inland drainage of north eastern Rajasthan indicates that these streams were formerly the part of the Ghaggar-Yamuna system and later on due to rise of the divide due to neotectonism and with the onset of aridity these streams were left as behaded streams. However, further field work is required for the confirmation of the hypothesis.

Studies of Basin Morphometry

The validity of climatic geomorphology depends on being able to establish a relationship between climate and process and to show that regional variations in climatic factors cause spatial differences in landforms. The significance of climate in relation to processes must be evaluated in order to make it beyond the descriptive morphogenetic studies. Similar to aeolian processes which are confined to the western part of Rajasthan, the fluvial systems are controlled by climatic regime and in particular to rainfall pattern.

In this study an attempt has been made to analyse the various parameters of fluvial morphometry and to show the relationship between them. In order to show the climatic control on the fluvial systems, morphometric parameters have been correlated with mean annual rainfall.

The analysis of 116 small basins of Rajasthan demonstrates that drainage network vary substentially with climate and particularly in relation to rainfall. Additional data are reguired to examine more thoroughly the hypothesis of whether drainage density, stream frequency, texture ratio and sinuósity index truely increase through arid to semiarid and then decrease in humid climatic zone. However, it can be ascertained from the little data available that the above hypothesis does not apply in the case of Rajasthan at the small basin scale. While at the meso scale (whole Rajasthan) morphometric properties like drainage density and stream frequency are related to mean annual rainfall.

Studies of Planer Surfaces

The study of planer surfaces has also been undertaken and four planer surfaces have been identified - Pre-cretaceous, mid-Miocene,post-Tertiary and Pleistocene. The remains of ancient cycles are lost in obscurity and the earliest of peneplane preserved in the relief of Rajasthan is that of following the Mesozoic. In the absence of any fossils in the pre-Cambrians of Rajasthan it had become very difficult to date the various planer surfaces.

In Rajasthan inselbergs are invariably occur in multicyclic landscapes. Hence they are the product of differential sub-surface deep weathering and subsequent removal of weathered material by fluvial erosion.

Morphogenetic Regions and Land Systems

The detailed analysis of morpho-climatic systems discussed, provides the basis for morphogenetic regionalization of Rajasthan . It has been felt that delimitation of morphogenetic regions of Rajasthan should be based on both present day phenomena and past morphogenetic features. Further, the classification can not solely be based on climatic data because climate affects morphogenetic processes and forms in both a direct and indirect way. The analysis of morphometric properties of drainage basins and spatial pattern of drainage density, stream frequency, slope, dissection index etc. indicate that the drainage density and stream frequency are spatially most coherently distributed and form the most manageable basis for regionalization.

Climate indirectly affects soil and vegetation pattern. Therefore, soil and vegetation in association with climate also have formed a basis of morphogenetic regionalization.

Thus on the basis of climatic, pedogenic, vegetation, drainage density and stream frequency the following five morphogenetic regions have been delineated:

- Arid Morphogenetic Region
- Transitional (between Arid and Semi-Arid) Morphogenetic Region
- Semi Arid Morphogenetic Region
- Sub-humid Morphogenetic Region
- Humid Morphogenetic Region

The five morphogenetic regions have been further divided into 45 landsystems.

The study of drainage density and stream frequency also help in identifying areas of erosion hazards. Such studies can be used as indices of water yield and condition of erosion. The high drainage density and stream frequency in south eastern Rajasthan indicate high runoff. Thus any clearance of vegetation and its replacement by other forms of landuse is likely to increase considerably the risk of floods. The hazards is naturally very high throughout the state. Great care is required in managing the development of the land in such areas otherwise danger of floods and erosion will be much increased with consequent effect with locally and the areas downstream.

The scheme of morphogenetic regionalization and land systems is essentially functional and express a certain relationship between climate, landforms, process and region.

Thus, it is condensed that morphogenetic study is a best developed as an energy system approach to geomorphology. Such an approach has much to offer planners and other workers requiring integrated approach to environmental studies.

Bibliography

Abbot, J. 1845: On Kunker Formations with Specimens. **Journal of the Asiatic Society of Bengal**, Vol 14, pp 442-4

Abichandani, C. T. and Roy, B B 1966: Rajasthan Desert - Its origin and Amelioration. **Indian Geographical Journal**, Vol 41 (3 & 4), pp 35-43

Abrahams, A.D. 1972: Drainage Density and Sediment Yields in Eastern Australia. **Australian Geographical Studies**, Vol 10, pp 19-41

Abrahams, A.D. 1980: Channel Link density and Ground Slope. **Annals of Association of American Geographers**, Vol 70, pp 80-93.

Agar, R.1960: The Post-glacial Erosion of the North Yorkshire Coast from the Tees Estuary to Ravenscar. **Proc. Yorks. Geol. Soc.**, Vol 32, pp 409-427.

Agrawal, D.P. and Kusumgar, S 1975: Radiocarbon Dates of Some Late Quaternary Samples. **Current Science**, Vol 44, pp 149-51.

Agrawal, D.P., Dhir, R.P. and Krishnamurthy, R.V. 1976: The Relevance of Kankar Studies to the Archaeology and Palaeoecology of Rajasthan, **Annual Archaeological Congress, University of Rajasthan, Jaipur.**

Agrawal, D.P., Dutta, P. S., Hussain, Z., Krishamurthy, R. V., Misra, V. N., Rajaguru, S.N.and Thomas, P.K. 1980: Palaeoclimate, Stratigraphy and pre-history in north and west Rajasthan. **Proceedings of the Indian Academy** of sciences (Earth and Planetary Sciences), Vol 89 (1), pp 51-66.

Agrawal, D.P., Gupta, S.K. and Kusumgar, S. 1970: Radiocarbon dates of some Quaternary Samples, **Current Science**, Vol 39,pp 219-22.

Agrawal, D.P., Kusumgar, S. and Pant, R.K.1975: Radio Carbon and Indian Archaeology, **Physics News**, Vol 64, pp 1-10.

Agrawal, D.P.and Roy,B. 1976: The Miliolite Problem: Some New S.E. and other Analytic Data, **Workshop on Pal-**

laeoclimate and Archaeology of Rajasthan, Gujarat. February 1976, p 10.

Agrawal, S. C. 1951: **The Samber Lake Salt Source** (Government of India Press), New Delhi, p 365.

Ahmed, E. 1968: Distribution and causes of gully erosion in India, Selected papers, 21st., IGU,Vol 1,pp1-13.

Ahmed, E. 1969: Origin and Geomorphology of the Thar Desert. **Annals of Arid Zone**, Vol 8, pp 171-80.

Ahmed, F. 1960: The Influence of Pleistocene Climates upon the Morphology of Cresta Scarps on the Colorado Plateau. **Annals of the Association of American Geographers**. Vol 50,pp 139-156.

Ahmed, F. 1986: Geological evidences bearing on the origin of the Rajasthan Desert (India). **Proceed. Indi-Nati. Sci. Aca.**, Vol 52(6), pp 1285-1306.

Alford, J.J. 1982: San Vicente arroyo. **Annals of the Association of American Geographers** Vol 72, pp 398-403.

Alllchin, B. 1958: The Late Stone Age of Ceylon. **Journal of the Royal Anthropological Institute**, Vol 88, pp 179-201.

Allchin, B. 1959: The Indian Middle Stone Age. **Bulletin of the Institute of Archaeology**, London. Vol 2, pp 1-35.

Allchin, B. and Goudie, A. 1971: Dunes, Aridity and Early Man in Gujarat, Western India. **Man**. Vol 6, pp 248-65.

Allchin, B., Goudie, A. Hegde, K. 1978: **The Pre-history and Paleogeography of the Great Indian Desert**, Academic Press, London.

Allchin, B., Hegde, K.T.M. and Goudie, A. 1972: Pre-history and Environmental change in Western India. A note on the Budha Pushkar Basin, Rajasthan, **Man.**, Vol 74,pp 542-64.

Allchin, B. and Goudie A., 1974: Pushkar: Pre-history and Climatic Change in Western India. **World Archaeology.**, Vol 5, pp 358-68.

Alllchin, B. 1976: The Discovery of palaeolithic Sites in the Plains of Sind and their Implicatioins. **The Geographical Journal.**, Vol 142, pp 471-89.

Allchin, B., Gondie, A., and Hegde, K. 1978: **The Pre-history and Palaeogeography of the Great Indian Deserts**, Academic Press, London.

Anderson, J.G. 1906: Solifluction, A component of sub-aerial Denudation. **J. Geol.**, Vol 14, pp 91-112.

Andres, W. 1980: On the palaeoclimatic significance of erosion and deposition in arid regions. **Zeit fur Geomorphie** (supliband), Vol 36, pp 112-122.

Antevs, E.V. 1952: Arroyo Cutting and Filling. **Jour. Geol.**, Vol 60,pp 375-385.

Arunachalam, B. 1981: Geomorphic Lineaments and the North Konkan Drainage. **Annals of the National Association of Geographers**, India., Vol 1 & 2, pp 1-9.

Ashwell, I.Y. 1966: Glacial Control of Wind and of Soil Erosion in Iceland. **Annals of the Association of American Geographers,** Vol 56, pp 529-540.

Auden, J.B. 1952: Some geological and chemical aspects of Rajasthan salt problem. **Bulletin of the National Institute of Sciences,** India, No. 1.

Bagnold, R.A. 1941: **The Physics of Blown Sand and Desert Dunes,** Methuen & Co Ltd. London, pp 265.

Bailey, R.W., Craddock, C.W. and Craft, A.R. 1947: Watershed management for summer flood control in Utah, US. Dept. of Agr. Miscall. Publi. 639, pp24.

Bailey, W., Forling, C.L. and Becraft, R.J. 1934: Floods and accelerated erosion in northern Utah, US. Dept. of Agri. Miscall. Publi. 196, pp 36.

Baker, V.R. 1974: Palateohydraulic interpretation of Quaternary alluvium near Golden, colorado, **Quate Research,** Vol 4, pp 94-112.

Baker, V.R. and Ritter, D.F. 1975: Competence of rivers to transport coarse bed load material, **Bull. Geol. Sci. America,**Vol 86, pp 1975-78.

Bakker, J.P. and Levett, Th. W. 1964: An Enquiry into the Problems of a Polyclimatic Development of Peneplains and Pediments Catch Plains in Europe during Senonian and Tertiary Period. **Publications services du Carte Geologique de** Luxemberg, Vol. 14 pp 27-75.

Bakker, J.P. 1967: Weathering of Granites in different climates in L° Evolution des Versants, P. Macar (Ed.) **Congres et Colloques de L°Universite de Liege.**, Vol 40, pp 51-68.

Barry, R.G. 1960: The Application of Synoptic Studies in Paleo-Climatology: A Case Study of Labrador-Ungava. **Geogr. Annlt.**, Vol 41A, pp 36-44.

Barry, R.G. 1960: The Application of Synoptic Studies in Paleo-climatology - A Case Study of Labrador-Ungava. **Geogr. Annlt.**, Vol 42A, pp 36-44.

Battle, W.R.B. and Lewis, W.V. 1951: Tempeprature Observations in Beryschrunds and their Relationship to Crique Erosion. **J. Geol.**, Vol 59, pp 537-545.

Beatty, C.B. 1972: The effect of Moisture on Slope Stability: A Classic Example from Southern Alberta, Canada. **J. Geol.**, Vol 80, pp 362-355.

Bedi, N. 1978: Geomorphology of Mahi River Basin in Gujarat, India. **Proceedings Symposium on Morphology and Evolution of Landforms,** pp 26-40.

Bedi, N. 1981: Applied Geomorphic Studies on Landsat Imagery in (H.S. Sharma (ed.) **Perspectives in Geomorphology,** Vol 3, Concept Publishing Company, New Delhi.

Bik, M.J. 1968: Morphoclimatic Observations on Praisie

Mounds. **Z. Geomorph.**, Vol 12, pp 409-469.

Bik, M.J.M. 1967: Structural Geomorphology and Morphoclimatic Zonation in the Central Highlands, Australian New Guinea. **Landform Studies.** Australian National University Press, Canberra. pp 26-47.

Birkland, P.W. 1968: Mean Velocities and boulder transport during Tahoe-age floods of the Truckee River, California, Naveda, **Bull. Geol. Soci. Ameri.**, Vol 79, pp 137-141.

Birot, P. 1968: The Cycle of Erosion in different Climates. **Batsford.**, Vol 133, pp 9-12.

Biswas, A.B. 1966: A Note on Certain Sand Dunes from the Kaliane Area, and Mahendragarh Area,Punjab. **Records of the Geological Survey of India.**, Vol 94, pp 221-8.

Biswas, R.K. and Ghose. R.N. 1978: Periglacial and Fluvioglacial Landscape in Jaisalmer District. **Proceedings International Symposium on Arid Zone Research and Development**, Ed. by H.S. Mann. CAZRI, Scientific Publishers, Jodhpur. 1980.

Blagoverchenskiy, E.N. 1968: The Dry Savannah of North West India (Thar Desert). **Soviet Geography,** Vol 6, pp 519-37.

Blandford, W.T. 1869: On the Geology of the Tapptee and Narbada Valleys, **Memoirs of the Geological Survey of India,** Vol 6, pp 163-385.

Blandford, W.T. 1877: Geological Notes on the Great Indian Desert between Sind and Rajputana, **Memoirs of the Geological Survey of India.**, Vol 10, pp 10-21.

Blandford,W.T. 1880: The Geology of Western Sind, **Memoirs of the Geological Survey of India,** Vol 17, pp 1-192.

Bloom, A.L. 1979: **Geomorphology.** Prentice Hall of India Pvt., New Delhi.

Bond, G.W. and Flint, R.F. 1968: Pleistocene Sand Ridges and Pans in Western Rhodesia. **Bulletin of the Geological-Society of America.**, Vol 79, pp 299-314.

Boocok, C. and Van Straten, O.J. 1962: Notes on the Geology and Hydrogeology of the Central Kalahari region, Bechuanaland Protectorate, **Trans. Geol. Soc. S. Afre.**, Vol 65, pp 125-71.

Bowler, J.M. 1970: **Late Quaternary Environments: A Study of Lakes and Associated Sediments in South Eastern Australia** (Ph.D dissertation, Australian National Universty, Canberra).

Bowler, J.M. 1976: Aridity in Australia: Age, Origins and Expression in Aeolian Landforms and Sediments. **Earth Science Reviews,** Vol 12, pp 279-310.

Bowler, J.M. and Polach, M.A. 1971: Radiocarbon Analysis of Soil Carbonates, An Evaluation from Paleosols in South-western Australia, Jerusalem. in D.H. Yaalon (ed)

Palaeopedology

Boye, M. and Fritsch, P. 1973: Degagement Artificial d'um dome, crystallin au Sud-Cameroun, **Travaux et Documents de Geog. Tropical**, Vol 8, pp 31-63.

Bradley, W.C. 1963: Large Scale Exfoliation in Massive Sandstones of the Colorado Plateau, **Bull. Geol. Soc. Am.**, Vol 74, pp 519-528.

Brady, L.F. 1936: The arroyo of the Rio-de Flag. a study of an erosion cycle. **Museum of Northern Arizona,** Museum Notes, Vol 6, pp 33-37.

Breed, C.S., Fryberger, S.G., McCawtey, C.K. and Lennartz **Desterts of Asia.** (in press). United States Geological Survey.

Brice, J.C. 1966: Erosion and Deposition in the Loess-mantled Great plains, Medicine Creek Drainage Basin, Nebraska, United States Geological Survey, **Professional Paper,** Vol 352 H, pp 255-339.

Browne, W.R. 1964: Grey Billy and the Age of Tor Topography in Monaro, NSW, **Proce. Linn. Soc. NSW,** Vol 89, pp 322-325.

Bryan, K. 1922: Erosion and sedimentation in Papago county. Arizona, **US Geol. Surv. Bulle,** Vol 730 B,pp19-90.

Bryan, K. 1925: Date of Channel Trenclhing (Arroyo cutting) in the Arid South West. **Science.**, Vol 62, pp 338-344.

Bryan, K. 1940: Erosion in the Valleys of the Southwest, **New Mexico Quartarly Review,** Vol 10, pp 227-232.

Bryan, K. 1941: Pre-columbian agriculture in the Southwest as conditioned by periods of alluviation, **Annals of the Associ. Ameri. Geogr.**, Vol 31, pp 219-242.

Bryan, K. 1954: The geology of chaco canyon, New Mexico in relation to the life and remains of the pre historic people of Pueblo Bonito, **Smithsonian Miscellaneous Collections.**, Vol 122 (7), pp 65.

Bryan, K. and Mason, S. 1932: Asymmatic Valleys and Climatic Boundaries. **Science.**, Vol 75, pp 215-216.

Bryan, R. and Yair,A. 1982: **Badland Geomorphology and Piping.** Edited volume, Norwich Geo. Books, pp 408.

Bryan, R.B. 1970: An improved Rainfall Simulation for Use in Erosion Research, Canada. **J. earth Sci.**, Vol 7, pp 1552-1561.

Bryson, R.A. and Barries, D.A. 1967: Possibilities of Major Climatic Modifications and their Implications: Northwest India, A case for study, **Bulletin of American Meteorological Society,** Vol 48, pp 136-42.

Budel, J. 1948: Das System der Klimatischeu Geomorphologie, Verhaudl. Deutscher Geographie., Vol 27, pp 65-100

Budel, J. 1957: Die Droppelten Einebnun-gsflachen in den feuchten Tropen. **Zeitschr. fur. Geomorph.**, Vol 1, pp

201-228.
Budel, J. 1959: The "Periglacial"- Morphologic effects of the Pleistocene Climate over the World. **International Geol. Rev.** Trans. by H.E. Wright and David Alt., Vol 1 & 3, pp 1-16.
Budel, J. 1963: Klima-genetische Geomorphologie, Geographische Rundschau reporduced in English as **Climatogenetic Geomorphology in Climatic Geomorphology.** E. Derbyshire (ed) Macmillan, London, 202-227 (1972), Vol 15, pp 269-285.
Burkham, D.E. 1981: Uncertainties resulting from changes in river form, **Jour. Hydraulic Divi., Anu. Soci. Civil Engi.**, Vol 107, Hy 5, pp 593-610.
Burnes, A. 1833: Description of the Salt Works at Panghpadder in Marwar. **Journal of the Asiatic Society of Bengal,** Vol 2, pp 365-66.
Butler, B.E., Blackburn, G., Bowler, J.M., Lawerence, C.R.,
Bewell, J.W. and Pels, S. 1973: **A geomorphic Map of the Riverine Plain of South-eastern Australia.** Australian National University Press. pp 33.
Butzer, K.W. 1963: Climatic-Geomorphologic Interpretation of Pleistocene Sediments in the Eurafrican Subtropics. Viking Fund Publ. **Antropology.**, Vol 36, pp 1-27.
Cailleux, A. and Romani, L. 1981: Modern Trends in Geomorphology, in H.S. Sharma (Ed.) **Perspectives in Geomorphology,** Vol 1, pp 9-20, Concept Publishing Company, New Delhi.
Caine, 1967: The Tors of Ben Lomond, Tasmania, **Z. Geomorph.**, Vol 11, pp 418-429.
Calkins, H.G. 1941: Man and Gullies, New Maxican, **Quarterly Review,** Vol 11, pp 69-70.
Campell, I.A. 1970: Climate and Overgrazing on the Shonto Plateau, Arizona. **The Professional Geographer,** Vol XXII, pp 132-141.
Carlston, C.W. 1963: Drainage density and streamflow, **US Geol. Survey Prof. Pap.** pp 8.
Carlston, C.W. 1966: The Effect of Climate on Drainage, Density and Streamflow. **Bulletin Inter. Assoc. Scient. Hydrol.**, Vol 11, pp 62-69.
Carroll, D. 1962: Rainwater as a Chemical Agent in Weathering a Review. **US Geol. Surv. Water-Supply. Irrigation Pap** 1535-9, pp 18.
Carson, M.A. and Tan, S.W. 1977: The land conservation conundrum of Eastern Barbado, **Annals of the Assoc. Ann. Geg.**, Vol 67, pp 185-203.
Carter, H.J. 1849: On Foraminitera, their Organisation and their Existence in a Fossilized State in Arabia, Sindh, Kutch and Kattyawari, **Journal of the Bombay Branch of**

the Royal Asiatic Society., Vol 3, p 158.

Chatterjee and Bagchi, K. 1957: Geomorphic Regions of East Rajasthan Pathar, geographical **Review of India,** Calcutta, Vol 19, No.4, pp 42-47.

Choubey, V.D. 1967: A Study of the Erosion Surfaces in the Sagar, Damoh, Jabalpur and Narsinghpur districts, Madhya Pradesh. **Proceedings of the seminar on Geomorphological Studies in India.** pp 164-173.

Chorley, R.J. 1958: Climate and Morphometry, **J. Geol.**, Vol 65, pp 628-638.

Chorley, R.J., Dunn, A.J., Backinsale, R.P. 1964: **The History of the study of Landforms - or the Development** of **geomorphology,** Vol I, Metheun.

Chorley, R.J. 1966: **The Application of Statistical Methods to Geomorphology, in G.H. Dury (Ed.) Essays in Geomorphology.** American Elsevier Publishing Company, New York, pp 275-387.

Christion, C.S. and Steward, G.A. 1953: General Report on Survey of the Katherine-Darwin Region, 1946. **CSIRO Australian Land Research Series** No. 1.

Clarke, J.I. 1966: **Morphometry from maps, in G.H. Dury (Ed.) essays in Geomorphology.** American Elsevier Publishing Company, New York, pp 235-274.

Clayton, K.M. 1960: The Landform Parts of Southern Essex. **Inst. of British Geographers. Translations and Papers.** Pub. No. 28, pp 55-74.

Clayton, K.M. 1971: Geomorphology - a study which spans the geology/geography interface. **J. Geol. Soc.** London, 127, pp 471-476.

Clayton, R.W. 1956: Linear Depressions in savanna landscapes, **Geog. Studies,** Vol 3, pp 102-126.

Colton, H.S. 1937: Some notes on the original condition of the little Colorado river: a side light on the problems of erosion. Museum of northern Arizona, **Museum Notes,** Vol 10, pp 17-20.

Common, R. 1966: Slope Failure and Morphogenetic Regions in G.H. Dury (Ed.) **Essays in Geomorphology.** Heinemann London, pp 53-81.

Cooke, R.W. 1974: The rainfall context of arroyo initiation in southern Arizona, **Zeit fur Geomorphology,** Vol 21, pp 63-75.

Cooley, M.E. 1979: Depths of channels in the area of the San Jhan basin regional uranium study, New Mexico, colorado, Arizona, and Utah, **US Geol. Suri. open file report**, 79, pp 1526-1538.

Cooper, W.S. 1958: Coastal Sand Dunes of Oregon and Washington. **Geol. Soc. Am. Mem.**, Vol 72, p 169.

Cooper, W.S. 1967: Coastal Sand Dunes of California. **Geol.**

Soc. Am. Mem., Vol 104, p 131.

Cosper, A.W. 1960: An Example of the Role of Micro-climate in Soil Genesis. **Soil Sci.**, Vol 90, pp 109-120.

Cottam, W.P. and Stewart, G. 1940: Plant succession as a result of grazing and Meadow desiccation since settlement in 1862, **Journal of Forestry,** Vol 38, pp 613-626.

Cotton, C.A. 1942: **Climatic Accidents in Landscape Making,** Whitcombe and Tombs, Christ Church, N.Z. p 354.

Cotton, C.A. 1958: **Climatic Accidents in Landscape Making.** Whitcombe and Tombs, Christwich,pp 354.

Crowther, E.M. 1930: The Relationship of Climate and Geological Factors to the Composition of Clay and the Distribution of Soil Types. **Proc. R. Soc.**, Vol 107, pp 10-30.

Cullingford, R.A., Davidson, D.A. and Lewis, J. 1980: **Time Scales in Geomorphology.** John Willey & Sons, New York.

Czudek, T. and Demek, J. 1970: Thermskarst in Siberia and its Influence on the Development of Lowland Relief. **Quaternary Res.**, Vol 1, pp 103-20.

Davies, J.J. 1969: **Landforms of Cold Climates.** Australian National University Press, Canberra, pp 200.

Davis, W.M. 1909: The Geographical cycle, Chapter 13 in **Geographical Essays,** Ginn and Co., New York.

Davis, W.M. 1913: Meandering valleys and underfrit rivers. **Annals of the Association of American Geographers.** AAAg, Vol 3.

Davitayor, F.F. 1969: Atmospheric Dust Content as a Factor Affecting Glaciation and Climatic Change. **Annals of the Association of American Geographers.**, Vol 59, pp 552-560.

De Terra, H. and Paterson, T.T. 1939: **Studies on the Ice Age in India and Associated Human Cultures,** Washington.

Demek, J. 1965: Slope Development in granite areas in Bohemian massif (Czechoslovakia), **Z. Geomorphology Suppli.**, Vol 5, pp 82-106.

Derbyshire, E. (Ed) 1973: **Climatic Geomorphology,** Macmillan, London.

Derbyshire, E. 1971: A Synoptic Approach to the Atmospheric Circulation of the Last Glacial Maximum in South Eastern Australia. **Palaeogeogr. Palaeo Clim. Palaeoecol.**, Vol pp 103-124.

Derbyshire, E. 1972: Tors, rock weathering and climate in southern victoria land, in R.J. Price and Sergden, C.E. (Eds) **Antarctica in Polar Geomorphology,** Inst. British Geog. Spec. Pub., Vol 4, pp 93-105.

Derbyshire, E. 1976: **Geomorphology and Climate** (an edited volume). A Wiley-Interscience Publication. John Wiley & Sons Ltd., London.

Deshpande, C.D. 1982: Understanding Indian Region, **National association of Geographers India**, pp 1-7.

Dey, A.K. 1956: The Shore Iines of India. **Quaternaria**, Vol 3, pp 95-100.

Dhinwa, P.S. 1982: Geomorphology and Human Settlement of the Berach Basin (Rajasthan). Unpublished Ph.D. thesis. University of Rajasthan, Jaipur.

Dhir, R.P. 1977: palaeoclimatic inferences from Quaternary Pedogenetic Processes in Arid Zone in D.P. Agrawal and B.M. Pande,(Ed) **Ecology and Archaeology of Western India**, Concept, New Delhi. pp 127-134.

Dikshit, K.R. 1970: Polycyclic Landscape and the Surfaces of Erosion in the Deccan Trap Country with Special Reference to Upland Maharashtra. **National Geographic Jour. of India.**, Vol 16, pp 236-252.

Dobyns, H.F. 1981: **From fire to flood: Historic human destruction of Sonaran desert, Riverina oases**, Ballano Press, Socorro, pp 222.

Dodge, R.E. 1902: **arroyo formation, science**, Vol 15, pp 746.

Dodge, R.E. 1910: The formation of arroyo in adobe filled valleys in the southwestern United States (Abstract) **Briti. Associ. Advance Scien. Rept.** 79, pp 531-532.

Dokuchaev, V.V. 1951: Soil Classification (Northern Hemisphere - Natural Soil Zones, 1900). **Acad Science**, USSR., Vol 6.

Douglas, I. 1976: Erosion rates and Climate: Geomorphological Implications. in E. Derbyshire (Ed.) **Geomorphology and Climate** Wiley, Chichester, pp 269-287.

Du Toit, A.L. 1908: Geological Survey of Parts of Mafeking and Vryburg. **Ann. Rept. Geol. Comm.** C.G.H., Vol 11, pp 125-92.

Du Toit, A.L. 1926: The Mier country S. Afri. **Geogr. J.**, Vol 9, pp21-26.

Dubey, R.S. 1968: Erosion surfaces in the Rewa Plateau, Madhya Pradesh, Selected papers, 21 IGU., Vol 1, pp 35-42.

Duce, J.T. 1918: The effect of Cattle on the erosion of Canyon bottoms. **Science**, Vol 67, pp 450-452.

Dunn, J.A. 1938: Post Mesozoic movements in the northern part of the Peninsula, **Memo. Geol. Surv. India**, Vol 73, pp 138-141.

Dury, G.H. 1970: A Resurvey of Part of the Hawkesbury river, NS Wales after 100 years. **Australian Geographical Studies**, Vol 8, pp 121-133.

Dutton, C.E. 1882: Tertiary history of the Grand Canyon district, **U.S. Geol. Surv. Monograph.**, 2, pp264.

Embleton, C and Thornes, J. 1979: **Process in Geomorphology.** (Edited). Edward Arnold publishers Ltd. London.

Evans, J.R. 1962: Falling and Climbing Sand Dunes in the

Cronese (Cat) Mountain Area. San Bernardino Country. California. **Journal of Geology.** Vol 70,pp 107-113.

Evans, S.L. and Meade, G.E. 1945: **Quaternary of the Texas High Plains.** University of Texas Publ. 4401, pp 485-507.

Fairbridge, R.W. (Ed.)1968: **The Encyclopadia of Geomorphology,** III, Reinhord Book Corporation, New York.

Fairbridge, R.W. 1970: **World palaeoclimatology of the Quaternary.** Revue de Geographie Physique et de Geologie Dynamique.

Falconer, J.D. 1911: **The Geology and Geography of Northern Nigeria,** Macmillan, London.

Fedden, F. 1884: The Geology of the Kathiawar Peninsula in Gujarat. **Memoirs of the Geological Survey of India,** Vol 21, pp 73-136.

Flint, R.F. 1957: **Glacial and Pleistocene Geology,** New York.

Flint, R.F. and Bond, G. 1968: Pleistocene Sand Ridges and Pans in Western Rhodesia. **Bulletin Geological Society of America.**, Vol 79, pp 299-313.

Folk, R.L. 1971: Longitudinal Dunes of the Northwestern Edge of the Simpson Desert, Northern Territory, Australia I-Geomorphology and Grain Size Relationship. **Sedimentology.**, Vol 16, pp 5-54.

Fourmier, F. 1960: Climate et erosion: la relation entre l' erosion du sol par l'eau et les preispitations atmospheriques (PUF, Paris). p 201.

Fox, C.S. 1923: The Bauxite and Aluminous Laterite Occurrence of India. **Mem. Geol. Surv. Ind.**, Vol 49, pp 186-187.

Frakes, L.A. and Kemp, E.M. 1973: palaeogene Continental Positions and Evolution of Climate. **Paleobot. Indica.**, Vol 1, pp 539-559.

Frye, J.C. 1961: Fluvial Deposits and Glacial Cycle. **Journal of Geology,** Vol 69, pp 600-3.

Gardiner, V. 1971: A drainage density map of Dartmoor. **Reports and Transactions of the Devonshire association for the Advancement of Science.**, Vol 103, pp 167-80.

Gardiner, V. 1972: A Morphometric Approach to Landform Aspects of Land Classification. Unpublished Ph.D. Thesis. University of Exeter., U.K.

Gardiner, V. 1974: Landform and Land Classification in North West Devon. **Reports and Transactions of the Devonshire Association for the Advancement of Science,** Vol 106, pp 141-53.

Gardiner, V. 1978: Redundancy and Spatial Organisation of Drainage Basin form indices. **Transactions of the Inst. of British Geographers.** New Series, Vol 3, pp 416-31.

Gardiner, V. 1981: The Impact of Climate on Fluvial System, **In H. S. Sharma (Ed.) Perspectives in Geomorphology.** Vol

2, pp 19-40, Concept Publishing Company, New Delhi.

Gerson, R. 1977: Sediment transport for desert Watersheds in erodible materials, **Earth Surface Processes**, Vol 2, pp 343-361.

Ghose, B. 1964: Geomorphological Aspects of the Formation of Salt Basin the Lower Luni Basin, Proceedings of Symposium of Problems of Indian Arid Zone, Jaipur, pp 169-78.

Ghose, B. and Singh S. 1965: Geomorphological Studies in Rajasthan-Analysis of Landforms of Saila block from Aerial Photographs. **Annals of Arid Zone**, Vol 4, (2), pp 207-216.

Ghose, B., Kar, A. and Zahid Husain, 1980: Comparative role of the Aravalli and the Himalayan river systems in the fluvial sedimentation of the Rajasthan Desert. **Man and Environment**, Vol 4, pp 8-12.

Ghose, B., Pandey, S. and Kar, A. 1977: Desertification Around the Thar - A Geomorphological Interpretation.**Annals of Arid Zone**, Vol 16, (3), pp 290-301.

Ghose, B., Pandey, S. and Kar, A. 1977: Geomorphology of the Rajasthan Desert, **Desertification and its control.** Indian Council of Agricultural Research, New Delhi.

Ghose, B., Pandey, S. and Singh, S. 1968: Processes and Extent of Erosion and its Effects on Land Use in the Central Luni basin, Western Rajasthan. **Annals of Arid Zone**, Vol 7, No.1, pp 15-30.

Ghose, B., Pandey, S., Singh, S. and Lal, G. 1967: Quantitative Geomorphology of the Drainage Basins in the Central Luni basin in western Rajasthan. **Z. Geomorph.** N.F., Vol 11, pp 146-160.

Ghosh, B., Pandey, S., Singh, S. and Lal, G. 1966: Geomorphology of the Central Luni Basin. Western Rajasthan. **Annals of the Arid Zone**, Vol 5, pp 10-25.

Gilbert, G.K. 1895: Lake Basins created by wind erosion. **J.Geol.**, Vol 3, pp 47-49.

Gilmour, D.A. and Bonell, M. 1979: Runoff processes in tropical Rainforest with Special Reference to a Study in Northeast Australian in Pitty, A.F. (Ed.) **Geographical Approachds to Fluvial Process**, Geoabstracts. Norwich.

Glassford, D.K. and Killigrew, L.P. 1976: Evidence for Quaternary Westward Extension of the Australian Desert into South Western Ausrtralia. **Search.**, Vol 7, pp 394-6

Glennie, K.W. and Evans, G. 1976: Reconnaissance of the Recent Sediments of the Ranns of Kutch. **Sedimentology**, Vol 23, pp 625-47.

Glennie, K.W. and Ewamy, B.D. 1968: Dikaka: Plants and Plant-Root Structures Associated with Aeolian Sand. **Palaeogeography, Palaeoclimatology and Palaeoecology.** pp

477-87.

Glennie, K.W. 1968: **Desert Sedimentary Environments.** Elsevier. Amsterdam, pp 222.

Godbole,N.N. 1972: Theories on the Origin of Salt Lakes in Rajasthan in India. **Proceedings of the 24th International Geological Congress,** Section. Vol 10, pp 354-7.

Gore, J.E. 1873: Note on a Bed of Fossiliferous, Kunkur in the Punjab. **Journal of the Royal Geological Society of Ireland,** (NS), Vol 4, pp 9-10.

Goudie, A.S. 1970a: Input and Output Considerations in Estimating rates of Denudation, **Earth Science Journal,** Vol 4, (2), pp 59-65.

Goudie, A.S. 1970b: Notes on some Major Dune Types in Southern Africa. **South African Geogrpahical Journal.** 52, pp 93-101.

Goudie, A.S. 1972a: The Concept of Post-Glacial Progressive Desiccation. Research Paper, School of Geography, Oxford, No. 4.

Goudie, A.S. 1972b: The Chemistry of World Calcrete Deposits, **Journal of Geology.**, Vol 80, No. 4, pp 449-63.

Goudie, A.S. 1972c: On the Definition of Calcrete Deposits. **Zeitschrist fur Geomorphologie.**, Vol 14, (4), pp 464-8.

Goudie, A.S. and Sperling, C.H.B. 1977: Long Distance Transport of Foraminiferal Tests by wind in the Thar Desert, Northwest India. **Journal of Sedimentary Petrology,** Vol 47, pp 630-33.

Goudie, A.S., Allchin, B. and Hedge, K.T.M. 1973: The Former Extensions of the Great Indian Sand Desert. **Geological Journal,** Vol 139, No.2, pp 243-57.

Govindan, P.R., Lele, V.S., Rajaguru, S.N. and Marathe, A.R. 1976: Dating of the Miliolite by Ca/Mg Ratio. Paper presented at Workshop on Palaeodimate and Archaeology of Rajasthan and Gujarat, PRL. pp 23-25.

Graf, W.L. 1979: Catastrophic theory as a model for change in fluvial systems, in Rhodes, DD. and Williames, G.P. (Eds.), **Adjustments of the Fluvial Systems,** Kendall/Hunt Publishers, Dubugue, pp 13-32.

Graf, W.L. 1983: The arroyo problem - Palaeohydrology and Palaeohydraulics in the short term, in Gregory,K.J. (Eds.), **Background to Palaeohydrology,** John Wiley and Sons Ltd.

Gregory, H.E. 1917: Geology of the Novajo Country, **US. Geol. Surv. Profess. Paper** 93, pp 161.

Gregory, H.E. 1938: The San Juan Country, **US Geol. Surv. Profess. Paper** 188,pp 123.

Gregory, H.E. 1950: Geology and Geography of the Zion Park region, Utah and Arizona, **US Geol. Surv. Profess. Paper** 220, pp 200.

Gregory, H.E. and Moore,R.C. 1931: The Kaiparowits region: A geographic and geologic reconnaissance of parts of Utah and Arizona, **US Geol. Surv. Profess., Paper 164**, pp 161.

Gregory, K.J. 1976: Drainage Networks and Climate in Derbyshire, E., (Ed.) **Geomorphology and Climate**, Wiley, London.

Gregory, K.J.and Gardiner, V. 1975: Drainage Density and Climate. **Zeitschrift fur Gecomorphologic**, Vol 19, pp 287--97.

Gregory, K.J. and Gardiner, V. 1979: Comment on Drainage Density and Streamflow. A CloserLook, in S.L. Dingman. **Water Resources Research**, Vol 15, pp 1662-4.

Grove, A.T. 1958: The Ancient Erg of Hàusaland and Similar Formations on the South Side of the Sahara. **Geographical Journal**, Vol 124, No. 4, pp 528-33.

Grove, A.T. 1969: Landforms and Climatic Change in the Kalahari and Ngamiland. **Geographical Journal.**, Vol 135, (2), pp 192-212.

Grove, A.T. and Goudie, A.S. 1971: Late Quaternary Lake Levels in the Rift Valley of Southern Ethiopia and Elsewhere in Tropical Africa. **Nature**, Vol 234,pp 403-5.

Grove, A.T. and Warren, A. 1968: Quaternary landforms and climate on the south side of the Sahara, **Geographical Journal**, Vol 134, (2), pp 194-208.

Gupta, B.L. and Singh, N.N. 1981: Some Findings of Rainfall Characteristics in Rajasthan. **Geographical Review of India**, Calcutta, Vol 43, (4), pp 326-341.

Gvozdetskiy, N.A., Gerenchuk, K.I., Isachenko, A.G. and Paebrazhoskiy, V.S. 1971: The Present State and Future Tasks of Physical Geography, **Soviet Geogr.**, Vol 12, pp 257-266.

Hack, J.T. 1941: Dunes of the western Navajo country, **Geographical Review**, Vol 31, pp 240-63.

Hack, J.T. 1960: Interpretation of Erosional Topography in Humid Temperate Regions. **American journal of Sc. Bradlay**, Vol 258A, pp 80-97.

Hadley, R.F. 1961: Some effects of Microclimate on Slope Morphology and Drainage Basin Development. **US Geol. Survey, Prof. Paper.**, Vol 424B, pp 832-833.

Hadley, R.F. 1967: Pediments and pediment forming processes. **Jour. Geol. Education**, Vol 15, pp 83-89.

Handley, J.R.F. 1952: The Geomorphology of the Nzega Area of Tanganika with special reference to the function of granite tors. **C.R. Congr. Geol. Inter. (Algiers)**, Vol 21, pp 201-10.

Harman, J.R. 1970: Forest and Climatic gradients along the Southeast Shoreline of Lake Michigan. **Annals of the Association of American Geographers.**, Vol 60, (3), pp 456-

465.

Harris, S.A. 1957: Mechanical Constitution of Certain Present day Egyptian Dune Sands. **Journal of Sedimentary Petrology,** Vol 27, (4), pp 421-34.

Hastenrath, S.L. 1967: The Barchgans of tghe Arequipa Region. Southern Peru. **Zeitschift fur Geomorphologie,** Vol 11, pp 300-331.

Hastings, J.R. 1959: Vegetation change and arroyo cutting in south eastern Arizona, **Jour. of Arizona Acad. Sciences,** Vol 1, pp 60-67.

Hastings, J.R. and Turner, R.M. 1965: **The Changing Mile.,** University Arizona Press, Tucson, pp 317.

Haynes, C.V. 1968: Geochronology of Late Quaternary alluvium, in Morrison, R.B. and Wright, H.E. (Eds.), **Means of Correlation of Quaternary Successions,** Univ. of Utah Press, Salt Lake City, pp 591-631.

Hegde, K.T.M. 1964: **The Pleistocene Deposits of the Lower Narmada River and an Early Stone age Industry from the River Chambal,** Baroda. In Wainwright, G.J.

Hegde, K.T.M. 1977: Use of Caliche as a Chronological and Environmental Tool in the Study of Quaternary Formatiions in Gujarat and Rajasthan. **Puratattva.,** Vol 8.

Hegde, K.T.M. and Switsur, V.R. 1973: Radiocarbon dates on the Buried Soil in the Lower Narmada Valley. **Current Science,** Vol 42, (17), pp 607-9.

Heron, A.M. 1917: The Geology of North Eastern Rajputana and Adjacent Districts. **Memoirs of the Geological Survey of India,** Vol 45, (1), p 128.

Heron, A.M. 1922: The Gwalior and Vindhyan System of South eastern Rajputana. **Memo. Geol. Surv. India.,** Vol 45, Part 2, pp 129-179.

Heron. A.M. 1923: Geology of Western Jaipur, **Reco. Geol. Surv. of India,** Vol 54 (4), pp 345-397.

Heron, A.M. 1925: Geology of Noroth Eastern Rajputana. **Memoirs Geol. Surv. of India,** Vol xiv.

Heron, A.M. 1936: The Geology of South Eastern Mewar, Rajputana.**Reco. Geol. Surv. of India,** Vol 68, Part I, pp 1-120.

Heron, A.M. 1938: The Physiography of Rajputana. **Proceeding 25th Indian Science Congress.,** Vol 25, pp 119-132.

Heron,. A.M. 1953: The Geology of Central Rajputana, **Memoirs of Geol. Surv. of India,** Vol 79.

Higgins, G.M., Baig, S. and Brinkmar, R. 1974: The sands of Thal: Wind regimes and sand ridge formations, **Zeitschrist fur Geomorphologie,** Vol 18, No. 3, pp 272-90.

Hjulstrom, F. 1949: Climatic changes and river patterns. **Geografiskar Annaler.,** Vol 31, pp 83-89.

Holland, T.H.and Christie, W.A.K. 1909: The origin of the salt deposits of Rajputana. **Records of the Geological Survey of India,** Vol 38, No. 2, pp 154-86.

Horton, R.E. 1945: Erosional Development of Streams and Drainage Basins-Hydrophysical Approach to Quantitative Morphology. **Bulletin of the Geological Society of America,** Vol 56, pp 275-370.

Hough, W. 1906a: Pueblo environments, **Ameri. Associ. for the Advance of Scie. Proceed.**, Vol 55, pp 447-454.

Hough, W. 1906b: Pueblo environment, **Science,** Vol 23, pp 865-869.

Hunt, C.B., Averitt, P. and Miller, R.L. 1953: Geology and Geography of the Henry Mountain region, Utah, **US Geol. Surv. Profess. Paper** 228,pp 234.

Huntington, E. 1914: **The Climatic Factor as illustrated in arid America Carnegic Institution of Washington,** Publication 192, pp 341.

Jacob, K., Sastry, M.V.A. and Sastry, V.V. 1952: A Note on the Micro fossils of Impure Gypsum from the Jamsar, mine Bikaner, and on the Possible Origin of the Gypsum. **Bulletin of the National Institute of Sciences of India,** Vol 1, pp.68-69.

Jaeger, F. 1939: Die Trochenseen de Erde, Petermanns Mitt. Erganzungssheft, 236.

Jagannathan, P. 1964: Climate, Environment and its Fluctuations in Rajasthan. **CSIR Symposium Problems of Arid Regions.** 1964.

Jakues, L. 1973: The role of Climate in the Quantitative and Qualitative Control of Karstic Corrosion. Symposium on Karst Moorphogenesis, Intul. Geogr. Union, European Regional Conf. Hungary. pp 122-152.

Jansen, J.M.L. and Painter, R.B. 1974: Predicting Sediment Yield from Climate and Topography. **J. Hydrol.**, Vol 21, pp 371-380.

Jennings, J.N. 1969: Desert dunes and Estuarine fill in the Fitsroy Estuary Kimberly Division, Western Australia. Paper presented to the Institute of British Geographers Conference, London, p 14.

Jenny, H. 1935: The Clay Content of the Soil as Related to Climatic Factors, Particularly Temperature. **Soil Sci.**, Vol 40, pp 111-128.

Johnson, W.D. 1900: The high plains and Their utilization, **US Geol. Surv. 21st Annual Report**, pp 609-741.

Joshi, R.V. 1970: The Characteristics of the Pleistocene Climatic Events in Indian Sub-continent - A land of Monsoon Climate. **Indian Antiquity,** Vol 4, pp 53-63.

Joshi, R.V. 1973: Quaternary Studies in the Sub-Himalayas, **Bulletian of the Deccan Collage, Research Institute,** Vol

33, pp 101-16.

Judson, S. 1952: Arroyos **Scientific American**, Vol 187, pp 71-76.

Karanth, K.R. 1970: Geohydrology of Budha Pushkar Internal Drinage Area, Ajmer District, Rajasthan. **Indian Geohydrology**, Vol 3, pp 58-78.

Karstem, R.A. and Juttle, S.D. 1971: **The Distribution of Climate Factors in Iowa and their Influence on Geomorphic Processes.**, Vol 77, pp 266-281.

Kerney, M.J., Brown, E.H. and Chaudhler, T.J. 1964: The Late-glacial and Post-glacial History of the Chalk Escarpment near book, Kent, **Phil. Trans. Roy. Soc. Series B,** Vol 248, pp 135-204.

Kesseli, J.F. 1946: A Neglected Field of Geomorphology. **Annals of Association of American Geographers.**, Vol 86, p 93.

Khan, A.A. and Banerjee, S.N. 1984: Geological and geomorphological studies around Tapti-Vaghur Confluence, district Jalgaon, Maharashtra, **Record. Geol. Surv. of India,** Vol 113 (6), pp 99-109.

Khandelwal, N.M. 1975 : On the occurrence of Gestropod Fossils in the Kanker deposits around Didwara, Rajasthan, **Current science**, Vol. 44(20), pp. 737-38.

King, C.A.M. 1966: **Techniques in Geomorphology.** Ernold E. London.

King, C.A.M. 1981: Climatogenetic Geomorphology. in H.S. Sharma (Ed.) **Perspectives in Geomorphology,** Vol IV, pp 37-62.

King, L.C. 1948: A Theory of bornhardts, **Geog. J.**, Vol 112, pp 83-86

King, L.C. 1950: The Study of World Plainlands: A New Approach in Geomorphology. **Quat. Journal Geol. Soci.** London, Vol 106, pp 101-131.

King, L.C. 1953: Canons of Landscape Evolution, **Geol. Soc. Am.Bull.**, Vol 64, pp 721-752.

King, L.C. 1962: **Morphology of the Earth.** Oliver and Boyd, pp 1-699.

King, L.C. 1963: **South African Scenery,** Oliver and Boyed. 3rd Edn.

Kirkby, M.J. 1969: **Infiltration, Throughflow and Overland Flow,** in Chorley, R.J. Water, Earth and Man Methuen, pp 215-227.

Kirkby, M.J. and Chorley, R.J. 1967: Throughflow, Overland Flow and Erosion. **Bull. Int. Assoc. Sci. Hydrol.**, Vol 12, No. 3, pp 5-21.

Knechtel, M.M. 1938: Geology and ground water resources of the valley of Gila River and San Simon - creek, Graham country, Arizona, **US Geol.Surv. Water Supply Paper,** 796-

F, pp 181-222.

Kondracki, J. and Richling, A. 1972: Synthetic Physico-geographical Research, **Geographica Polinica**, Vol 22, pp 13-25.

Koppen, W. 1923: **Die Kilmate der Erde, Grundrissder, Klimakunde, De Gruyter,** Berlin, p. 369.

Koppen, W. and Geiger, 1936: **Hand buch der Klimatologie,** Vol 1, Part C, Gebruder Borntraeger, Berlin.

Krishnan, M.A. 1952: Geological history of Rajasthan and its relation to present day conditions. **Proceedings of Symposium on the Rajputana Desert,** pp 19-31.

Kumar, V. and Laul, V.P. 1981: Records of glacmation in Jaisalmer Basin, District Jaisalmer, Rajasthan, Paper presented at 3rd Convention of Indian Association of Sedimentologists.

La Touche, T.H.D. 1902: Geology of western Rajputana, **Memoir of the Geological Survey of India,** Vol 35, No. 1, pp 1-110.

La Touche, T.H.D. 1910: Relics of the Great Ice Age in the Plains of Northern India, **Geological Magazine,** Vol 7, No. 5, pp 193-201.

La Touche, T.H.D. 1919: The Submerged Forest at Bombay, **Records of the Geological Survey of India,** Vol 49, No. 4, pp 214-9.

Lakhanpal, R.N. and Rose, M.N. 1951: Some Tertiary Leaves and Fruits of Guttiferae from Rajasthan. **Journal of Indian Botanical Society,** Vol 30, pp 131-9.

Lamb, H.H. and Woodroff, A. 1970: Atmospheric Circulation during the Last Icr age. **Quat. Res.,** Vol 1, pp 29-58.

Lancaster, I.N. 1981: The Pans of the Southern Kalahari Botswana, **Journal of Geography,** London, pp 82-98.

Langbein, W.B. and Leopold, L.B. 1966: River Meanders - Theory of minimum variance. **United States Geological Survey, Professional Paper,** 422-H.

Langbein, W.B. and Schumm, S.A. 1958: Yield of Sediment in Relation to Mean Annual Precipitation. **Transactions American Geophycisal Union.,** Vol 39, pp 1076-1084.

Leopold, A. 1921: A plea for recognition of artifical works in forest erosion and control policy, **Journal of Forestry,** Vol 19, pp 267-273.

Leopold, L.B. 1951 a : Rainfall Frequency: An aspect of climatic variation. **Trans. Americ. Geophysi. Uin.,** Vol 32, pp 347-357.

Leopold, L.B., Emmett, N.W. and Myrick, R.M. 1966: Channel and Hillslope Processes in a Semi Arid Area, New Mexico. **US Geol. Surv. Prof. Paper,** Vol 3526.

Leopold, L.B., Wolman, M.G. and Miller, J.P. 1964: **Fluvial Processes in Geomorphology.** Freeman, San Franciso, pp

522.

Leshnik, L.S. 1968: Prehistoric Explorations in North Gujarat and Parts of Rajasthan. **East and West (N.S.),** Vol 1183 and 4, pp 295-310.

Linton,D.L. 1955: The Problem of Tors. **Geog. J.,** Vol 121, pp470-87.

Love, D.W. 1979: Quaternary fluvial geomorphic adjustments in Chaco Canyon, New Mexico. In Rhodes, D.D. and Williams, G.P. (Eds.) **Adjustments of the Fluvial systems**, Kendall/Hunt Publishers, Dubuque, pp 227-308.

Mabbutt, J.A. 1952: A study of granite relief from South West Africa, **Geol. Mag.**, Vol 89, pp 87-96.

Mabbutt, J.A. 1966: Mantle - Controlled planation of pediments, **Ann. J. Sci.**, Vol 264, pp 78-91.

Mabbutt, J.A. 1971: The Australian Arid Zone as a Prehistoric environment. (E.J. Mulvaney, and J. Colson). (Ed.) **Aboriginal Man and Environment in Australia.**

Madhav, C., Bhatnagar, P.S., Dassarma, D.C., Dasgupta, A.K. and Sinha, S. 1979: Environmental Apppraisal of Jaipur urban complex. International Conference on Jaipur and Environs, (Abstract) University of Rajasthan, Jaipur.

Malde, N.E. and Scott, A.G. 1977: Observations of contemporary arroyo cuttings near Santa Fe, New Mexico, USA, **Earth Surfance Processes,** Vol 2, pp 39-54.

Mann, H.S., Malhotra, S.P. and Kalla, J.C. 1974: Desert Spread - A Quantitative Analysis in the Arid Zone of Rajasthan. **Annals of the Arid Zone,** Vol 13, pp 103-13.

Marston, R.B. 1952: Ground cover requirements for summer storm runoff control on aspen sites in northern Utah, **Journal of Forestry,** Vol 50, pp 303-307.

Martin, P.S. 1963: The last 10,000 years: A Fossil Pollen Record of the American Southwest, University of Arizona Press, Tucgon, pp 87.

Martin, P.S., Schoenwetter, J. and Arms, B.C. 1961: Palynology and Prehistory, the last 10,000 years. Geochronology Laboratories, University of Arizona.

Marxwell, J.C. 1960: Quantitative Geomorphology of the San Dimas Experimental forest, California. Office of Naval Research, Geography Branch, Project, N.R. 389-042. Technical Report 19.

Masannat, Y.M. 1980: Development of piping erosion conditions in the Banson area, Arizona, U.S.A. **Quarterly Journal of Engineering Geology**, Vol. 13, pp. 53-61.

Mathur, C.M. and Shankaranarayana, H.S. 1972: **Soils of Thar Desert, Personal Communication.**

Mathur, C.M., Ganu, S.N., Moghe, V.G. and Jain, S.V. 1974-75: **Soils of Rajasthan.** Survey and classification in Retrospect and Prospect. Soil Survey Organization,

Department of Agriculture, Rajasthan, Jaipur.

Mauley, G. 1959: The Late Glacial Climate of North West England, **Geol. J.**, Vol 2, pp 188-215.

Mc Craw, J.D. 1965: Landscape of central Otago, N.Z., **Geog. Soc. Misc. Series, No.5.** pp 30-45.

Medlicott, H.B. and Bladford, W.T., 1879: **A Manual of the Geology of India**, Calcutta. Government of India Publication.

Meher-Homji, V.M., 1973: Is the Sind-Rajasthan Desert the result of a recent climatic change **Geogorum.**, Vol. 15, pp 47-57.

Meigs, P., 1952: Arid and semi-arid climates of the world, **procee. 17th Internat. Geog. Cong. Washington**, pp. 135-138

Melton, M.A. 1957: An Analysis of the relations among Elements of Climate, surface properties and geomorphology. O.N.R. Tech Rep 11, Office of Naval research Geography Branch, Project NR 389-042 : Technical Report 11, Columbia

Melton, M.A., 1958: Correlation Structure of Morphometric Properties of Drainage Systems and their Controlling Agents. **J. Geol.**, Vol. 66, pp 442-460.

Melton, M.A., 1960: Intra-valley Variation in Slope Angles Related to Microclimate and Erosional Environment. **Bull Geol Sec. Am.**, Vol 71, pp 133-144.

Melton, M.A., 1965: The geomorphic and Palaeoclimatic significance of alluvial deposits in southern Arizona, **Jour. of Geology.**, Vol. 73, pp.1-38.

Miller, A.A., 1961: Climate and the Geomorphic cycle. **Geography.**, Vol.. 46, Part 3, No.3, pp 185-197.

Miller, V.C., 1953: Relation of Quantitative Geomorphic Study of Drainage Basin-characteristics in the linch Mountain Area, Virginia and Tennesse. Technical report 3, Office of the Naval research, Department of Geology, columbia.

Miller, J.P. and Wendrof, F. 1958: Alluvial chronology of the Tesuque valley, New Mexico, **Journal of Geology**, Vol. 66, pp. 177-194.

Miser, H.D. 1924: The Soin Juan Canyon, southern utah, U.S. Geological survey, water supply paper 538, pp 80.

Mishra, V.N., 1961: Palaeolithic Cultures of western Rajputana. **Bulletin of the Deccan College, Research Institute.**, Vol 21, pp 85-156.

Misra, V.N. 1967: **Preand Proto History of Berach Basin south Rajasthan.** Poona.

Misra, V.N. 1971: Two Microlithic Sites in Rajasthan - a Preliminary Investigation. **The Eastern Anthropologist**, Vol 24, No.3, pp 237-88.

Misra, V.N., 1973: Bagor- A late Mesolithic Settlement in north west India, **World Archaeology**, Vol 5, No. 1, pp 92-110.

Misra, V.N., Rajaguru, S.N., Agrawal, D.P., Thomas, P.K., Husain, Z, and Dutta, P.S. 1980: Pre-history and Palaeoenvironment of Jayal, western Rajasthan. **Man and Environment**, Vol. IV, pp 18-30.

Morgan R.P.C. 1970: Climatic Geomorphology, Its Scope and future, **Geographica**, Vol 6, pp 26-35.

Morgan, R.P.C. 1971a: A Test of Climatic Geomorphology by Morphometric Analysis in West Malaysia, unpublished Ph.D. thesis, University of Malaya.

Morgan, R.P.C. 1972: Observations on Factors Affecting the Behaviour of a First Order Stream. **Transanctions, Institute of British Geographers.**, Vol 56, pp 171-86.

Morgan, R.P.C. 1973: The Influence of scale in Climatic Geomorphology: A Case study of Drainage Density in West Malaysia. **Geogr. Annuli.**, Vol 55A, pp 107-115.

Morgan, R.P.C. 1976: The Role of Climate in the Denudation System : A Case Study from West Malaysia in Derbyshire E. (Ed). **Geomorphology and Climate,** Willey: London.

Morisawa, M.E 1962: Relation of Quantitative Geomorphology of Streams Flow in Representative Watershed of Appalachian Plateau Province. **Bulletin and Geological Society of America.**, Vol 73, pp 1025-1046.

Mosley, M.P. 1972 : Evolution of discontinuous gully system, **Annals of the Association of American Geography**, Vol. 62, pp. 655-663.

Mukhopadhyay, A.K. and Ghose, R.N. 1976: The Problem of the Stratigraphic Position of the Bap Boulders, Rajasthan. **Indian Journal of Earth Science,** Vol 3, No.2, pp 220-228.

Nair, R.R. 1974: Holocene Sea Levels on the Western Continental Shelf of India, **Proceedings of the Indian Academy of Science.**, Vol B 79, pp 197-203.

Oldham, C.F. 1893: The Saraswati and the Lost River of the Indian Desert, **Journal of the Royal Asiatic Society** (NS), Vol 34, pp 49-76.

Oldham, R.D. 1926: The Cutch (Kachh) Earthquake of 16th June 1819 with a Revision of the Great Earthquake of 12th June 1897. **Memoir of the Geological Survey of India,** Vol 46, No.2, pp 71-147.

Ollier, C.D., 1960: The Inselbergs of Uganda, **Zeit. f. Geomorph**, Vol.4, pp. 43-52.

Ollier, C.D., 1965: Some features of granite weathering in Australia, **Zeit, f. Geomorph.**, Vol. 9, pp. 285-304.

Ollier, C.D., 1981: **Tectonics and Landforms**, Longmans, London.

Otvos, E.G. 1971: Relict Eolian Dunes and the Age of the Prairie Coastwise Terrace, South eastern Louisiana. **Bulletin of the Geological Society of America**, Vol 82, pp. 1753-8.

Padmaja, G. 1976: Geomorphology of the Mej Basin, Rajasthan. Unpublished Ph.D. theis. University of Rajasthan, Jaipur.

Padmaja, G. Rao and Sharma, H.S. 1978: Some morphometric techniques with relation to discharge to Musi River basin (AP). **Proceedings Symposium on morphology and Evolution of landforms**, pp. 168-176.

Pal, S.K. 1973: Quantitative Geomorphology of Drainage Basins of Himalayas, **Geographical Review of India**, Vol 35, pp. 81-101.

Palmer, J. and Neilson, R.A. 1962: The origin of granite Tors on Dartmoor, Devonshire, **Procee York Geol. Soc.**, Vol. 33, pp. 315-340

Pandey, S. and Chatterji, P.C. 1970: Genesis of Mitha Ranns, Kharia Rann, and Kanodwala Ranns in the Great Indian Desert.**Annals of Arid Zone**, Vol 9, No. 3, pp. 175-80.

Pandey, S. Singh, S. & Ghosh, B. 1964-71: Orientation, Distribution and origin of sanddunes in the central Luni basin. **Proc. Symp. Problems of Indian Arid Zone**, Jodhpur, pp.84-91.

Passarge, S. 1904: **Die Kalahari**, Berlin, Raimer.

Peel, R.F. 1960 : some Aspects of Desert Geomorphology. **Geography**, Vol 45, Part 4, pp. 241-260

Peel, R.F. 1967: Geomorphology: Trends and Problems. **The Advancement of Science**, Vol.24, No. 124.

Pels, 1966: Late Quaternary Chronology of the Riverine Plain of Southeastern Australia. **J. Geol. Soc.** Aust., Vol. 13, pp. 27-40.

Penck, A. 1910: Versuch einer klimakhassifikation auf physiographischer Grendlage. Sitzungsbericht d Preuss. Akad. d. Wissensch. Math-Phys. **Klasse.**, Vol. 12, pp. 236-246

Penck, A. 1973: Attempt at a classification of climate on a physiographic Basis, **Climatic Geomorphology**, Macmillan & Co Ltd., pp. 51-60.

Peterson, H.V., 1950: The problem of gullying in western valleys, in Trask, B.P. (Edn.) **Applied sedimentation**, wiley, New York, pp 407-434.

Pickup, G., 1975: Downstream variations in morphology, flow conditions and sediment transport in an eroding channel, **Zeit fur Geomorphologic** (Supplement band), Vol.19, pp. 443-459.

Pley, J. de. 1972: A Quantitative Comparison Between Rainfall erosion capacity in a tropical and a middle latitude region, **Geog. Polinica**, Vol. 23, pp. 141-150.

Ploy, V. de, 1977: Mechanical properties of hillslopes and their relation to gullying in central semi-arid Tunisia, **zeit. fur Geomorphologie** (Supplement band), Vol.21, pp. 177-190.

Pomerol, C. 1964: Influence du climat periglaciaire sur le moddole des versants crayeus de la vallee de la science a l'aval de Mantes. **Ann Geogr.**, Vol. 73, pp. 704-707.

Poser, H. 1948: Boden- Und Klimaverhattmisse in Mittel und Westeuropa Wahrend der wurmeiszeit, **Erudkunde,** Vol. 8, pp. 53-68.

Pottlier, L.C. 1950: The Geographic Cycle in Periglacial Regions as it is Related to Climatic Geomorphology. **Ann. Assoc. Am. Geog.**, Vol. 40, pp. 214-236.

Pramila Kumar and Rai, R.K., 1972: Has Deccan Foreland Rejuvenated ? **The Geographer**, vol. 19, pp. 34-36.

Pramila Kumar and R.K. Rai, 1981: Has Deccan Foreland Rejuvenated? in H. S. Sharma (Ed.) **Perspectives in Geomorphology.**, Vol. IV, pp. 101-114.

Price, W.A.. 1958: Sedimentary and Quaternary Geomorphology of south Texas. Transactions of the Gulf Coast, **Association of Geological Societies,** Vol. 8, pp. 41-75.

Quinn, J.H., 1957: Paired river terraces and Pleistocene glaciation, **Journal of Geology**, Vol. 65, pp. 149-166.

Radha Krishnan, B.P., 1967: The Western Ghats of the Indian Peninsula, Proceedings of Seminar on Geomorphological studies in India, Sagar, pp. 4-14.

Ragnon, p. 1967: Climatic Influences on the African Hoggar During the Quaternary-Based on Geomorphologic observations. **Annals of the Association of American Geographer**, Vol. 57, pp. 115-127.

Rahu, P.H. 1967: Sheetfloods, Streamfloods and the Formation of Pediments. **Annals of the Association of American Geographers**, Vol 57, No. 3, pp. 593-604.

Raikes, R. L. 1967: **Water, Weather and Prehistory.**

Raikes, R.L. 1965: The Mohenjo-Daro Floods, **Antiquity,**Vol. 39, pp. 196-203.

Raikes, R.L. and Dyson, R.H. 1961: The Prehistoric Climate of Baluchistan and the Indus Valley. **American Anthropologist**, Vol. 63, pp. 265-282.

Rajaguru, S.N. and Marathe, A.R. 1976: On the origin of the Milliolite Formation in Hiran villey, Saurashtra - A Sedimentologic Approach. Paper presented to the workshop on Palaeoclimate and Archaeology of Rajasthan Gujarat PRL, 22-26 Feb, pp. 1-10.

Ramanujam, C.G.K. 1968: some Observations on the Flora of the Cuddalore Sandstone Series, Cretaceous, Tertiary formations of south India. **Geol soc Ind.**, Vol. 2, pp.

271-285.

Rathjens, S. (Ed). 1971: **Klimatsche Geomorphologie.** Wissenschoftliche Buchgesellschalf. Dramstadt. p. 485.

Ravi Shankar and Dubey, R. 1984: Landforms, neotechonism and geothermal activity in part of Tapi Basin, district Jalgaon, Maharashtra. A study. Record **Geol. Surv. India**, Vol. 113 (6), pp. 70-85.

Reagon, A.B. 1922: Archaeological notes on Pine River Valley, Colorado and the Kayenta - Tuba region, Arizona, **Transactions of the Kansas Academy of Science**, Vol. 30, pp. 262-267.

Reagon, A.B. 1924 a : Stream aggradation through irrigation, **Pan Amserican Geologist**, Vol. 42, pp. 335-344.

Reagon, A.B. 1924 b : Recent changes in the Plateau region, **Science**, Vol. 60, pp. 283-285.

Reeves, C.C. 1966: Pluvial lake basins of west Texas. **J. Geol.**, Vol. 74, pp. 269-91.

Rich, J.L. 1911: Recent stream trenching in the semi-arid portion of south western New Mexico, a result of removal of vegetation cover, **American Journal of Science**, Vol. 32, pp. 237-245.

Richardson, H.L. 1945: Discussion: the significance of terraces due to climatic oscillations, **Geological Magazine**, Vol. 82, pp. 16-18.

Rohdenburg, H. 1969: Slope pedimentation and climatic change as principal factors of planation and scarp development in tropical Africa. **Giesseuer Geogr Schrift.**, Vol. 20, pp. 57-152.

Rohdenburg, H. 1971: Einfuhrung in die klimagenetische Geomorphologie, Lenz-Verhag, Giesson, p. 350.

Ross, C.P. 1923: The Lower Gila Region, Arizona, **U.S. Geological Survey Water Supply Paper** 498, pp. 237.

Roth, E.S. 1965: Temperature and water content as factors in desert weathering, **J. Geol.**, Vol. 73, pp. 454-468.

Roy Choudhari, S.P., Agrawal, R.P., Datta Biswas, N.R., Gupta, S.P. and Thomas, P.K., 1963: **Soils of India**, I.C.A.R., New Delhi, pp. 284-310.

Roy Choudhary, S.P. 1952: Certain Geomorphological aspects of Rajasthan district. **Bull. Nat. Inst. Sci. India.**, Vol. 251.

Roy, B.B. and Sen, A.K. 1968: Soil Map of Rajasthan. **Annals of Arid Zone**, Vol. 7, pp. 1-14.

Roy, B.B., and Pandey, S. 1970: Expansion or Contraction of the Great Indian Desert. **Proceedings of the Indian National Science Academy**, Vol.36, B-No. 6, pp. 331-344.

Roy, B.B., Chatterji, P.C. and Pandey, S. 1969: Genesis of Carbonate Pan in Arid region of Rajasthan. **Annals of Arid Zone.**, Vol.8, pp. 181-87.

Rudbergs, S. 1961: Geomorphological Processes in a cold semi-aird Region. **Axed Heiberg Island research Reports.** McGill Uniersity, pp. 139-150.

Ruzumova, V.N. and Kheraskov, N.P. 1963: Geologic types of weathering crusts, Doklady Akad **Nauk SSSR,** Vol. 148, pp. 87-89.

Sahni, A and Kumar, V. 1974: Palaeogeography of the Indian sub-continent. **Journal, Paleogeogra, Paleo climatology, Paleoecology,** Vol. 15, No. 3, pp. 209-226.

Sarenthein, M. 1972: Sediments and history of the Post Glacial Transgression in the Persian Gulf and North West Gulf of Oman, **Marine Geology,** Vol. 12, pp. 245-66.

Satyanarayan, Y. 1963: Ecology of the central Luni Basin, Rajasthan. **Annals of the Arid Zone,** Vol. 2, No. 1, pp. 83-94.

Savigear, R.A.G. 1965: A Technique for Morphological Mapping, **Annals of the Association of American Geographers**, Vol. 55, pp. 514-38.

Saxena, S.K. and Singh S. 1976: Some Observations on the Sand Dunes and Vegetation and Bikaner District in western Rajasthan. **Annals of Arid zone,** Vol. 15, No.4, pp. 313-322.

Saxena, S.K., Gaur, Y.D. and Satyanarayan, Y. 1964: Studies on Dune Ecology, **Agricultural Research,** Vol. (4), pp. 128-29.

Scheideggor, A.E. 1961: **Theoretical Geomorphology,** Springer-Verlag, Berlin.

Schoenwetter, J. 1962: The Pollen analysis of eighteen archeological sites in Arizona and New Mexico. In chapters in the Prehistory of EASTERN Arizona, **Fieldiana, Anthropology,** Vol. 53, pp. 168-209.

Schoenwetter, J and Eddy, F.W. 1964: Alluvial and polynological reconstruction of environments, Navejo Reservation. District, **Museum of New Mexico - Papers in Anthropology,** Vol. 13, pp. 155.

Schumm, S.A. 1956: Evolution of Drainage Systems and Slopes in Badlands at Perth Amboy, New Jersey, **Geol. Soc. Am. Bull,** Vol. 67, pp. 597-646.

Schumm, S.A. 1965: Quaternary Paleohydrology. in H.E. Wright and D.G. Frey. (Ed.) **The Quaternary of the United States.**

Schumm, S.A. 1968 a : Speculations concerning palaeohydrologic control of terrestrial sedimentation, **Bulletin of the Geological Society of America,** Vol. 79, pp.1573-1588.

Schumm, S.A. 1968b: River adjustment to altered hydrologic regimen Murrumbidgee River and Palaeochannels, Australia, **US Geological Survey Professional Paper,**598, pp. 65.

Schumm, S.A. 1969: Geomorphic Implications of Climatic Changes in R.J. chorley Eyre (Ed.) **Water, Earth and Man,** pp. 525-534, Mathuen Ltd.

Schumm, S.A. 1969: Geomorphic Implications of Climatic Changes. **Water, Earth and Man,** Ed. by R.J. Chorley, Eyre Mathuen Ltd., pp. 525-534.

Schumm, S.A. and Hadley, R.F. 1961: Progress in the Application of Landforms Analysis in Studies of Semi-arid Erosion, **US Geol Surv. Cir.**, Vol. 437.

Schumm, S.A. and Litchy, R.W. 1963: Channel widening and Flood plain construction along Cimarron river in south western Kansas, **U.S. Geological Survey Professional Paper**, 352, pp. 71-88.

Sen, Depawati, 1972 : Geomorphology of the Aravalli Range (Raj)-a re-interpretation of residual erosion surfaces. **Geog. Rev. of India**, Vol. 34, No.3, pp. 262-87.

Senter, D. 1937: Tree rings, valley floor deposition, and erosion in chaco canyon, New Mexico, **American Antiguity,** Vol. 3, pp. 68-75.

Shankarnarayan, H.S. and Hirakerur, L.R. 1972: Characterization of some soils of North Indian Plains. **J. Indian Soc. Soil Sci.**, Vol.20, p.157.

Sharma, G.R. 1973b: Stone Age in the Vindhyas and the Ganga Valley, Bombay, in D.P. Agarwal and A Ghosh (Ed.) **Radiocarbon and Indian Archaeology** pp 106-110.

Sharma H.S. 1968: Genesis and Pattern of Ravines in the Lower Chambal Valley, **Geog. Review of India,** Vol. 30, No.4 pp. 14-25.

Sharma, H.S. 1968: Genesis of Ravines of the Lower Chambal Valley, India, Selected Papers, 21st I.G.U., Vol. i, pp. 114-118.

Sharma, H.S. 1972: Erosion surfaces in the Lower Chambal Valley, India, **International Geography, Canada,** Vol. 1, pp. 61-63.

Sharma, H.S. 1976: Morphology of Revines of the Morel Basin, India, **International Geography,** Moscow, Vol. 1, pp. 219-223.

Sharma, H.S. 1979: **Physiography of the lower Chambal Valley,**Concept Publishing Company, New Delhi.

Sharma, H.S. 1980: **Ravine erosion in India.** Concept Publishing Company, New Delhi.

Sharma, H.S. 1981: **Perspectives in Geomorphology,** 4 Vols, (Ed.). Concept Publishing Company, New Delhi.

Sharma, H.S. 1983: Inland Drainage of North-eastern Rajasthan and its Possible Link with the Ghaggar Yammuna system. **Transaction's of Institute of Indian Geographers,** Vol 5, No. 1, pp. 19-22.

Sharma, H.S. and Banêrjee, R.C. 1984: A natural event and its aftermath. A study of Jaipur city and its environs, **Annals of the National Association of Geograpahers**, India, Vol. IV (1), pp. 1-9.

Sharma, H.S. and G. Padmaja Rao, 1978: Evolution of Hillside Slopes of the Morel Basin (Rajasthan). **Proceedings of Symposium on morphology and evolution of landforms**, pp. 177-188.

Sharma, M.L. 1982: Geomorphology of the Gambhir Basin. Unpublished Ph.D. thesis. University of Rajasthan, Jaipur.

Sharma, V. 1968: Quantitative Terrain Types in Chaur-Rajgarh Tract of the Lesser Himalayas, **Geographical Review of India**, Vol. 30, No. 1, pp. 31-42.

Singh, G. 1969: Palaeobotanical features of the Thar Desert. **Annals of the Arid Zone,** Vol. 8, No. 2, pp. 188-95.

Singh, G. 1971: The Indus Valley culture seen in the Context of Post-glacial Climate and Ecological Studies in Northwest India. **Archaeology and Physical Anthropology in Oceania,** Vol. 42, pp. 178-89.

Singh, G. 1977: Stratigraphical and palynological evidence for desertification in the Great Indian Desert, **Annals of Arid Zone, Vol. 16, No. 3, pp. 310-320.**

Singh, G. and Agrawal, D.P. 1976: Radiocarbon Evidences for Deglaciation in North-western Himalaya, India, **Nature,** Vol. 260, p.232.

Singh, G., Joshi, R.D., Chopra, S.K. and Singh, A.B. 1974: Late Quaternary History of Vegetation and Climate of the Rajasthan Desert, India. **Philosophical Transactions and of the Royal Society,** 267B, pp. 467-501.

Singh, G., Joshi, R.D. and Singh, A.P. 1972 : Stratigraphic and Radio-carbon evidence for the age and development of three salt lake deposits in Rajasthan, India, **Quaternary Research,** Vol 2, pp. 496-505.

Singh, S. 1977: Geomorphological Investigation of the Rajasthan desert. Central Arid zone Research Institute, Jodhpur. Monograph No. 7.

Singh, S. 1977: Geomorphology and Climatic Changes During Quaternary period in Rajasthan Desert, **Recent Researches in Geology,** Vol. 9, pp. 121-130.

Singh, S. 1977: Sand Dunes and Palaeoclimate in Jodhpur District, Western Rajasthan, **Man and Environment,** Vol. 1, pp. 7-15.

Singh, S. and Ghose, B. 1977: Geomorphology of the Luni Basin and its Palaeoclimatic Inferences in D.P. Agrawal and Panday (Ed.) **Ecology and Archaeology of western India,** Concept Publishing company, New Delhi, pp. 135-146.

Singh, S., Pandey, S. and Ghose, B. 1966: Geomorphology of Siwana Development Block in Barmer district, **Annals of**

Arid Zone, Vol. 5, pp. 117-26.

Sinha, S. 1976: A Review of the Studies on the Origin of Salinity of Rajasthan Salt Lakes. Paper presented by workshop on palaeoclimate and Archaeology of Rajasthan and Gujarat, PRL, February 1976, pp. 23-6.

Smith, E. 1833: Notes on the Specimens of the Kanker formation, and on Fossil Bones collected on the Jamwa. **Journal of the Asiatic Society of Bengal**, Vol.2, pp. 622-31.

Smith, H.T.U. 1965 : Dune Morphology and chronology in central and western Nebraska, **Journal of Geology**, Vol. 73, pp. 557-558.

Smith, D.I. and Atkinson, T.C. 1976 : Process, land forms and climate in limestone regions in E. Derbyshire. **Geomorphology and climate**, wiley, London, pp. 367-409.

Smithson, F. 1939: Statistical methods in Sedimentary Petrology. **Geol. Mag.**, Vol. 76, pp. 417-427

Snead, R.E. and Frishman, S.A. 1968: Origin of Sands on the east side of the Las Dela valley. West Pakistan. **Bulletin of the Geological Society of America.**, Vol. 79, pp. 1671-6.

Sood, R.K. and Pathan, S.K. 1982: Palaeoenvironment of Rajasthan-Observation from Landsat Imagery. Abstracts of papers Fourth Indian Geography Congress Bombay.

Spate, O.H.K 1954: **India and Pakistan**, Metheren, London.

Sperling, C.H.B. and Goudie, A.S. 1975: The miliolite of Western India: A Discussion of the Aeolian and Marine Hypotheses. **Sedimentary Geology**, Vol. 13, No. 1, pp. 71-5.

Starkel, L. 1972: The Role of Catastrophic Rainfall in the shaping of the relief of the lower Himalaya, **Geography Polinica**, Vol. 21, pp. 103-147.

Stein, 1942: A Survey of Ancient Cities along the 'Lost' Saraswati River. **Geographical Journal**, Vol. 99, pp. 173-82.

Stoddart, D.R. 1969: Climatic Geomorphology: review and reassessment, **Progress in Geography**, Vol. 1, pp. 161-222.

Stoddart, D.R. and Pillai, C.S.G. 1972: Raised Reefs of Ramanathapuram, South India. **Transactions of the Institute of British Geographers.**, Vol. 56, pp. 111-125.

Strahler, A.N. 1950: Equilibrium theory of Erosional Slo Approached by Frequency, Distribution Analysis. **American Journal of Science**, Vol. 248, pp. 673-696.

Strahler, A.N. 1952: Hypsometric. Area Altitude Analysis and Erosional topography. **Bulletin of Geological Society of America.**, Vol. 63, pp. 1117-1142.

Strahler, A.N. 1952: Quantitative Geomorphology of Erosional Landscapes. **19th International Geological Congress**, Algirs. Section 3, Vol. 3, pp. 341-354.

Strahler, A.N. 1954: Statistical Analysis in Geomorphic Research. **Journal of Geology,** Vol. 62, pp. 1-25.

Strahler, A.N. 1957: Quantitative analysis of watershed **Geom. Trans Am Geophysical Union,** Vol. 38, pp. 913-920.

Strahler, A.N. 1958: Dimmensional Analysis Applied to Fluvially Eroded Landforms. **Bulletin of Geol Soc of Am**, Vol. 69, pp. 279-300

Strahler, A.N. 1959: Quantitative Drainage Basins, **Bull Geol Soc Amer**, Vol. 70.

Strahler, A.N. 1964: Quantitative Geology of Drainage Basins and Channel Networks. **Handbook of applied hydrology.** (Chow V.T. Editor). McGraw-Hill Book Co., pp. 39-76.

Strahler, A.N. 1971: Quantitative analysis of Erosional Landforms. **Physical Geography,** Willey Eastern Pvt Ltd in New Delhi, pp. 481-500.

Subbarao, B. 1952: Archaeological Explorations in the Mahi valley. **Journal of the MS University Baroda,** Vol. 1, pp. 33-72.

Subbarao, B.P. 1958: **Personality of India.** 2nd Edn. Baroda.

Subramanian, K.S. and Mani, G. 1978: Implications of Letrites in High Landforms in Tamilnadu. Proceedings of symposium on morphology and evolution of landforms, pp. 208-216.

Subramanian K.S. and Mani, G. 1978: Geomorphic Significance of Bauxites below Mio-phiocene Sediments in the west coast. Proceedings of Symposium on morphology and evolution of landforms, 83-87.

Subramanian, A.R. and Prasad Rao, G.S.L.H.V. 1981: Eco-Climates of Rajasthan State Moisture Regime, **Geographical Review of India,** Vol. 43, pp. 68-74.

Subramanian, K.S. and Murleedharan, M.P. 1985: Origin of Palghat Gap in South India, A synthesis, **Journ. Geol. Soc. Ind.**, Vol. 26(1), pp. 28-37.

Swift, T.T. 1926: Rate of Channel Trenching in the South-west, **Science,** Vol. 63, pp.70-71.

Tanner, W.F. 1961: An alternate Approach to Morphogenetic Climates. **South-eastern Geologist,** Vol. 2, pp. 252-257.

Tardy, Y. Bocquier, Paquet, H. and Millet, G. 1973: Formation of day from Granite and its Distribution in relation to Climate and Topography. **Geoderma**, Vol. 10, pp. 271-284.

Thomas, M.F. 1965: Some aspects of the geomorphology of tors and domes in Nigeria, **Zeit. f. Geomorph**, Vol. 9, pp.63-81.

Thornber, J.J. 1910: The grazing ranges of Arizona, **Arizona Agricultural Experiment Station Bulletin,** Vol. 65, pp. 245-360.

Thornbury, W.D. 1954 : **Principles of Geomorphology**, John wiley, Newyork.

Thornes, J.B. 1980: Structural instability and ephermeral channel behaviour, **Zeit fur Geomorphologic**, Vol. 36, pp. 233-244.

Thornthwaite, C.W. 1933: The climate of the Earth. **Geog. Rev.**, Vol. 23, pp. 433-440.

Thornthwaite, C.W. 1948: An Approach Toward a Rational Classification of Climate. **Geographical Review**, Vol.38, pp. 55-94.

Tindale, B. 1941: The Handaxe Used in the Western Desert of Australia, **Mankind**, Vol. 3

Tod, J. 1829: **Annals and antiquities of Rajasthan.**

Todd, K.R.U. 1939: The Palaeolithic Industries of Bombay. **Journal of the Royal Anthropological Institute.**, Vol. 69.

Tricart, J. 1974: Existence de P'eriodes seches au quateraireen Amazonie et dans les regions voisines. **Reuve de Geomorphologie Dynamique**, Vol. 23, pp. 145-158.

Tricart, J. 1981: Climatic Influence on Geomorphology. in H.S. Sharma (Ed.) **Perspectives in geomorphology.**, Vol. 1, pp. 21-36.

Tricart, J. and Cailleux, A. 1965: **Introduction a la g'eomorphologie climatique**, SEDES, Paris.

Tricart, J. and Cailleux, A. 1972: **Introduction to Climatic Geomorphology**, Trans. C.J.K. de Jonge, Longmans, London.

Tricart, J. and Mainguet, M. 1965: Characteristiques Granulomeriques de quelques sables eoliens du desert peruvien, Aspects de la dynamique des Barchanes, **Revue de Geomorphologie Dynamique.**, Vol. 15, No. 7-9, pp. 110-121.

Tuen, Y.F. 1966: New Mexican gullies: A critical review and some recent observations, **Annals of the Association of American Geographers**, Vol. 56, pp. 573-597.

Twidale, C.R. 1972: Landforms development in the lake Eyre Region. **Geographical Review**, Vol. 62, No. 1, pp. 40-70.

Twidale, C.R. 1978: Bornhardts, **Zeit. Geomorph. Supple. Band**,Vol. 31, pp.111-137.

Twidale, C.R. 1981: Granitic Inselbergs. in H.S. Sharma (Ed.) **Perspectives in Geomorphology**, Vol. I, pp. 127-168.

Twidale, C.R. 1981: Planation Surfaces: Theory and Reality. in H.S. Sharma (Ed.) **Perspectives in Geomorphology.**, Vol. I, pp.63.83.

Twidale, C.R. 1981b: **Origin and age of inselbergs** (Serton Memorial lecture) Unit. Stellenbosch.

Vaidyanadhan, R. 1962a: Effect of Uplift and Structure on

Drainage in the Southern part of Cuddapah basin, **Journal Geol. Soc. of India.**, Vol 3, pp 70-85.

Vaidyanadhan, R. 1964a: **Recognition and Correlation of Erosion Surfaces in the Southern part of Cuddapah basin, Jown Geological Society of India.**, Vol 5, pp 121-127.

Vaidyanathan, R. 1964b: Geomorphology of the Cuddapah Basin. **Jour Indi. Geosci. Ass.**, Vol 4, pp 29-36.

Vaidyanadhan, R. 1967: An Outline of the Geomoorphic History of India. South of N. Latitude 18, **Proccedings of Geomorphological studies in India.** pp 121-130.

Vaidyanadhan, R. 1971: Lineaments and their Importance in Landform Studies. **Journal of the Geological Society of India**, Vol 12, 3, pp 299-302.

Vaidyanadhan, R. 1979: Climatic Geomorphology. All India Summer Institute on recent development in applied geography, University of Rajasthan, Jaipur, 16 July - 9 August.

Valdia, K.S., Joshi D.D., Sanwal, R. and Tandon, S.K. 1984 : Geomorphologic Development across the active Main Boundary Thrust -an example from the Nainital hills in Kumaon Himalaya, **Jour. Geol. Soc. India**, Vol 25 (12), pp. 761-774.

Van der Hammer, T. 1972: Changes in Vegetation and Climate in Amazon Basin and Surrounding Areas During the Pleistocene. **Geologie en Mijnbouw.**, Vol 46, pp 79-95.

Varma, P. 1960: **Udaipur region-A study in geomorphology.** Indian Geographer, Vol. 5.

Vats, P.C., Singh, S. Ghose, B. and Kaith, D.S. 1976: Types, orientation and distribution of sanddunes in Bikaner district. **Geographical Observer**, Vol 12, pp 67-75.

Verma, B.L. 1980: Droughts in Rajasthan. Unpublished Ph.D. thesis, University of Rajasthan. Jaipur.

Verma, K.K. and Mathur, U.B. 1976: Data from Field and Laboratory Analyses on the Problem of the Origin of Saurashtra Miliolite. Papers presented at workshop on Palaeclimate and Archaeology of Rajasthan and Gujarat PRL. 23-26 Feb, pp. 1-18.

Verma, R.P. and Roy, P.K. 1981: Palaeo-drainage studies and its utility in Flood Problems. A case history of Lower Luni Basin in H.S. Sharma (Ed.) **Perspectives in Geomorphology** Concept, New Delhi, Vol. IV, pp. 137-143.

Verma, V.K. and Tandon, S.K. 1971: Stream Tortuosity: A new method of visual estimation. **Z. Geomorph N.F.**, Vol. 12, pp. 165-171. Berlin-Stuttgarh.

Verstappen, H. T. and Delft, I.T.C. 1964: Karst Morphology of the Star Mountains (central New Guinea) and its Relation to Lithology and Climate, **Zeitschrift fur Geomorphologie.**, Vol 8, pp 40-49.

Verstappen, H.T. 1968: On the origin of longitudinal (seif) dunes. **Zeitschrift fur Geomorphologie**, Vol 12, pp 200-20.

Verstappen, H.T. 1970: Aeolian Geomorphology of the Thar Desert and Palaeoclimates. **Zeitschrift fur Geomoraphische Abhandlungen**, Vol. 60, pp.341-353.

Verstappen, H.T., Ghose, B. and Pandey, S. 1969: Landforms and resources in central Rajasthan (India) Result of Jalor Pilot Survey, Publi. Inst. **Aerial Survey and Earth Sci.**, Vol 51, pp 1-20.

Virendra Kumar and Laul, V.P. 1981: Records of glaciation in Jaisalmer Basin, District Jaisalmer, Rajasthan, 3rd convention, Indian Association of sedimentologists.

Vishnu Mittre and Savithri, R. 1973: Supposed remains of Rice (Oryza sp.) in the Terracotta cakes and Pai of Kalibangan, Rajasthan **The Palaeobotanist**, Vol. 22, No.2, pp. 124-126.

Vishnu-Mittre, 1974: Plant Remains and Climate from the Late Harappan and other Chalcolithic Cultures of India- A study in inter-relationships. **Geophytology.**, Vol. 4, pp. 46-53.

Vishnu-Mittre, 1978: Palaeoecology of the Rajasthan Desert during the last 10,000 years. **The Palaeobotanist**, Vol. 25, pp. 549-558.

Wadia, D.N. 1960: **The Post-glacial Desiccation of Central Asia: Evolution of the Arid Zone of Asia.** Monograph No. 1, National Institute of Sciences, India.

Wayland, E.J. 1953: More about the Kalahari **Geog. J.**, Vol. 119, Part 1, pp. 49-56.

Weeks, A.G. 1969: The stability of Natural Slopes in South-east England as Affected by Periglacial Activity. **Quart J. Eng. Geol.**, Vol. 2, pp. 49-61.

Weinert, H.H. 1961: Climate and weathered Karroodolerite.**Nature.**, Vol. 191, pp.325-329.

Weinert, H.H. 1965: Climatic Factors Affecting the Weathering of Igneous Rocks. **Agric Met.**, Vol. 2, pp. 27-42.

Weischet, W. 1969: Klimatologische Regeln zur Vertikalverteilung der Niederschlage in Tropengebirgan, **Die Erde.** Vol. 100, pp. 287-306.

Wertz, J.B. 1966: The Flood Cycle of Ephemeral Mountain streams in the Southwestern United states. **Annals of the Association of American Geographers.**, Vol. 56, pp. 598-633.

Wertz, J.B. 1970: The start of an ephemeral stream, **Zeit fur Geomorphologic** (Supplement band), Vol. 14, pp. 96-102.

West, W.D. 1962: The Line of the Narmada and son Valley, **Current Science.**, Vol. 31, pp. 143-144.

West, W.D. and Choubey, V.D. 1964 : The Geomorphology of the country around sagar and Katangi, M.P.—an example of superimposed drainage, **Journ. Geol. Soci. of India**, Vol. 5, pp. 41-55.

Weyman, D.R. 1974: Runoff Processes, Contributing area and Streamflow in a Small Upland Catchment. Institute of British Geographers, **Special Publication.**, Vol. 6, pp. 33-43

Wilhelmy, H. 1958: **Klimamorphologie der Massengesteine.** Georg. Westermann Verlag. Braunschweig., pp. 238.

Wilhelmy, H. 1968: Indusdelta und Rann of Kutch. **Erdkunde,** Vol. 177-91.

Wilhelmy, H., 1969: Das Vrstromtal am Ostrand der Indusebene und dar Sarasvati-problem. **Zeitochrift fur geomorphologie, Supplement band,** Vol. 8, pp. 76-83.

Williams, G.E., and Polach, H.A. 1971: Radiocarbon dating of Arid Zone Paleosols, **Bulletin of the Geological Society of America**, Vol. 82, pp. 3069-3086.

Williams, M.A.J. and Adamson, D.A. 1974: Late Pleistocene Desiccation along the White Nile, **Nature,** Vol. 248, pp. 584-586.

Wilson, L. 1968: Morphogenetic Classification. **Encyclopedia of Geomorphology.** R.W. Fairbridge (Ed) Reinhold, New York, pp. 717-728.

Wilson, L. 1973 a : Relationships between Geomorphic Processes and Modern Climates as a Method in Palaeoclimatology in E. derbyshire (Ed) **climatic Geomorph°loogy,** Macmillan, London, pp. 269-284.

Wilson, L. 1973: Variations in Mean Annual sediment Yield as a Function of Mean Annual Precipitation. **Am. J. Sci.**, Vol. 273, pp. 335-349.

Winn, F. 1926: The west Fork of the Gila river, **Science,** Vol. 64, pp. 16-17

Wolman, M.G. and Gerson, R. 1978: Relative scales of Time and Effectiveness of Climate in watershed Geomorphology. **Earth Surface Process**, Vol. 3, pp. 189-208.

Womack, W.R. and Schumm, S.A. 1977: Terraces of Douglas creek, northwestern Colorado, and example of episodic erosion, **Geology,** Vol. 5, pp. 72-76.

Wooley, R.R. 1946: Cloudburst in Floods in Utoh 1850-1938, **U.S. Geolo, Surv. water Supply paper**, 994.

Wooley, R.R. and Alter, J.C. 1938: Precipitation and Vegetation, **Transections American Geophysical Union**, Vol. 19, pp. 604-607.

Woolnough, W.G. 1930: The Influence of Climate and Topography in the Formation and Distribution of Products of-Weathering. **Geol. Mag.**, Vol. 67.

Yaalon, D.H. and Ganor, E. 1966: The Climatic Factor of Wind Erodibility and Dust Blowing in Israel. **Israel Journal**